Statistics in Economics

Statistics in Economics

JOHN D. HEY

 Praeger Publishers • New York

To Marlene

Published in the United States of America in 1977
by Praeger Publishers, A Division of Holt, Rinehart and Winston,
200 Park Avenue, New York, N.Y. 10017

Library of Congress Cataloging in Publication Data

Hey, John Denis.
    Statistics in economics.

    Includes index.
    1. Statistics. I. Title.
HA29.H616 1977   519.5'02'433   76-5863
ISBN 0-275-64930-X

Printed in the United States of America

789 074 987654321

Contents

I am indebted to the Literary Executor of the late Sir Ronald A. Fisher, F.R.S., to Dr Frank Yates, F.R.S., and to Longman Group Ltd., London, for permission to reprint Table III from their book *Statistical Tables for Biological, Agricultural and Medical Research.*

Preface

This book is the result of the conviction, gained over several years of teaching statistics to students of economics, that there existed a need for an introductory text on statistics which presented the material required by these students for them to appreciate the contents of their economics course. This book, therefore, contains the statistical material that the student is likely to encounter when he examines the empirical literature in economics. In order for him to be able to evaluate critically the way economic theories have been subjected to test, he must be aware of the underlying methodology of the statistical analysis. In my experience, students find statistics conceptually difficult, and often relapse into regarding the subject as consisting of the memorisation of certain formulae. Many existing texts encourage the reader to do this; either by producing results out of thin air, or by glossing over conceptually difficult pieces of analysis. It is my belief that, unless the student has a solid base of conceptual understanding of statistical methods, he will never be in a position to criticise intelligently the confrontation of theory with evidence. I have tried to provide this base of understanding by giving 'illustrations' of all major results, and by concentrating on elucidating difficult concepts.

The typical student of economics often has a fairly limited mathematical background. I have therefore limited the mathematical techniques to those that the student already needs for his economics courses. Limiting the level of mathematical sophistication has another important benefit: it allows the reader to concentrate on the statistical concepts under discussion, rather than worry about the mathematical proofs.

The book is designed for a one-year course in statistics for students with no previous statistical knowledge, but with some knowledge of economics and some experience of algebraic manipulation and elementary differential calculus. The book could be used as an introduction for students intending to study econometrics, but is basically designed as a self-contained text for those students of economics whose only study of statistics is a course of this nature. It should provide material for a lecture course of forty to fifty hours, supplemented by tutorials or classes of some ten to fifteen hours for discussion of the exercises.

The material, style and format of this book owe much to the many students who have moulded my ideas, and helped me realise where the conceptual difficulties of statistics lie. I am also greatly indebted to many of my colleagues in the Economics Department of Durham University for helpful comments and suggestions, in particular to Joost van Doorn who suffered for two years in giving tutorials to students I had taught. Several of the chapters were read and commented on by John Creedy to whom I am most grateful. Any remaining deficiencies lie in the chapters he did not read, or in the suggestions of his that I did not take up.

A particularly large debt of acknowledgement is owed to Mrs Sheila Drake who, in an extremely short period of time, managed to convert my scribbled drafts into an aesthetically pleasing manuscript.

Finally, I must thank my wife who, although totally unacquainted with statistics, read through the entire manuscript, correcting my grammar and attempting to totally eradicate my split infinitives. An even greater debt is owed to her for her constant help and encouragement throughout the writing of this book.

J.D.H.   *York, January 1977*

1

Introduction

Economics is conventionally divided into two broad areas: economic theory and applied economics. The methodology of economics (in common with other sciences) involves a continuing interplay between these two areas. Without such an interplay, progress in economics as a science is impossible. The role of statistics in economics is to provide a 'bridge' between economic theory and applied economics, so that this interplay may take place, and so that progress in economic knowledge is possible. We see, therefore, that statistics plays a crucial role in economics, and that knowledge of statistical analysis constitutes an essential requirement for any serious student of economics.

This book is designed to give students of economics the statistical knowledge necessary for them to appreciate fully the material in their economics courses. In keeping with the methodology of economics, any course in economics requires that students read both the theoretical and the applied literature in that field. Only by understanding the statistical concepts and techniques used will the student be able to evaluate critically the empirical evidence, and to use the evidence to evaluate the validity of competing economic theories.

The contents of this book have been chosen to cover, as far as possible, the main statistical techniques commonly used in economics, and which the student is likely to encounter in the course of his reading in economics. We make no pretence that we have included all techniques that have been used in economics, or are likely to be of use to economists. However, although we have concentrated on the particular techniques used in the majority of the applied literature, we deliberately stress throughout the *methodology* of statistics. This concentration on the importance of statistical method is designed to enable the student to evaluate critically a specific statistical technique even though such a technique may not have been explicitly discussed in this book.

Some of the ideas and concepts discussed above may appear rather abstract at this stage, but some insight into these concepts may be gained by a fuller discussion of them at this point. However, full appreciation of their importance may not come until the entire book has been read. First, we discuss in more detail the methodology of economics.

As we remarked in our introductory paragraph, the methodology of economics requires a continuing interplay between economic theory and applied economics. In order to examine any problem in economics the first step is to advance an appropriate theory for its analysis. Theories are based on a set of 'plausible' assumptions, usually concerning the behaviour of economic agents and the environment in which they behave. For example, the theory of demand assumes that the agents (either individuals or individual households) are 'rational', and that their behaviour is characterised by the assumption that they allocate their income in such a way as to maximise their level of utility; the environment in which the agents operate is assumed to be characterised by known and fixed prices for all the goods in their utility function. The theory of the perfectly competitive firm assumes that the agents (the firms) behave so as to maximise their profits; their environment is assumed to consist of known production functions, known and fixed input prices and known and fixed output prices. In general, the assumptions on which economic theories are based, although appearing plausible, may not be directly testable. How, for example, would we test whether individuals did in fact attempt to maximise their utility? Many individuals would be unaware of what 'utility' is, and certainly very few of us have an explicit utility function which we carefully maximise (subject to constraints) every time we go into a shop.

However, the recognition that the direct testing of assumptions is difficult, if not impossible, and that it is unimportant (for example) whether people explicitly differentiate their utility function every time they go into a shop or whether they merely act 'as if' they do, leads economists to concentrate on testing the *conclusions* of their theories.

It is at this stage that theory is confronted with evidence, and this is where the crucial role of statistics arises. If the conclusions of the theory and the evidence are contradictory (and we leave for the moment the problem of determining when a contradiction exists), then three possibilities arise: either the theory is wrong, or the evidence is wrong, or both are wrong. The second possibility, that the evidence is wrong, is not as absurd as it may sound. The evidence could be wrong in two ways: the relevant information may not have been collected accurately; or the evidence that has been used does not correspond to the concepts employed in the theory. As an illustration of the first point, the Central Statistical Office in the U.K. is prepared only to assert that its figures for Gross National Product are likely to be within ±3 per cent of the 'correct' figures, while their figures for stock changes are likely only to be within ±10 per cent of the correct figures. As an illustration of the second point we need look only as far as Friedman's theory of consumption, which is framed in terms of 'permanent income' and 'permanent consumption' — concepts which, although useful in theory, have no direct observable counterparts.

If we are convinced that a contradiction between the conclusions of a theory and the evidence is not due to incorrect evidence, we must deduce that the conclusions of the theory are wrong. This implies (excluding the trivial case of a logical flaw in the argument deducing the conclusions from the assumptions) that the assumptions on which the theory is based must be wrong. Thus we are led to a modification of the assumptions, the way in which the assumptions need to be modified hopefully being indicated by the kind of contradiction apparent between the theory and the evidence. Upon these new assumptions a revised theory is constructed, and, once again the conclusions of this revised theory are confronted with the evidence. This process continues until a theory is devised that is consistent with the evidence. However, we must be careful in interpreting this procedure; a faulty interpretation might lead us to allow the evidence alone to determine the theory. This would be dangerous in that it would imply a different 'theory' for every observed phenomenon; not only would this make theory worthless, but it would also mean that we would be unable to apply the knowledge gained from one body of evidence to the analysis of a similar problem in a different setting.

Such considerations lead to certain desiderata for theories; as a general rule we would like each theory to explain as wide a range of observed phenomena as possible. However, we also prefer simplicity of explanation to complexity; this objective may well conflict with that of generality noted above. We would also prefer that theories explaining different evidence (in different areas of economics) had at least mutually consistent assumptions; we would be unhappy if the theory of demand for durable goods had assumptions about individual behaviour that conflicted with those used in the theory of demand for non-durable goods. This criterion of mutually consistent sets of assumptions in a sense recognises that economic theory is only partially developed, and that in time several different theories may become special cases of a more general theory which encompasses them all.

We return to consider in more detail the idea that, the larger the body of evidence that a theory can explain, the better is the theory. Obviously this statement needs to be qualified. Consider the 'theory': 'if the price of a good goes up, either more of the good will be bought, or less of the good will be bought, or the same amount of the good will be bought'; this theory 'explains' a tremendous range of phenomena! That this 'theory' is useless is apparent. It is useless because it *cannot* be wrong; no amount of evidence can possibly refute it. This example naturally leads us to consider how a theory may be proved or disproved. If a theory is to be of use, it must be capable of predicting what will happen (or what has happened) in an unknown situation. If a theory has been constructed merely to explain a specific set of evidence, then its worth is restricted to that specific set. However if a theory is constructed to explain a whole range of possible phenomena (some of which necessarily have not been observed) then *until the theory is disproved* it is of value. Obviously, as the theory is tested against more and more evidence, and is found to be consistent with it, we can place more confidence in its validity. However, this argument necessarily implies that, if a theory is to be of value, it must be capable of predicting phenomena as yet unobserved, and there remains the possibility that this new evidence will be inconsistent with the theory. Thus, if a theory is to be of value, its validity must necessarily remain in doubt; we can therefore never *prove* a theory, we can only *disprove* it. As more and more evidence consistent with a theory accumulates, then the confidence we can place in its validity increases, but we can never be 100 per cent sure that a theory is correct.

The above argument is crucial to the methodology of economics, and to the role of statistics in economics. Since we can never prove a theory, but can only disprove it, then the only way to strengthen our confidence in the validity of a theory is to confront it with as much evidence as possible *that could conceptually disprove it*. The more tests of this nature that the theory survives, the more confidence we can place in it. It would be pointless either to construct a theory that cannot possibly be refuted by any evidence, or to test a theory against a set of evidence that cannot possibly refute it.

Statistics recognises that the evidence that is available to test a particular theory is only a part of all the possible evidence of relevance to that theory. Statistics therefore attempts to quantify the degree of confidence one can place in a theory after its confrontation with partial evidence. In keeping with the methodology of economics discussed above, in the application of statistics to economics we will be trying to show that a hypothesis (or theory) is not disproved, rather than that it is proved. For example, if our theory leads to the conclusion that there should be a relationship between consumption and income, our statistical methodology is to try and disprove that there is no relationship, rather than to try and prove that there is a relationship. However, since our tests are necessarily based on partial evidence, we can never disprove in absolute terms that no relationship exists. We can only conclude that it is extremely unlikely that no relationship exists. Statistics attempts to quantify statements like 'extremely unlikely'; a typical conclusion of a test concerning a hypothesised relationship would be: 'it is extremely unlikely, if no relationship exists, that we would obtain the evidence that we have, but there remains a 1 per cent (or 5 per cent) chance that the evidence we have could arise even though no relationship does exist'.

A connected role of statistics is to estimate the parameters of the economic theory. We are not only interested in whether a relationship exists between consumption and income, but also in the form of the relationship and the values of the parameters of the relationship. For example, we might wish to know whether the consumption function is linear and, if so, what is the value of the marginal propensity to consume. If decisions are to be taken on the basis of estimates of the relevant parameters, then we wish also to know how accurate these estimates are likely to be. (If they are based only on partial evidence, they are unlikely to be completely accurate.) We wish also to know whether and how their accuracy might be improved. Statistics attempts to quantify the 'accuracy' of estimates; a typical estimation procedure would result in a statement like: 'on the basis of the available evidence, we can be 95 per cent (or 99 per cent) confident that the true value of the marginal propensity to consume lies between 0.59 and 0.63'.

Intuitively, we can see that by gathering more evidence we should be able to improve the accuracy of our estimates. However, as economists we should be able to see that this does not necessarily mean that we should collect *all* the available information. Collecting evidence is costly, both in time and money, and the increased accuracy may not be worth the increased cost. Obviously, the question of how much evidence to collect depends upon the use to which we are to put our estimates. For the government deciding on a tax cut to reduce unemployment, accuracy in the estimation of the aggregate marginal propensity to consume is obviously very crucial, as the costs of mis-estimation are high. On the other hand, for a private individual wondering whether to invest a spare $10 (out of a large fortune) in consumption goods industries, accuracy in the estimation of the aggregate marginal propensity to consume is not so crucial.

So far our discussion of the methodology of economics, and of the role of statistics in economics, has much in common with a similar discussion concerning any other science. However, unlike the physical sciences, economics has particular problems in that, in many situations, it is unable to obtain *experimental data*, and it is unable to carry out *controlled* experiments. The physical scientist can set up a laboratory experiment and control to a certain extent those variables whose influence he does not wish to observe (on earth, of course, he is unable to remove the effects of gravity, though even this difficulty is being overcome by experiments carried out in orbiting satellites). However, the economist, although he too frames his *theory* keeping irrelevant variables constant (the famous *ceteris paribus* clause), often has difficulty in keeping these variables constant when he wishes to collect evidence on his theory. The data available for use by economists are, in many cases, generated by the actual working of the economic system. The economist would certainly not be popular if he were allowed to control the economy of the country so that he might obtain experimental data (this is not to say that the country might not be better off in the long run if economists were allowed to do so!). This creates two problems: first, in many cases the economist is *unable* to obtain all relevant information, and therefore must *necessarily* base his applied analysis on only partial evidence; secondly, he must be aware that the data he does have were not obtained experimentally, and the *ceteris paribus* clause he invoked in his theoretical analysis may not hold with respect to the evidence.

Such considerations naturally lead us to discuss the problem of investigating the validity of the assumptions used when we apply statistical analysis to economic theory using economic data. We will see in the course of this book that the statistical techniques used for testing hypotheses and for obtaining estimates of economic parameters rely on certain assumptions concerning the way that the evidence was obtained. The physical scientist may be able to ensure that these assumptions do hold by organising his experiment in such a way that they do. In some cases, the economist may also be able to ensure the validity of his assumptions by controlling the way he collects his evidence. However, as we have already noted, the economist is often unable to control the way that the evidence is obtained. In such cases it is crucial that the validity of the assumptions (which underlie the statistical techniques) is itself subject to investigation. Therefore, in the later stages of this book we deliberately stress the assumptions used in the analysis, and discuss at length how we may test the assumptions to determine their validity. We deliberately stress the importance of the assumptions, not only because the conclusions of the statistical analysis depend crucially on their validity, but also so that the reader may gain a solid foundation of the concepts of statistics, and of statistical methodology, and is thereby enabled to investigate statistical problems not explicitly covered in this book.

Now we discuss in more detail the organisation of the book. We have already noted that the contents have been basically determined by the needs of economic students; shortly, we will discuss the actual material in each chapter.

First, we must turn our attention to the thorny problem of *mathematics*. Partly because we recognise that many economics students have a limited mathematical background, we deliberately restrict the mathematical techniques used in this book. The only mathematical 'techniques' used in this book are elementary differential calculus (up to constrained maximisation) and algebraic manipulation. These are, in fact, exactly the same parts of mathematics that are required in (undergraduate) economic theory; thus you are not required to know any more mathematics for the understanding of this statistics book than you are expected to know for the successful completion of your economic theory courses. However, the limited mathematical background of many economics students is only part of the reason for the restriction of the number of mathematical techniques used in this book. Perhaps a stronger reason lies in the nature of statistics itself: statistics is *conceptually* difficult; it is completely different from most other subjects with which you will be familiar, since it is concerned with *inference* rather than *deduction*. Economic theory is of a *deductive* nature; it deduces conclusions from given assumptions. Statistics, however, is of an *inferential* nature; it infers something about a totality from a limited body of evidence. In a sense, economics deduces particularities from the general, while statistics infers generalities from the particular. These differences should become apparent as you proceed through the book. Our aim therefore, given the conceptually difficult nature of the subject, is to try to present it in the clearest possible way, and to avoid making conceptual understanding difficult by using excessive amounts of abstract mathematics. You may find some of the mathematics tedious, but you should not find it difficult. Above all, you should concentrate your attention on the underlying concepts, rather than worry about the algebraic derivations. We deliberately prove as many of the major results as we can (within the limitations of the mathematical techniques employed), so that we may achieve our previously expressed aim of stressing the importance of the assumptions, and of showing clearly which assumptions are used at each stage of the analysis. However, since we appreciate that many students have difficulty in interpreting abstract mathematical results, we include, after each major result, an 'illustration' of the result. We also include 'illustrations' of results stated without proof (where a proof would necessitate too high a level of mathematical sophistication) so that, once again, intuitive insight into, and conceptual understanding of, the meaning of the results may be obtained.

Some students may regard learning statistics as consisting of memorising formulae. This is clearly a dangerous procedure; it may well be better to know no statistics than simply to memorise 'magic formulae'. Any formula is based on a set of assumptions; if those assumptions are not valid, then the use of the formula is invalid, and this may lead not only to misleading conclusions, but also to dangerously wrong conclusions. Memorising the assumptions as well as the formulae is probably an improvement; better still is to know the methodology of statistics, so that one may *derive* the appropriate formula for any given problem.

At the end of each chapter (excluding this one) we provide a set of exercises designed to reinforce knowledge of the material presented in that chapter. It is all too easy to fall into the trap of looking at an exercise, saying 'that looks easy', and passing on without tackling it. Avoid this temptation at all costs. It is only when you tackle exercises that you learn whether you fully understand the material you have read. You must continually verify your understanding by working through the exercises. Each chapter builds on the foundation laid in the earlier chapters: you do not want to find yourself halfway through chapter seven only to find that the edifice of what you thought was your understanding has come tumbling down.

Finally we discuss the material to be covered in the rest of the book. Chapter 2 is, in a sense, preliminary to our basic theme of statistical inference, but it deals with the important problem of organising the raw data (or evidence) into a meaningful form. It shows how we can present a set of data in tabular or diagrammatical form so that the main characteristics of the data may be observed. It also discusses how we may summarise a whole set of data by a few summary measures.

Chapter 3 lays the foundation for the study of statistical inference by dealing with the theory of probability. Since inference from partial evidence always entails the possibility of making incorrect inferences, we need to be able to express, in a meaningful form, how probable such incorrect inferences are. Thus the study of probability is an essential prerequisite for any statistical analysis.

Chapters 4 to 9 are all concerned with statistical inference in different situations. The central concepts and methodology of statistical inference are introduced in chapter 4, with reference to the problems of testing hypotheses about, and estimation of, the proportion of members of some population having a certain characteristic (of an 'all-or-nothing' nature). Although introduced in this relatively simple context, the ideas of chapter 4 are repeated throughout the remainder of the book. Thus, chapters 5 to 9 use exactly the same concepts as chapter 4, albeit in increasingly complex applications. Chapter 5 is concerned with inference about the mean and variance of a particular characteristic in some population. Although we will define our terms more precisely in the relevant chapters, it may be helpful, in relating this paragraph to the preceding discussion, to think of the 'population' as equivalent to the totality of all possible evidence, while our inferences are based on a 'sample' from the population, in other words, partial evidence.

Chapter 6 uses the material in the preceding two chapters to illustrate the problems of choosing the 'best' way of estimating parameters. As might be expected, we see in this chapter that there is no unique 'best' method of estimation, and in general the 'best' method depends on the use to which we put our estimates. However we discuss certain properties of estimators that may be considered desirable in fairly general contexts.

The remaining three chapters of the book contain the material of most relevance in applied economics. These last three chapters apply the statistical concepts and methodology introduced in the earlier chapters to the study of relationships between variables. The majority of economic theories present their conclusions in the form of relationships between variables (quantity demanded of a good is related to the price of the good and the income of consumers, aggregate consumption is related to aggregate income and the rate of interest, and so on). Chapter 7 considers the simplest of such hypothesised relationships, where one variable is hypothesised to be dependent on just one other variable. Chapter 7 considers how we may test such hypotheses, and how we may estimate the relevant parameters.

Chapter 8 broadens the analysis of chapter 7 to allow us to investigate theories where a variable is postulated to depend on several other variables. Chapter 8 also investigates how we may test for the validity of the assumptions used in chapter 7, and how our results may be altered (and what modifications we may have to make in our analysis) if it appears that these assumptions do not, in fact, hold.

Finally, chapter 9 broadens our analysis still further to consider cases where the conclusions of theories are cast in terms of a series of simultaneous equations, and thus one set of variables is jointly dependent on another set of variables. For example, the theory of price determination in a perfectly competitive market involves three 'relationships': the supply curve, the demand curve and the market-clearing condition. Two variables, price and quantity, are jointly determined by the interaction of these three 'relationships', and are thus dependent on the variables that shift the demand and supply curves. Particular problems arise in the investigation of such simultaneous systems. However, these problems are discussed only partially; a full treatment would take us into the realms of econometrics, which is outside the scope of this book.

By the time you have completed the book, you should be in a position to evaluate critically the majority of the applied literature relevant to your economics courses. You should also be able to undertake your own empirical analyses, and to subject to test your own theories of economics. Finally, and perhaps more importantly, you should have gained a whole new insight into the subject matter and methodology of economics.

2

Data Presentation

2.1 Introduction

This book is concerned with the application of statistical analysis to economic data. These data may have been collected by government departments, by research agencies or by the economist himself. Before any sophisticated analyses can be performed on them, certain preliminaries of data organisation need to be carried out. These preliminaries achieve two basic objectives: first, to arrange the data into a form suitable for analysis, and secondly, and perhaps more importantly at this stage, to allow the analyst to 'get a feel' for the data. This second objective, although rather nebulous, is very crucial, as it gives an opportunity to spot any regularities or irregularities in the data, to see if any obvious mistakes have been made in the transcription of the numbers, and to note whether there are any deficiencies in the method of collection of the data.

2.2 Frequency Distributions

Organisation of the data basically means arranging them in some orderly fashion so that the basic structure of the data and their key characteristics can be observed. We illustrate these ideas by means of a simple example. Suppose we are interested in household incomes in a particular area, and suppose we have selected a sample of 50 households in that area. (At this stage in our discussion the method of selection of the 50 households does not concern us; later we will see that if we wish to make general inferences about households' incomes from the evidence of these 50 households we need to be careful about the method of selection.) Suppose we have contacted these 50 households and found out what their incomes were in a particular week. We might find the results of this investigation looked like table 2.1.

Inspection of table 2.1 shows immediately that we are unlikely to deduce very much from the data in their present form. One fact that may be apparent, though even this is difficult to spot, is that household number 21 has an income ($376.50) considerably higher than any other household. This should be a signal to check this figure to see if any mistake had been made in recording it or transcribing it. Suppose however that we have satisfied ourselves that this figure is in fact correct. What else can we get out of the data in their present form? We can note that the lowest income is $20.65 and the highest $376.50; but the latter, as we have

Table 2.1   The incomes of 50 households in a particular week

Household number	Household income ($)	Household number	Household income ($)
1	80.35	26	168.00
2	130.50	27	131.85
3	162.90	28	87.20
4	186.95	29	164.35
5	37.95	30	27.55
6	183.05	31	144.35
7	140.45	32	99.00
8	160.60	33	138.50
9	113.50	34	109.95
10	219.50	35	149.55
11	211.30	36	231.15
12	20.65	37	195.50
13	155.05	38	46.15
14	177.30	39	173.90
15	200.40	40	229.35
16	103.95	41	103.95
17	137.05	42	127.60
18	240.35	43	66.10
19	250.00	44	221.80
20	121.05	45	194.40
21	376.50	46	283.60
22	296.05	47	134.40
23	256.30	48	151.30
24	59.35	49	273.90
25	265.00	50	120.65

already noted, is considerably higher than any other, and thus gives rather misleading information about the range of incomes.

Obviously, then, we need to arrange the data in some way before we can get much useful information from them about the structure of the incomes of the 50 households. One thing we could do is to arrange the households in ascending (or descending) order by the magnitude of their incomes. You should try this yourself to see that, although it does help a little in getting a feel for the data, it still does not provide much more insight into the pattern of these 50 households' incomes (though it is a useful preliminary for further analyses).

A more illuminating way of arranging the data is to divide the households into classes according to their incomes. Thus we could say, for example, that there are 15 households with incomes between $100 and $150 (households numbered 2, 7, 9, 16, 17, 20, 27, 31, 33, 34, 35, 41, 42, 47 and 50), and only 12 households with incomes between $150 and $200. (We have to be careful about the dividing point between these classes, so that a household with an income of exactly $150 is not in-cluded in both classes.) If we continued in this way, dividing the households into classes—$0 to under $50, $50 to under $100, $100 to under $150, etc.—we would get table 2.2.

Table 2.2   Information of table 2.1 grouped by income

Income class	Household's numbers	Number of households
$ 0 to under $ 50	5, 12, 30, 38	4
$ 50 to under $100	1, 24, 28, 32, 43	5
$100 to under $150	2, 7, 9, 16, 17, 20, 27, 31, 33, 34, 35, 41, 42, 47, 50	15
$150 to under $200	3, 4, 6, 8, 13, 14, 26, 29, 37, 39, 45, 48	12
$200 to under $250	10, 11, 15, 18, 23, 36, 40, 44	8
$250 to under $300	19, 22, 25, 46, 49	5
$300 and over	21	1
Total		50

Table 2.2 gives us information on household income in the form of a *frequency distribution*. We define the *frequency* of income in a particular class as the number of households whose income is contained within that class. Thus the frequency of income in the income class $0 to under $50 is 4, as there are four households with incomes below $50. Similarly the frequency in the class $100 to under $150 is 15. Note that our extreme observation (household number 21) causes problems in the choice of classes; we could have used a class $300 to under $350, and another $350 to under $400, but this would have meant a frequency of zero in the former class. Not that a zero frequency causes any difficulties in itself, but providing two separate classes instead of the '$300 and over' class would have complicated the interpretation of table 2.2 without giving much compensating gain in information.

This last sentence provides a clue as to the choice of classes. Obviously the choice made in order to condense the data of table 2.1 into the frequency distribution of table 2.2 was arbitrary. We could have chosen a completely different set of classes. However, our choice of classes should meet certain criteria. First, we note that table 2.2 is an improvement on table 2.1 in the sense that the former conveys more information about the pattern of household incomes in the sample. We can get a better general picture of the distribution of household income from the frequency distribution of table 2.2 than from the raw data of table 2.1. However in another sense table 2.2 conveys less information than table 2.1: table 2.2 tells us that there are four households with income less than $50, but it does not tell us whether these four households have an income of $49.99 or of $0.01. We have lost this information by grouping the data into a frequency distribution, and obviously the fewer the classes and the larger their size the more information we lose.

Table 2.3   Alternative income
frequency distribution

Class number i	Income class	Number of households f_i
1	$ 0 to under $ 25	1
2	$ 25 to under $ 50	3
3	$ 50 to under $ 75	2
4	$ 75 to under $100	3
5	$100 to under $125	6
6	$125 to under $150	9
7	$150 to under $175	7
8	$175 to under $200	5
9	$200 to under $225	4
10	$225 to under $250	4
11	$250 to under $275	3
12	$275 to under $300	2
13	$300 and over	1
Total		50

Consider for example choosing classes only $25 in width rather than $50; we thus get table 2.3, where f_i denotes the frequency in the ith class. Table 2.3 gives more detailed information than table 2.2; for example it shows that, of the four households with incomes less than $50, only one of them has an income below $25. However, table 2.3 still provides less detailed information than the raw data.

Whether the smaller width classes of table 2.3 give a frequency distribution providing a clearer picture of the *pattern* of the income distribution than do the larger width classes of table 2.2 is a matter open for individual interpretation. Inspection of the two tables suggests that a choice of classes giving a frequency distribution somewhere between the two may be best.

Obviously no general rules can be laid down for the choice of classes. The final choice must depend on the trade-off between loss of detailed information and gain in understanding of the pattern of the data. We must be careful however not to choose the classes in such a way as to present a misleading picture of the distribution; the purpose of these tables is to give insight, not to confuse, obscure or mislead.

In both of our examples all except the final class are the same width. This is obviously not necessary, and we may well find that the optimal choice of classes to achieve the trade-off mentioned above gives classes of varying widths.

At this stage the introduction of some terminology will facilitate the future discussion. For each class we define the *lower limit* as being the minimum value in that class, and the *upper limit* as the maximum value.

The class *mid-point* is naturally halfway between the lower and upper limits, while the *width* of the class is the difference between the upper and lower limits. Some classes may be open — that is, they have no upper or no lower limit. For example, class number 13 in table 2.3 is open — it has no upper limit. For such open classes the width and mid-point must be chosen arbitrarily unless further information (for example the raw data of table 2.1) is available.

Finally, for each class we define the *relative frequency* (of observations in that class) as the proportion of the total observations which are in that class. For example in class 1 of table 2.3 the relative frequency is $1/50 = 0.02$; i.e. 2 per cent of the (total) households are in class 1.

In general if we have n observations, and we divide the data up into m classes, $i = 1, 2, \ldots m$, with the frequency in the ith class denoted by f_i, then we must have*

$$\sum_{1}^{m} f_i = f_1 + f_2 + \ldots + f_m = n \tag{2.1}$$

since each of the n observations must be allocated to one and only one of the m classes. The relative frequency in the ith class is

$$f_i/n.$$

From (2.1) it follows that the sum of the relative frequencies over all classes must be unity; i.e.

$$\sum_{1}^{m} f_i/n = f_1/n + f_2/n + \ldots + f_m/n = 1. \tag{2.2}$$

Before proceeding further we must introduce an important distinction which will clarify the interpretation of upper and lower limits. This distinction concerns the type of variable we are observing. Suppose the characteristic of the households we are interested in is the number of persons in the household. This variable can only take integral values $(1,2,3, \ldots)$ — we cannot conceive of a household with 2½ persons in it! Similarly the values of household income can take only certain values; if income is paid in cash it must be some multiple of 1 cent. It is a physical impossibility to find \$149.9875 in a pay packet. However some variables are not restricted to certain values. Consider height for example: it is perfectly possible to find someone 69.3421 in. tall, or 69.3422 in. tall, or even 69.34215 in. tall. We might not be capable of measuring this accurately, but we know that, if someone who is now six feet tall was once two feet tall, at some point in time he attained every single conceivable height between two and six feet. Variables such as height, which are not

*Readers unfamiliar with the sigma notation (Σ) are referred to Appendix A1.

limited to certain values, are termed *continuous variables*; those like number of persons in a household or income in a pay packet, which can only take certain values, are termed *discrete variables*. Note that discrete variables need not be only integral-valued as our example on income shows.

Sometimes we may be interested in a continuous variable which we can observe only in a discrete form. Suppose we wish to measure height, but have a ruler measured only in inches; our observations on height will thus be to the nearest inch — the actual observations are discrete, although the original variable we were trying to measure was continuous.

Similarly, our observations on household income may be a discrete form of a continuous variable. Suppose, for example, a person worked 40 hours at a basic hourly rate of $3.625 and then did half-an-hour of overtime at a rate of $9.99 per hour. This would entitle him to an income of $149.995. However he would never actually be paid this — he would either get $149.99 or $150. Suppose firms' policy was actually to pay someone only $149.99 whenever the income due was above $149.99 but less than $150. Then the upper limit of our class 6 ($125 and under $150) includes all people whose income was anything up to $149.9999 . . . . The dividing line between class 6 and class 7 is $150, with an income of exactly $150 being in class 7 and anything slightly below $150 being in class 6. $150 thus represents the lower limit of class 7 and the upper limit of class 6.

If however firms' policy was to pay $150 to anyone whose income was just above $149.995, and $149.99 to those whose income was just below $149.995, then this figure would represent the dividing line between the two classes and would be the upper limit of class 6 and the lower limit of class 7.

If we take the former interpretation, class 6 has a lower limit of $125, an upper limit of $150, a width of $25 and a mid-point of $137.50. On the latter interpretation of firms' policy, class 6 has a lower limit of $124.995, an upper limit of $149.995, a width of $25 and a mid-point of $137.495. In order to decide on the dividing lines between the classes we need to know something about the nature of the data, and the way they were collected.

After this sizeable digression we now return to our main theme, that of organising the raw data of table 2.1 into a form suitable for analysis. We have already seen that arranging the data into a frequency distribution enables us to get some insight into the pattern of the data; we now consider how the presentation of the data in a visual form may improve our comprehension of them further.

2.3 Visual Presentation

We consider the presentation of the frequency distribution as given in tabular form by table 2.3 in a diagrammatical form. The usual method is by use of a *histogram*. A frequency histogram is obtained by showing the classes on the horizontal axis, and the frequency of observations on the vertical axis in the form of a horizontal bar extending over the relevant class. Thus, to portray the fact that there are 9 households with incomes between $125 and $150, we place a horizontal bar at a height of 9 over the range $125 to $150. For reasons that will become obvious, we will slightly modify the classes in table 2.3 so that class 13 becomes four classes: '$300 and under $325', '$325 and under $350', '$350 and under $375' and '$375 and under $400', with frequencies of 0, 0, 0 and 1 respectively. The frequency histogram portraying the information in table 2.3 is shown in figure 2.1.

The advantages of this visual presentation should be immediately apparent; a comparison of table 2.3 with figure 2.1, although both convey exactly the same information, shows the superiority of the latter form for immediate assimilation of the main characteristics of the data. Of course we could also present the frequency distribution of table 2.2 in the form of a histogram. The choice of which particular histogram is 'best' depends on the outcome of the same discussion concerning the tabular presentation.

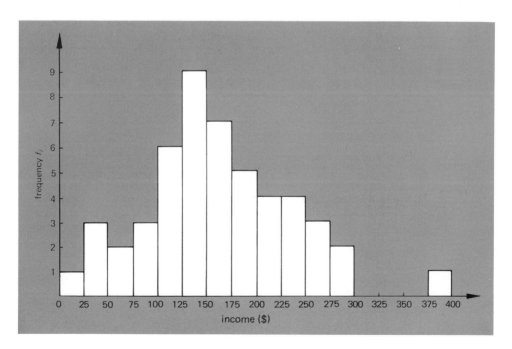

Fig. 2.1   Frequency histogram of data in table 2.3

Instead of presenting the data in a frequency histogram we could have used a *relative frequency histogram*, plotting relative frequencies against income. Obviously the shapes of the two histograms would be the same; since to find relative frequencies we divide all frequencies by the same number n (in this case 50), so only the scale of the vertical axis would be altered. However relative frequency histograms are useful when we wish to compare two different distributions based on different numbers of observations.

As far as possible we want our visual presentation to portray the main characteristics of the data without any 'distortions'. Distortions would be introduced if the classes had different widths. Suppose the data of table 2.1 had been presented to us in the frequency distribution of table 2.4. (Ignore for the moment the last two columns.) If we plotted the frequency histogram we would get figure 2.2. Comparing this with figure 2.1 we see clearly that figure 2.2 presents a distorted picture of the distribution of incomes. Table 2.4 itself is perfectly accurate, but the portrayal of it in the frequency histogram of figure 2.2 is clearly misleading. The problem lies in the differing class widths: the frequencies in classes 2, 3 and 6 of table 2.4 are the same, but the classes are $50, $25 and $75 wide respectively. We can correct for differing class widths by expressing the frequency in each class relative to its width. Thus for every one dollar of width of class 2 there are 9/50 households, but for every one dollar of width of class 3 there are 9/25 households. The ratio of frequency to width is termed the *frequency density*. For the ith class the frequency density is f_i/w_i where w_i is the width of the ith class. The penultimate column of table 2.4 gives the frequency densities (all

Table 2.4   Frequency distribution with differing class widths

Class number i	Income Class	Frequency f_i	Width ($) w_i	Frequency density ($\times 150$) f_i/w_i	Relative frequency density ($\times 7,500$) f_i/nw_i
1	$ 0 to under $ 75	6	75	12	12
2	$ 75 to under $125	9	50	27	27
3	$125 to under $150	9	25	54	54
4	$150 to under $175	7	25	42	42
5	$175 to under $225	9	50	27	27
6	$225 to under $300	9	75	18	18
7	$300 and over*	1	150	1	1
Total		50			

*estimated upper limit $450

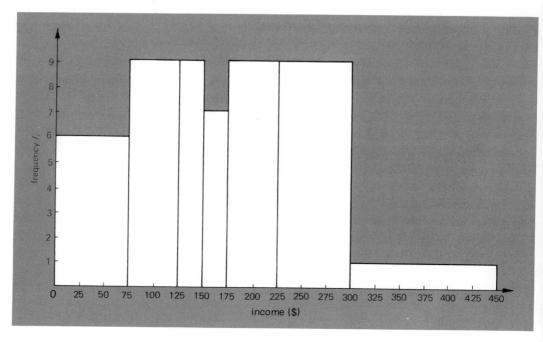

Fig. 2.2   Distorted visual presentation

multiplied by 150 to avoid awkward decimals). The unit of frequency density (in this example) is numbers per dollar. Similarly we define the *relative frequency density* as the relative frequency divided by the class width, i.e. f_i/nw_i for the ith class. The relative frequency densities (all multiplied by 7,500) are given in the last column of table 2.4. Note that the frequency densities and the relative frequency densities differ only by a common multiple of n (in this case, 50). Thus, presenting the data in a frequency density histogram or a relative frequency density histogram will give the same shaped histogram — only the scale on the vertical axis will differ. Plotting the relative frequency density histogram for table 2.4 gives figure 2.3. This presents a much less distorted picture of the data than does figure 2.2. Obviously figure 2.1 presents a more *detailed* picture than figure 2.3, but this is inevitable since the former was based on a more detailed frequency distribution.

Besides providing an undistorted visual picture, the representation of table 2.4 by a relative frequency density histogram has some interesting properties. Consider the rectangular block marked A in figure 2.3, covering the income range \$75 to under \$125. The height of this block is 27/7,500 (the relative frequency density in class 2) and its width is 50; the area of block A is thus $27/150 = 9/50$, the relative frequency of observations in this class. This is no coincidence, since the height of the block over the ith class is f_i/nw_i (the relative frequency density) and the

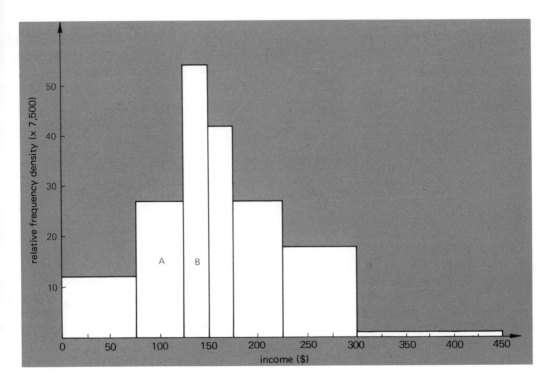

Fig. 2.3   Relative frequency density histogram

width is w_i, thus giving an area of $(f_i/nw_i) \times w_i = f_i/n$, the relative frequency in the ith class. Returning to our specific example, block B has area $(54/7,500) \times (25) = 9/50$, the relative frequency in class 3.

Since the sum of the relative frequencies must be unity (see (2.2)), it follows that the total area under a relative frequency density histogram must also be unity. This, together with the result already shown, that the area under this histogram between any two (income) values gives the relative frequency of observations between these two values, are two very important properties of the relative frequency density histogram. Their importance, however, may not become apparent until the next chapter.

Histograms are not the only ways of visually presenting data, and for some purposes there are clearly superior alternatives. I intend to discuss only one alternative, which is very closely linked to the histogram method; there are many texts that consider other alternatives. The frequency *polygon* is derived from the framework of the frequency histogram by plotting the mid-points of the horizontal histogram bars and joining the points together. For illustration the frequency polygon representation of table 2.3 is given in figure 2.4, and should be compared with the histogram in figure 2.1.

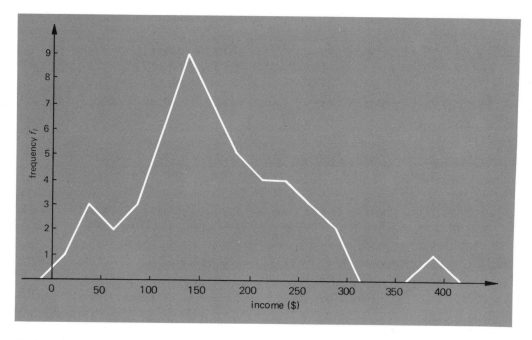

Fig. 2.4   Frequency polygon of data in table 2.3

We can similarly define relative frequency polygons, frequency density polygons and relative frequency density polygons in the framework of the corresponding histograms. Of course the earlier discussion of distortions applies with equal force to polygon presentation as it does to histogram presentation. You should verify for yourself that the interesting area property of the relative frequency density histogram (that the total area under the 'curve' is unity) carries over to the relative frequency density polygon if the classes are all the same width, and carries over approximately if the classes are of varying widths.

For a histogram presentation of a frequency distribution of discrete data a small problem arises. Although there are necessarily 'gaps' in discrete data, a portrayal by histogram should not have gaps between the blocks as this would make visual interpretation difficult. The blocks should be centred on the mid-points of the classes. Such a problem does not arise for the *polygon* presentation of discrete data.

2.4 Joint Frequency Distributions
Often we are interested in the values of two variables, rather than in just a single variable, where the two variables may be related to each other. An obvious example is household consumption and household income.

Table 2.5   Household income and consumption in a particular week

Household number	Household income ($)	Household consumption ($)	Household number	Household income ($)	Household consumption ($)
1	80.35	78.05	26	168.00	167.15
2	130.50	125.05	27	131.85	120.05
3	162.90	172.10	28	87.20	92.35
4	186.95	187.05	29	164.35	143.25
5	37.95	58.10	30	27.55	51.25
6	183.05	144.40	31	144.35	132.75
7	140.45	114.50	32	99.00	133.55
8	160.60	146.80	33	138.50	138.10
9	113.50	127.25	34	109.95	106.95
10	219.50	167.05	35	149.55	162.50
11	211.30	201.45	36	231.15	204.45
12	20.65	58.05	37	195.50	157.40
13	155.05	158.55	38	46.15	78.20
14	177.30	141.40	39	173.90	173.95
15	200.40	180.45	40	229.35	187.30
16	103.95	96.55	41	103.95	132.20
17	137.05	143.10	42	127.60	117.60
18	240.35	214.55	43	66.10	88.55
19	250.00	226.15	44	221.80	197.70
20	121.05	122.00	45	194.40	165.25
21	376.50	300.95	46	283.60	211.80
22	296.05	247.05	47	134.40	137.10
23	256.30	228.50	48	151.30	145.40
24	59.35	90.95	49	273.90	238.15
25	265.00	204.95	50	120.65	116.75

Suppose, for the 50 households sampled, we had collected data not only on their income in a particular week but also on their consumption in that particular week. These data in their raw form would look like table 2.5, and in this raw form little information of much value can be extracted. If we are interested only in household consumption by itself we could use the methods already discussed to present the consumption data in a meaningful form. If however we are interested in the relationship between income and consumption, we could extend these methods to construction of a *joint frequency distribution*. We could, for example, construct the joint frequency distribution in the tabular form of table 2.6.

Table 2.6 shows for example that, of the five households with an income between $50 and $100, four of these (households numbered 1, 24, 28 and 43) spent between $50 and $100, while the remaining household (number 32) spent between $100 and $150. The last column shows the totals in each income class, and is of course just the frequency distribution of income as given by table 2.2. Similarly, the last row shows the

Table 2.6   Joint frequency distribution of data of table 2.5

Income class	$0 to under $50	$50 to under $100	$100 to under $150	$150 to under $200	$200 to under $250	$250 and over	Totals
$ 0 to under $ 50		4					4
$ 50 to under $100		4	1				5
$100 to under $150		1	13	1			15
$150 to under $200			2	10			12
$200 to under $250				4	4		8
$250 to under $300					5		5
$300 and over						1	1
Totals		9	16	15	9	1	50

frequency distribution of consumption. Table 2.6 shows very clearly the association between income and consumption, as one would expect.

There are many ways we could demonstrate this association in a diagrammatical form. We could extend the histogram method to illustrate this joint distribution. Unfortunately this would involve a three-dimensional diagram, which is rather difficult to draw and interpret on a two-dimensional page. A more usual, and useful, presentation is by means of a *scatter diagram*. This involves the use of the raw data (of table 2.5) and would be of no use if we had the data only in the form of table 2.6. A scatter diagram merely involves plotting each observation (household) on a graph in which the axes represent the two variables (consumption and income). Figure 2.5 is the scatter diagram of the data of table 2.5.

2.5 Measures of Location
Although the presentation of data in a frequency distribution is a useful first step in organising a set of data, in that it provides some insight into the pattern of the data, it is still somewhat unwieldy for analysis. Often it is useful to be able to summarise a whole set of data by just one summary measure. A natural kind of measure would be one that is 'typical' or 'average'. Obviously there are many ways of compressing a whole set of observations into a single summary statistic; and just as obvious, there is no unique 'best' way. In particular, we will see that 'best' can only be defined relative to the use to which the summary measure is to be put and the form of the distribution. Some of these *measures of location* in common use are the mean, the median and the mode, though there are many others which may prove more useful in some applications. The idea, as mentioned above, of a measure of location is to provide some idea of the 'average' observation; the mean does this by answering the question: 'What value, if repeated n times, has the same total as all n observations

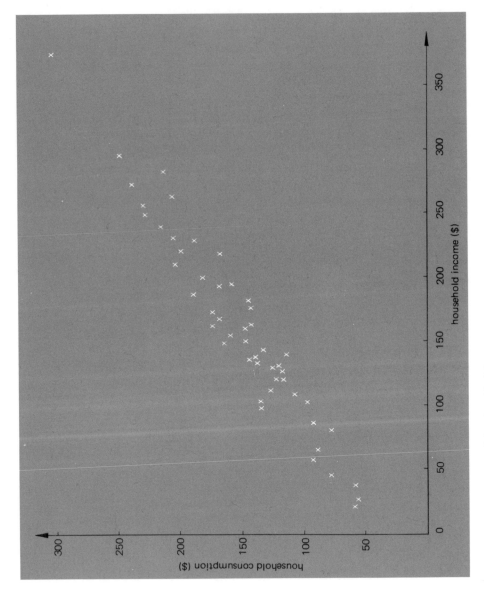

Fig. 2.5   Scatter diagram of data in table 2.5

added together?'; the median answers: 'What is the value below (above) which half the observations lie?'; and the mode answers: 'What is the most common observation?'. More formal definitions are provided below, but it must be stressed that these three are only a subset of all possible measures of location, and none of them are necessarily 'best' in any sense.

Suppose we denote our variable of interest by X, and suppose we have n observations $X_1, X_2, \ldots, X_n$ on this variable. Denoting the (arithmetic) mean of the observations by $\overline{X}$, it is defined by

$$n\overline{X} = X_1 + X_2 + \ldots + X_n = \sum_1^n X_i$$

that is

$$\overline{X} = \frac{1}{n} \sum_1^n X_i. \tag{2.3}$$

For our observations on household income we have, from table 2.1, $n = 50$, $X_1 = 80.35$, $X_2 = 130.50$, etc, and the total income of all 50 households $(\sum_1^n X_i)$ is thus \$8,059.50 giving a mean income of

$$\overline{X} = \frac{1}{n} \sum_1^n X_i = \frac{1}{50}\, 8{,}059.50 = 161.19.$$

The value of the mean has a simple interpretation: if all 50 households were to pool their incomes and then divide the pooled income equally between them, all households would get an income of \$161.19.

An interesting property of the mean emerges if we consider the variables x_i defined for all i by

$$x_i = X_i - \overline{X} \tag{2.4}$$

then it follows that

$$\sum_1^n x_i = \sum_1^n X_i - \sum_1^n \overline{X}$$
$$= n\overline{X} - n\overline{X} \text{ from } (2.3)$$

that is

$$\sum_1^n x_i = 0. \tag{2.5}$$

The variable x_i measures the deviation of the ith observation from the mean, and equation (2.5) merely says that the sum of all such deviations is zero.

If data are not available in raw form, but only in the form of a frequency distribution, certain assumptions have to be made in order to calculate the mean. Suppose the data are grouped in m classes, with the mid-point of the ith class being y_i and the frequency in the ith class being f_i. Without further information we do not know the particular values taken by the f_i observations in the ith class; all we know is that they take some value between the lower and the upper limits of that class. For simplicity we could assume that all f_i observations lie at the mid-point of the class, or what is equivalent, the mean value of the f_i observations is y_i. Whichever assumption we make, the sum of the observations lying in the ith class is $f_i y_i$. Thus the total sum of all observations is

$$f_1 y_1 + f_2 y_2 + \ldots + f_m y_m = \sum_1^m f_i y_i.$$

Now the total number of observations is

$$f_1 + f_2 + \ldots + f_m = \sum_1^m f_i = n,$$

and thus the mean is given by

$$\frac{1}{n} \sum_1^m f_i y_i = \sum_1^m f_i y_i / \sum_1^m f_i. \qquad (2.6)$$

To illustrate, suppose the only information we have on the incomes of the 50 households is that given by table 2.2. We carry out the necessary computations in table 2.7. The first row shows that we assume the four

Table 2.7   Calculation of mean of frequency distribution

Class number i	Income class	Frequency f_i	Mid-point y_i	$f_i y_i$
1	\$ 0 to under \$ 50	4	25	100
2	\$ 50 to under \$100	5	75	375
3	\$100 to under \$150	15	125	1875
4	\$150 to under \$200	12	175	2100
5	\$200 to under \$250	8	225	1800
6	\$250 to under \$300	5	275	1375
7	\$300 and over	1	375*	375
Totals		50		8,000

*see text

households in the first income class have a total income of \$100. A problem arises with the last class as we are not told its upper limit. Unless we have some futher information, we need to guess its likely value. Here we have assumed an upper limit of \$450 which gives a mid-point of \$375.

We have therefore $\Sigma f_i y_i = 8,000$, and this is the approximate total income, giving a mean income (using (2.6)) of $8,000/50 = \$160$. Comparison with the mean calculated from the raw data (161.19) shows a small error induced by the partial information of the frequency distribution.

Sometimes we may wish to attach more weight to some observations than to others. Suppose we have observations $X_i (i = 1, \ldots, n)$ and wish to attach weight W_i to the ith observation. Then the *weighted* (arithmetic) *mean* of the Xs is given by

$$\sum_1^n W_i X_i / \sum_1^n W_i. \qquad (2.7)$$

Note that equation (2.3) is just a particular case of (2.7) with $W_i = 1$ for all i, that is equal weight given to all observations. Note too that equation (2.6) is equivalent to a weighted mean of the ys (the mid-points) where $W_i = f_i$ for all i; that is, the mid-points are weighted by the frequencies. You should convince yourself that this is intuitively reasonable.

As most of our subsequent analysis uses the mean as 'the' measure of location (not because it is the 'best' measure in general, but because it is the most suitable for our purposes), we will not devote much space to other measures of location. We will limit our discussion to two others — the median and the mode — although there are many others, including the geometric mean and the harmonic mean.

For data in their raw form, the *mode* is simply defined as the most common observation. In a frequency distribution, the mode can be taken as the mid-point of the modal class defined as the class with the highest frequency or frequency density. More sophisticated definitions of the mode are available.

For our income data in their raw form of table 2.1, only two households have the same income ($\$103.95$); thus the 'most common observation' is $\$103.95$. However, even though this is the mode, there are fairly obvious reasons for not taking it to be a 'typical' or 'average' income. Consulting table 2.3, the modal class is '$\$125$ to under $\$150$' (since it has the highest frequency), and the mode can thus be taken as $\$137.50$. This appears to give a better picture of the 'average' income.

The *median*, as mentioned earlier, is the 'middle' observation. For raw data, the median is thus the $(n + 1)/2$th observation when the data are placed in ascending order if n is odd, or the value halfway between the $(n/2)$th observation and the $(n + 2)/2$th observation if n is even.

For data in a frequency distribution the actual calculation is rather messier, though the idea is the same. Again, certain assumptions have to be made, and you should try to find the assumptions implicit in the approach outlined below:

(a) call the total frequency n;

(b) find the class in which the cumulative frequency reaches $(n + 1)/2$ — call this the jth class;

(c) subtract the cumulative frequency up to the bottom of this class from $(n + 1)/2$; i.e. find

$$\frac{n + 1}{2} - \sum_{1}^{j-1} f_i;$$

(d) divide this by the frequency in the jth class, f_j, to give the proportion of observations in the class up to the median, and multiply by the class width, w_j, to give the distance from the lower class limit to the median; i.e. find

$$\frac{w_j}{f_j} \left(\frac{n + 1}{2} - \sum_{1}^{j-1} f_i \right);$$

(e) then, if l_j is the lower limit of the jth class, the median is

$$l_j + \frac{w_j}{f_j} \left(\frac{n + 1}{2} - \sum_{1}^{j-1} f_i \right).$$

Inspecting table 2.1, we see that 25 households have an income less than or equal to \$151.30, while 25 households have an income greater than or equal to \$155.05. Thus the median income (\$153.175) divides the households into two groups of equal size.

Which of the three measures of location (the mean = \$161.19, the mode = \$137.50, the median = \$153.175) is the 'best' measure of the 'average' income is open to interpretation, and much depends upon the use to which the 'average' is to be put. However, in the particular case of a unimodal symmetrical distribution (that is, a distribution which has a single mode or 'peak', and which looks the same whether viewed from the right or from the left) we have no problem of choice, since all three measures are the same. (You should check the validity of this assertion yourself.)

2.6 Measures of Dispersion

The various measures of location represent only one particular aspect of the distribution, that is the 'average' value. Naturally the representation of a whole set of data by one measure, although providing a useful indicator, does mean the loss of a great deal of information. In many applications it is found useful to consider, in conjunction with the 'average', a measure to indicate the 'spread' of the data about the 'average'. This need becomes apparent when we consider the following two sets of data:

Set 1:   1, 4, 7, 10, 13, 16, 19, 22, 25, 28, 31.
Set 2:   11, 12, 13, 14, 15, 16, 17, 18, 19, 20, 21.

Both sets of data have the same mean (16) and the same median (also 16, owing to the symmetry). The mode is undefined for both sets. However it is obvious that the 'spread' of the observations about the mean (or median) is much greater for set 1 than for set 2. A method for measuring 'spread' or 'dispersion' can fairly easily be defined for comparison of these two particular sets of data, but in general we can see that there is no unique 'best' method to measure dispersion. Once again the 'best' measure of dispersion is only defined relative to the use to which the measure will be put and the form of the distribution.

Although there are many measures of dispersion that can be used in different applications, we will mention only a few, concentrating on those of most use in our subsequent analysis. The simplest and most obvious measure is the *range*, which is the difference between the highest and the lowest observations. This may be useful, but it suffers from the fact that it ignores all observations in between the highest and the lowest, and in particular the distribution of these intermediate observations. It may also give a misleading picture of the spread if one of these two extreme values is 'out of line' with the rest of the observations. A measure of spread which gets round this last criticism is the *interquartile range*, defined as the difference between the lower quartile and the upper quartile. The lower quartile is the value below which one-quarter of the observations lie; the upper quartile is the value below which three-quarters of the observations lie. However the interquartile range still suffers from the defect of not taking into account all the observations.

One way of arriving at a more satisfactory definition of spread is to consider the values taken by the xs, the deviations of the observations from their mean, defined earlier in (2.4) by

$$x_i = X_i - \bar{X} \quad (i = 1, 2, \ldots, n)$$

where X_i is the ith observation and $\bar{X}$ is the mean of the n observations.

As x_i measures how far the ith observation is from the mean, it would seem reasonable to use some kind of average of the x_i as a measure of how widely the distribution is spread around its mean. Obviously a straightforward (arithmetic) mean is no use as a measure since $1/n(\Sigma_1^n x_i) = 0$ (from (2.5)). We can avoid this cancelling out of positive and negative values (of the xs) by simply ignoring the signs, and considering the mean value of the *magnitude* of the x_i as a possible measure. This measure is known as the *mean deviation*, and is defined formally by

$$\frac{1}{n}\sum_{1}^{n}|x_i| = \frac{1}{n}\sum_{1}^{n}|X_i - \bar{X}| \tag{2.8}$$

where $|x|$ means the magnitude of x (e.g. $|-6| = |6| = 6$).

As is apparent the mean deviation takes into account all observations, and thus is likely to be a better measure than the range or the interquartile range. However the mean deviation suffers from two defects, the significance of which may not be immediately apparent at this stage. The first is that it is difficult to manipulate. The second is that it attaches the same weight to large and to small deviations.

A further measure of dispersion defined in terms of the x_i is the mean of the *squared* deviations from the mean. This is termed the *variance* and is denoted by var X or σ_X^2. Thus

$$\sigma_X^2 \equiv \text{var } X \equiv \frac{1}{n}\sum_{1}^{n}x_i^2 = \frac{1}{n}\sum_{1}^{n}(X_i - \bar{X})^2. \tag{2.9}$$

The variance is necessarily positive, and gives more weight to large deviations than to small. It is more mathematically tractable than the mean deviation. We notice however that the variance is measured in squared units; for example, if X is measured in pounds the variance is in pounds squared. To obtain a measure of spread based on the variance but in the same units as X, we define the *standard deviation* of the observations, denoted by s.d.(X) or σ_X, as the (positive) square root of the variance. Thus

$$\sigma_X = \text{s.d.}(X) = \sqrt{\left(\frac{1}{n}\sum_{1}^{n}x_i^2\right)} = \sqrt{\left[\frac{1}{n}\sum_{1}^{n}(X_i - \bar{X})^2\right]}. \tag{2.10}$$

Our discussion so far has been concerned with measures of dispersion for data in their raw form. If the data are available only in frequency distribution form we need once again to make some assumptions to evaluate these measures. Consider for example the variance. If we assume that the f_i observations in the ith class are all at the mid-point y_i of that class, then for each of these f_i observations the squared deviation from the mean is $(y_i - \bar{X})^2$ where $\bar{X}$ is the mean as given by (2.6). Thus the sum of the squared deviations for the f_i observations in the ith class is $f_i(y_i - \bar{X})^2$, and thus the mean squared deviation over all m classes (i.e. the variance) is

$$\sigma_X^2 \equiv \text{var } X \equiv \frac{1}{n}\sum_{1}^{m}f_i(y_i - \bar{X})^2 = \sum_{1}^{m}f_i(y_i - \bar{X})^2 \Big/ \sum_{1}^{m}f_i.$$

To illustrate, consider the data as given in table 2.1. We have already shown that the mean income $\bar{X}$ is 161.19. Now $X_1 = 80.35$ and so

$x_1 = X_1 - \overline{X} = -80.84$; that is the first household has an income \$80.84 below the mean income. Similarly $x_2 = 130.50 - 161.19 = -30.69$, etc. Thus $x_1^2 = 6535.1056$, $x_2^2 = 941.8761$, etc., and the sum of all these squared deviations is (see table 2.9):

$$\sum_1^{50} x_i^2 = 273{,}340.645.$$

Thus the variance of income is

$$\sigma_X^2 \equiv \operatorname{var} X = \frac{1}{n}\sum_1^n x_i^2 = 5{,}466.8129$$

and the standard deviation is

$$\sigma_X \equiv \sqrt{\operatorname{var} X} = 73.938 \quad \text{(to three decimal places)}.$$

It is useful to note that all except one of the observations lie within two standard deviations of the mean (i.e. between $161.19 \pm 2 \times 73.938$, i.e. between 13.314 and 309.066), while 34 of the 50 households (i.e. 68 per cent) have incomes within one standard deviation of the mean (i.e. between 161.19 ± 73.938, i.e. between 87.252 and 235.128). These results help to give intuitive content to the standard deviation as a measure of spread.

If the raw data of table 2.1 were not available, and we had only the data on household incomes in the form of the frequency distribution of table 2.2, we would use the results in (2.6) and (2.11). Table 2.8 shows the necessary computations. Hence

$$\sigma_X^2 \equiv \operatorname{var} X = \sum_1^m f_i(y_i - \overline{X})^2 \Big/ \sum_1^m f_i = 276{,}250/50 = 5{,}525$$

giving the standard deviation

$$\sigma_X = \sqrt{\operatorname{var} X} = 74.330.$$

Comparing this with the standard deviation (73,938) calculated from the raw data shows again a small error as a consequence of the loss of information when the raw data are condensed into a frequency distribution.

We do not intend to discuss any other measures of dispersion. Throughout the remainder of this book we shall mainly be using the standard deviation (or its square, the variance) as our measure of dispersion. As mentioned before, this is not because the standard deviation is the 'best' measure of spread in all applications, but because, as will become apparent, it can be considered the best in most of our applications. For similar reasons we will tend to concentrate on the (arithmetic) mean as our measure of location.

Table 2.8   Calculation of mean and variance of frequency distribution*

Class number i	Frequency f_i	Mid-point y_i	$f_i y_i$	$y_i - \bar{X}$	$(y_i - \bar{X})^2$	$f_i(y_i - \bar{X})^2$
1	4	25	100	−135	18,225	72,900
2	5	75	375	− 85	7,225	36,125
3	15	125	1,875	− 35	1,225	18,375
4	12	175	2,100	15	225	2,700
5	8	225	1,800	65	4,225	33,800
6	5	275	1,375	115	13,225	66,125
7	1	375	375	215	46,225	46,225
Totals	50		8,000			276,250

$\bar{X} = \sum_1^7 f_i y_i / \sum_1^7 f_i = 8000/50 = 160$ *Data from tables 2.2 and 2.7

2.7 Other Measures

In the last two sections we have been considering how we may summarise a whole set of data by two indicators, one indicating the 'average' or 'typical' value in the data and one indicating the 'spread' or 'dispersion' of the data around the 'average'. Although these measures of location and of dispersion are useful indicators, we obviously lose a great deal of information contained in the raw data when we compress all n observations into just two summary measures. In general therefore it seems reasonable to say that these two measures alone may not be sufficient adequately to describe the data. (However, we will see cases where a whole set of data *can* be completely described by two such measures alone.) Further general measures may thus be needed to describe and compare sets of data.

We shall not be referring to these other measures in the rest of this book, but it is important to be aware of their existence and their usefulness in describing and comparing sets of data.

These measures are designed to give some indication of the 'shape' of the distribution: whether it is skewed or symmetrical, whether it has a sharp peak or a flat peak, and so on. For example a natural third measure, after the 'average' and the 'dispersion', is the degree of 'skewness' in the data. Such a measure should also indicate whether the distribution is skewed to the right or to the left, or whether it is symmetric (i.e. not skewed). Consider for example the two distributions pictured in figure 2.6. Both distributions have the same mean (2½); they also have the same variance. Thus the mean and variance are unable to indicate the differences between the two distributions. Obviously the difference lies in their skewness: distribution A is skewed to the right, distribution B to the left. A measure of skewness should indicate this difference. The reader may like to construct such an indicator of skewness.

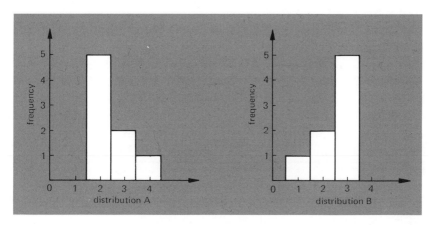

Fig. 2.6   Illustration of skewness

Finally we return briefly to the study of joint frequency distributions which we left at the end of section 2.4. Here the tabular presentation of a joint frequency distribution and the visual presentation by means of a scatter diagram helped to give a general picture of the way two variables are associated. Obviously it would help to have some numerical summary measure of the degree of association. If the two variables are denoted by X and Y, and we have n pairs of observations on X and $Y - (X_1, Y_1)$, $(X_2, Y_2), \ldots, (X_n, Y_n)$ — one such measure of association, termed the *covariance* between X and Y and denoted by cov(X, Y), is defined by

$$\text{cov}(X, Y) = \frac{1}{n}\sum_1^n x_i y_i = \frac{1}{n}\sum_1^n (X_i - \overline{X})(Y_i - \overline{Y}). \qquad (2.12)$$

We interpret this measure by use of the scatter diagrams of figure 2.7. In each scatter diagram, the horizontal line is at $\overline{Y}$ — the mean of the Y values — and the vertical line at $\overline{X}$ — the mean of the X values. Note that

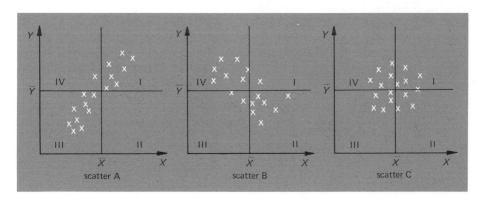

Fig. 2.7   Illustration of association

to the right of $\overline{X}$ all $x_i > 0$ since $X_i > \overline{X}$, and to the left all $x_i < 0$. Similarly, above $\overline{Y}$ all $y_i > 0$ and below $\overline{Y}$ all $y_i < 0$. Consider then the four quadrants marked I to IV. In I all observations have $x_i > 0$ and $y_i > 0$ and thus the product $x_i y_i > 0$. In II all observations have $x_i > 0$ but $y_i < 0$, giving a product $x_i y_i < 0$. Similarly in III $x_i y_i > 0$ and in IV $x_i y_i < 0$. Consider scatter A: in this all observations except one are in quadrants I and III, thus all products $x_i y_i$ except one are positive. Thus the covariance will be positive. In scatter B the covariance will be negative since all except three points lie in quadrants II and IV. In scatter C we have roughly equal numbers of observations in all four quadrants, with the negative products $x_i y_i$ in II and IV cancelling out with the positive products in I and III, giving a covariance of approximately zero.

Thus for positively associated variables (scatter A) the covariance is positive, for negatively associated variables (scatter B) the covariance is negative, and for variables that are not associated (scatter C) the covariance is small. Thus the covariance is indeed useful as a measure of association. However, two qualifications about it must be made. First, it is only a measure of a certain kind of association (viz. linear association). Consider figure 2.8: obviously there is some relation between Y and X, but as we can see by applying the above reasoning, the covariance is zero. Covariance is not designed to detect non-linear association.

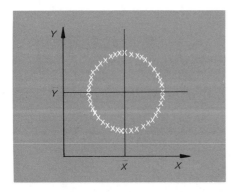

Fig. 2.8   Example of non-linear association

Secondly, as a measure of association it is not invariant with respect to the units in which X and Y are measured. We can correct for this problem by defining the *correlation coefficient* between X and Y (denoted by r) in terms of the covariance between X and Y by

$$r = \frac{\text{cov}\,(X,\,Y)}{\sqrt{(\text{var}\,X \cdot \text{var}\,Y)}} = \frac{\text{cov}\,(X,\,Y)}{\sigma_X \sigma_Y}. \tag{2.13}$$

Naturally r preserves all the qualities of the covariance as a measure of association. However it has other properties which the covariance does not possess. First, it does not depend on the units in which X or Y are measured; since if all X observations are multiplied by two, the covariance is doubled but so is the standard deviation of X ($\sqrt{\text{var } X}$), and thus r is unchanged. Secondly, r has the important property that its magnitude is always less than unity, that is

$$-1 \leq r \leq 1.$$

We do not intend to provide a.proof of this until chapter 7. However exercise 2.9, which shows that r takes the values ± 1 where the observations lie exactly along a straight line, should provide intuitive understanding of this result.

Table 2.9   Calculation of covariance and correlation coefficient for data of table 2.5

i	X_i	Y_i	$x_i = X_i - \bar{X}$	$y_i = Y_i - \bar{Y}$	x_i^2	y_i^2	$x_i y_i$
1	80.35	78.05	−80.84	−72.644	6535.1056	5277.1507	5872.5409
2	130.50	125.05	−30.69	−25.644	941.8761	657.6147	787.0144
3	162.90	172.10	1.71	21.406	2.9241	458.2168	36.6043
·	·	·	·	·	·	·	·
·	·	·	·	·	·	·	·
·	·	·	·	·	·	·	·
50	120.65	116.75	−40.54	−33.944	1643.4916	1152.1951	1376.0897
Σ	8059.50	7534.70	0.0	0.0	27,3340.645	13,6599.672	186,919.235

$$\bar{X} = \frac{1}{n}\Sigma X_i = \frac{1}{50}\,(8{,}059.50) = 161.19$$

$$\bar{Y} = \frac{1}{n}\Sigma Y_i = \frac{1}{50}\,(7{,}534.70) = 150.694$$

We illustrate by calculating the covariance between X and Y and their correlation coefficient where X and Y are household income and household consumption as given in table 2.5. Table 2.9 indicates part of the calculations used to derive the result. We have

$$\sigma_X^2 \equiv \text{var } X = \frac{1}{n}\Sigma x_i^2 = \frac{1}{50}\,(273{,}340.645) = 5{,}466.8129$$

$$\therefore \quad \sigma_X = \sqrt{\text{var } X} = 73.938$$

$$\sigma_Y^2 \equiv \text{var } Y = \frac{1}{n}\Sigma y_i^2 = \frac{1}{50}\,(136{,}599.672) = 2{,}731.9934$$

$$\therefore \quad \sigma_Y = \sqrt{\text{var } Y} = 52.268$$

$$\text{cov }(X,\,Y) = \frac{1}{n}\Sigma x_i y_i = \frac{1}{50}\,(186{,}919.235) = 3{,}738.3846$$

and so

$$r = \frac{\text{cov}(X, Y)}{\sqrt{(\text{var } X \cdot \text{var } Y)}} = \frac{3{,}738.3846}{(73.938)(52.268)} = 0.967$$

The value of the correlation coefficient (0.967) indicates a high but not perfect positive association between household income and consumption.

2.8 Transformations

Suppose we have n observations $X_1 X_2, \ldots, X_n$ on a variable X. Suppose a second variable Y is related to X in such a way that

$$Y_i = a + X_i \quad (i = 1, 2, \ldots, n)$$

where a is a constant.

Then $\bar{Y}$, the mean of the Y observations, is related to $\bar{X}$, the mean of the X observations, as follows:

$$\bar{Y} = \frac{1}{n} \sum_1^n Y_i = \frac{1}{n} \sum_1^n (a + X_i) = \frac{1}{n} \left(\sum_1^n a + \sum_1^n X_i \right)$$

$$= \frac{1}{n} \left(na + \sum_1^n X_i \right) = a + \frac{1}{n} \sum_1^n X_i$$

that is

$$\bar{Y} = a + \bar{X}. \tag{2.15}$$

Also the deviation of the ith Y observation from its mean

$$y_i = Y_i - \bar{Y} = (a + X_i) - (a + \bar{X}), \text{ from (2.14) and (2.15)}$$

thus

$$y_i = X_i - \bar{X} = x_i$$

and so

$$\sigma_Y^2 \equiv \text{var } Y = \frac{1}{n} \sum_1^n y_i^2 = \frac{1}{n} \sum_1^n x_i^2 = \text{var } X \equiv \sigma_X^2 \tag{2.16}$$

thus

$$\sigma_Y = \sigma_X.$$

From (2.15) and (2.16) we see that the mean of Y is a plus the mean of X, while the variances of the two variables are the same. This should be intuitively reasonable since the distribution of Y is the same as the distribution of X 'pushed to the right' by a distance a.

Consider now the relation

$$Y_i = bX_i \quad (i = 1, 2, \ldots, n) \tag{2.17}$$

then

$$\bar{Y} = \frac{1}{n}\sum_1^n Y_i = \frac{1}{n}\sum_1^n (bX_i) = b\,\frac{1}{n}\sum_1^n X_i = b\bar{X}. \tag{2.18}$$

Hence

$$\sigma_Y^2 \equiv \text{var } Y = \frac{1}{n}\sum_1^n y_i^2 = \frac{1}{n}\sum_1^n (bx_i)^2 = b^2\,\frac{1}{n}\sum_1^n x_i^2$$

that is

$$\sigma_Y^2 \equiv \text{var } Y = b^2 \text{ var } X \equiv b^2 \sigma_X^2 \tag{2.19}$$

Thus if one variable is a constant multiple (b) of another, its mean is that multiple of the other's mean, and its variance is the square of the multiple times the other's variance. This last result (2.19) should not surprise us when we remember that variance is a squared quantity. Taking the (positive) square root of both sides of (2.19) gives

$$\sigma_Y = |b|\sigma_X$$

which is intuitively sensible, since the transformation (2.17) has 'spread' the distribution of X by the factor b to give the distribution of Y.

Combining the results above we see that if

$$Y_i = a + bX_i \quad (i = 1, 2, \ldots, n) \tag{2.20}$$

then

$$\bar{Y} = a + b\bar{X} \tag{2.21}$$

and

$$\text{var } Y = b^2 \text{ var } X$$

or

$$\sigma_Y = |b|\sigma_X. \tag{2.22}$$

We consider a particularly important example of the linear transformation (2.20). Consider the n observations $X_1, X_2, \ldots, X_n$ on a variable X. Denote the mean and variance by $\bar{X}$ and σ_X^2 as usual. Consider the transformation

$$Z_i = a + bX_i \quad (i = 1, 2, \ldots, n)$$

where

$$a = -\frac{\bar{X}}{\sigma_X} \text{ and } b = \frac{1}{\sigma_X}$$

that is

$$Z_i = \frac{X_i - \bar{X}}{\sigma_X} \quad (i = 1, 2, \ldots, n).$$ (2.23)

Applying the general results (2.21) and (2.22) gives

$$\bar{Z} = -\frac{\bar{X}}{\sigma_X} + \frac{1}{\sigma_X}\bar{X} = 0$$

and

$$\sigma_Z = \frac{1}{\sigma_X}\sigma_X = 1$$

that is, Z has mean zero and standard deviation (and thus variance) one.

The variable Z, as given in terms of X by (2.23), is termed the *standardisation* of X, and Z_i is the standardised value of X_i ($i = 1, 2, \ldots, n$). Note that the value of Z_i expresses the value of X_i in terms of the number of standard deviations it is from the mean. Note also that Z is dimensionless.

Standardisation of variables is very useful when we wish to compare a particular value of a variable in one distribution with a particular value of another variable in another distribution. For example we might want to know whether a teacher's salary of $6,000 in America is in a relatively different position in the American income distribution than a teacher's salary of £1,500 in Britain is in the British distribution, or whether a pupil's history mark of 50 was worse relative to the rest of his class than his 85 in mathematics. These comparisons are difficult to make directly, not only because in the first example the units of measurement are different, but also because the position and spread of the distributions are different. We thus need to 'standardise' the raw data to facilitate comparison.

Suppose that in a particular class the mean history mark was 40 and the standard deviation of the history marks was 10; suppose that the mathematics marks had mean 85 and standard deviation 5. We see therefore that the maths marks are more narrowly distributed around a higher mean than the history marks. This obviously makes direct comparison of individual marks in different subjects meaningless as an indicator of relative performance. A pupil with a mark of 50 in history is 10 marks above the mean, which is one standard deviation above the mean. The same pupil with a mark of 90 in maths also is one standard deviation above the mean in the maths distribution. Thus a pupil with 50 in history and 90 in maths has done equally well, *relative to the class*, in both subjects. A pupil with 20 in history and 75 in maths has standardised marks of -2 (i.e. $(20 - 40)/10$) in history and -2 (i.e. $(75 - 85)/5$) in maths; that is, he has done equally badly (relative to the class) in both subjects. Finally, a pupil with 70 in both subjects has standardised marks of 3 for history and -3 in maths, and so has performed relatively much better in history.

Table 2.10   Illustration of standardisation

i	X_i	$x_i = X_i - \overline{X}$	x_i^2	$Z_i = (X_i - \overline{X})/\sigma_x$	$z_i = Z_i - \overline{Z}$	z_i^2
1	7	3	9	1.5	1.5	2.25
2	2	-2	4	-1.0	-1.0	1.00
3	5	1	1	0.5	0.5	0.25
4	1	-3	9	-1.5	-1.5	2.25
5	6	2	4	1.0	1.0	1.00
6	3	-1	1	-0.5	-0.5	0.25
7	4	0	0	0.0	0.0	0.00
Σ	28	0	28	0.0	0.0	7.00

$$\overline{X} = \frac{1}{n}\Sigma X_i = \frac{1}{7}(28) = 4$$

$$\sigma_X^2 = \frac{1}{n}\Sigma x_i^2 = \frac{1}{7}(28) = 4 \qquad \sigma_X = 2$$

To close this section, we give one complete example of the procedure used to standardise a variable. Consider the variable X with values as given in the second column of table 2.10. The third and fourth columns provide the calculations necessary to find the mean $\overline{X}$ (=4) and the standard deviation σ_X (= 2). The fifth column calculates Z from X according to our standardisation procedure (2.23). The sum of the Z values is zero, showing that $\overline{Z} = 0$. Finally the variance of Z given by $\sigma_Z = (1/n)\Sigma z_i^2 = (1/7)(7) = 1$. Thus Z has zero mean and unit variance, confirming our general result.

Using the transformation (2.23) we can standardise any variable into a standardised variable with zero mean and unit variance. The importance of this should become apparent in the later chapters.

2.9   Summary
In this chapter we have considered ways of presenting data both in tabular and diagrammatical form. We noted that the relative frequency density histogram method of visual presentation has some interesting area properties, the importance of which should become apparent in subsequent chapters. Ways of summarising a whole set of data by a few summary measures were discussed, with the mean, median and mode given as examples of measures of location and the standard deviation (amongst others) as an example of a measure of dispersion. Methods of illustrating

and summarising joint frequency distributions were also discussed, and the covariance was introduced as a measure of association between two variables. Finally, by examining the relations between the summary measures of variables related linearly, we showed how we may transform any variable into a standardised variable, that is one with mean zero and variance unity. This standardisation was shown to be useful for purposes of comparison.

2.10   Exercises

2.1 For the data given below, construct:

> either the frequency histogram
> or the relative frequency histogram
> or the frequency density histogram
> or the relative frequency density histogram

choosing that histogram which you think gives the 'best' representation of the data. Give reasons for your choice. Construct also the associated polygon.

Age distribution of the population of the U.S. 1970 (thous)

Age	Number	Age	Number
under 5 years	17,184	35 and under 45	23,126
5 and under 10	19,876	45 and under 55	23,269
10 and under 15	20,805	55 and under 65	18,648
15 and under 20	19,285	65 and over	20,156
20 and under 25	17,176		
25 and under 35	25,278	Total	204,800

Source: *Current Population Reports*, U.S. Bureau of the Census.

2.2 Find a copy of *Current Population Reports—Consumer Income* (series P-60). Plot the distribution of incomes before and after tax in the U.S. for any recent year. Comment.

2.3 The *Current Population Reports—Consumer Income* (series P-60) gives the distribution of household income for the states and for certain regions. Choose any two states or regions, and represent the distributions by some suitable visual presentation. Comment. For each of the two states or regions, find the mean, median and modal income. Is one particular measure best for comparison purposes? Calculate also the standard deviation of incomes in the two states or regions and compare.

2.4. Refer to exercise 2.2. Find the mean and median income before tax. Detail any problems you encountered in the calculation and mention how (if at all) you overcame them. Do you think that one measure of location is better than the other in this context?

2.5 In the *Current Population Reports* you will also find joint frequency distributions of household income and household consumption. Choose one such table, and discuss how you might present the information in it in a visual form. Also adapt the covariance formula as given by (2.12) for use with a joint frequency distribution. Calculate the covariance and correlation between household income and consumption.

2.6. Show that the variance as given by (2.9) can be expressed in the form:

$$\text{var } X = \frac{1}{n} \sum_1^n X_i^2 - \bar{X}^2 .$$

This form often facilitates calculation of the variance.

2.7. In an examination in economics the mean mark was 72 and the standard deviation of marks was 15.

(i) determine the standardised marks of students receiving marks of:

(a) 60   (b) 93   (c) 72

(ii) find the marks corresponding to the standardised marks of:

(a) 1.0   (b) −1.8   (c) 0.8.

2.8. A student received a mark of 84 in an economics examination for which the mean mark was 76 and the standard deviation of marks was 10. In sociology, for which the mean was 82 and variance 256, he got a mark of 90. In which subject was his relative standing higher? A second student gained a mark of 66 for both economics and sociology. In which subject was *his* relative standing higher?

2.9. Show that if two variables are related by

$$Y_i = a + bX_i \quad (i = 1, 2, \ldots, n)$$

where a and b are constants, then

$$\text{cov}(X, Y) = b \text{ var } X$$

and so the correlation coefficient r between X and Y is either plus one (if b is positive) or minus one (if b is negative). Interpret this result.

2.10. Show that $\text{cov}(X, X) = \text{var } X$. Comment.

3

Elementary Probability Theory

3.1 Introduction

The last chapter was concerned with the presentation and summarisation of data. As we remarked there, the organisation of data into a meaningful form is an important first step in abstracting the main essentials of the data. However our interest in the data does not stop there; we are also interested in comparing sets of data, and making general inferences from a sample of data about the population from which they were drawn. For example we may be interested in household income in a particular area (e.g. New York State). However, we may not be able (or, as we shall see, may not need) to contact every single household in that area and find out its income. We therefore contact a sample of households and find out the incomes of the households in the sample. We want to know if, and under what circumstances, we can infer the characteristics of the incomes of the whole population of households in that area from the characteristics of household incomes in the sample. We also want to know how accurate such inferences are likely to be. Intuitively we can see that the larger the sample the more accurate the inferences will be. However, increasing the sample size will increase the cost of obtaining the inferences. We thus have a classic 'trade-off' between increasing accuracy and increasing cost. If accuracy is very important (and this of course depends on the use to which we put our results) then a larger sample would be needed to achieve the optimum trade-off than if accuracy is relatively unimportant. If sampling cost was high and the importance of accuracy was low a relatively small sample may be optimal. If sampling cost was low and the importance of extreme accuracy was crucial then a relatively large sample would be optimal, and we might even be forced to the extreme case where we sample the whole population (this would obviously be the case where the sampling cost was zero, and any increase in accuracy resulted in an increase in benefits).

As the above discussion shows, before we can decide on the size of the sample to collect (or whether we ought to 'sample' the whole population) we need to know how to calculate the accuracy of inferences made from sample evidence, and how the accuracy changes as the sample size changes.

In order to determine the accuracy of inferences based on sample evidence we need to study how the characteristics of samples drawn from a known population vary as different samples are drawn. For example, suppose we are interested in mean household income in a particular area; the mean income in a sample of size 50 (say) will of course vary depending on which particular households are in the sample. However if we know *how* the sample mean income varies, that is if we know the distribution of the sample mean income, then we have some idea of how far a particular sample's mean income is likely to be from the population mean income.

To find the distribution of some sample characteristic over all samples we need to know how likely it is that the sample characteristic takes certain values. In more formal terms, we need to know the *probability distribution* of the sample characteristic. This chapter provides the basic material necessary for determining the probability distribution of certain sample characteristics (we define the terms used here in more formal terms shortly). The first step is the study of probability theory.

3.2 Probability

We introduce the basic ideas of probability with some very simple examples, avoiding for the moment the complexities of real-life economic problems. The terminology too may appear somewhat abstract, but, as we will see, the concepts introduced in this chapter provide a useful framework for the application of probability and statistics in economics.

We wish to study situations in which several outcomes are possible and where the particular outcome at any moment is determined purely by chance. Familiar examples include tossing a coin, rolling a die and dealing a hand of cards. We use the term *experiment* to describe such situations. Each experiment has several different outcomes; the outcome on any particular repetition is determined purely by chance. Consider the set of all possible outcomes of an experiment broken down into their simplest distinct form. Suppose there are n such outcomes in this set, and denote them by E_1, E_2 ..., E_n. We call these n distinct outcomes *simple events*. These simple events have the property that if we carry out the experiment once only then one and only one of the simple events can result. We illustrate by means of the following experiments:

Experiment 1. Toss a coin once and note which face is uppermost. There are two simple events $(n = 2)$; E_1 : head, E_2 : tail. (We ignore the possibility of the coin landing on its edge.) The actual numbering of the events is of course arbitrary.

Experiment 2. Roll a (six-sided) die once, and note which side lies uppermost. Here there are six simple events, which can be conveniently denoted by: E_1 : '1' uppermost, E_2 : '2' uppermost, ... , E_6 : '6' uppermost.

Experiment 3. Toss a coin twice, and note which face lies uppermost on each toss. The four simple events are: E_1 : both tosses heads; E_2 : first toss head, second tail; E_3 : first toss tail, second head; E_4 : both tosses tails. Notice carefully that the three outcomes, 'two heads', 'one head and one tail' and 'two tails', do not constitute the list of all simple events of this experiment, since the middle of these three ('one head and one tail') can be broken down further into E_2 and E_3 as defined above. However these three *would* be the simple events of an experiment in which two indistinguishable coins were tossed jointly, or one in which the order of the outcomes of two tosses of a single coin was not recorded.

The last remark should indicate that it may not be easy to decide on the set of simple events, and that this set may crucially depend on the way the experiment is organised and the way the results of the experiment are presented. However the simple events will always satisfy the property that on a single repetition of the experiment one and only one of the simple events can result.

For intuitive understanding of some of the results that we will prove later in this chapter, it is often helpful to represent each simple event by a point in a diagram. We then define the *sample space* of the experiment as the set of n points representing the n simple events. The actual position of the n points may well not have any geometrical meaning; in such cases the representation is useful only as an aid to intuitive understanding of subsequent results.

For the three experiments described above, possible representations in sample spaces are given in figure 3.1. Also given in this figure is the sample space of a fourth experiment, defined by:

Experiment 4. Roll two distinguishable dice and note the number uppermost on each die. There are 36 simple events, represented by the 36 points in the sample space in figure 3.1. The labelling of the points has been omitted for clarity.

We define a *compound event* as a set, or collection, of simple events. For example, in experiment 2 (rolling a die once), the compound event 'getting a '4' or above' is the collection of simple events E_4, E_5 and E_6. If we denote this compound event by A_1 then we can write

$$A_1 = (E_4, E_5, E_6).$$

Thus if the outcome of rolling a die once is either a '4' (E_4) or a '5' (E_5) or a '6' (E_6) then we have got a '4' or above, and so event A_1 has occurred. Similarly, the compound event $A_2 = (E_2, E_4, E_6)$ is the event 'rolling an even number'.

For experiment 3 (tossing a coin twice) the compound event 'getting at

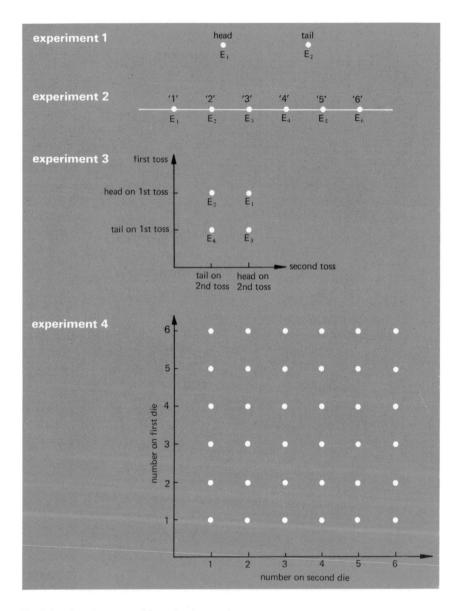

Fig. 3.1   Sample spaces of four simple experiments

least one head in the two tosses' is: the set $A_1 = (E_1, E_2, E_3)$ while $A_2 = (E_3, E_4)$ can be described as 'getting a tail on the first toss'. Any compound event can be expressed as a set of simple events.

We can represent compound events in the sample space by simply drawing a closed curve around the simple events from which it is compounded. The two compound events defined in the last paragraph are illustrated in figure 3.2.

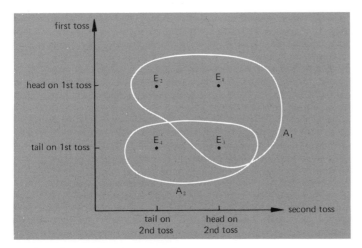

Fig. 3.2   Illustration of compound events

We define two events to be *mutually exclusive* if the occurrence of one precludes the occurrence of the other. Consider once again experiment 3 (tossing a coin twice) and consider the two compound events $A_2 = (E_3, E_4)$ and $A_3 = (E_1, E_2)$. A_2 is the event 'getting a tail on the first toss' and A_3 is the event 'getting a head on the first toss'. Obviously A_2 and A_3 cannot happen simultaneously — we cannot get both a head and a tail on the same toss! Thus A_2 and A_3 are mutually exclusive. However the events $A_1 = (E_1, E_2, E_3)$ ('getting at least one head in the two tosses') and $A_2 = (E_3, E_4)$ are not mutually exclusive, since if we get first a tail and then a head both A_1 and A_2 have occurred. Notice that A_1 and A_2, which are not mutually exclusive, have a simple event (E_3) in common, while A_2 and A_3, which are mutually exclusive, have no simple event in common. This leads us to the general result, which is a consequence of the fact that two simple events are by definition mutually exclusive, that two compound events are mutually exclusive if and only if they have no simple event in common. This result is easily proved. Consider two compound events A_1 and A_2 composed of h and k simple events respectively. Number the n simple events in such a way that we can write

$$A_1 = (E_1, E_2, \ldots, E_h)$$
$$A_2 = (E_g, E_{g+1}, \ldots, E_{g+k}) \quad (\text{where } g + k \leqslant n)$$

Suppose that $h < g$, that is that there is no simple event contained both in A_1 and A_2. Now we know that on a single repetition of the experiment one and only one of the n simple events E_1, E_2, ..., E_n can result. Denote the outcome of a particular repetition by E_i. If $1 \leqslant i \leqslant h$ then A_1 has occurred, but A_2 can not have occurred since, for $1 \leqslant i \leqslant h$, E_i is not contained in A_2. Similarly if $g \leqslant i \leqslant g + k$, then A_2 has occurred but A_1

cannot have occurred. Thus A_1 and A_2 cannot occur simultaneously and thus are mutually exclusive. (Of course if $h < i < g$ or $g + k < i \leqslant n$ then neither A_1 nor A_2 have occurred.) Conversely, suppose that $g \leqslant h$, that is simple events $E_g, \ldots, E_h$ are contained in both A_1 and A_2. Then if any one of $E_g, \ldots, E_h$ occurs both A_1 and A_2 occur simultaneously, and thus A_1 and A_2 are not mutually exclusive.

In our diagrammatic representation we can see immediately whether two compound events are mutually exclusive by seeing whether they have any simple events in common. More formally, two compound events are mutually exclusive if we can draw two closed curves (containing their respective simple events) that do not intersect each other. In figure 3.2 we see that A_1 and A_2 are not mutually exclusive since E_3 is contained in both, and there is no way of drawing the two closed curves without their intersecting.

As the question of common simple events is crucial to whether events are mutually exclusive or not, it is obviously useful to use some shorthand way of expressing the set of events (if any) that are common to any two compound events. This set is termed the *intersection* of the two events (a glance at figure 3.2 should provide the reason for this term). We shall denote the intersection of two events A_1 and A_2 very simply by (A_1 and A_2). By definition (A_1 and A_2) consists of the simple events that are contained in *both* A_1 *and* A_2. Thus the event (A_1 and A_2) only occurs if *both* A_1 *and* A_2 occur.

Reverting to experiment 3 and events $A_1 = (E_1, E_2, E_3)$ and $A_2 = (E_3, E_4)$, we have

$$(A_1 \text{ and } A_2) = (E_3).$$

Thus, if E_3 ('tail followed by head') occurs then both A_1 ('at least one head') and A_2 ('tail first') occur.

Mutually exclusive events have no simple events in common and so their intersection is empty. We thus have:

$$(A_1 \text{ and } A_2) = (\phi) \text{ for } A_1, A_2 \text{ mutually exclusive} \qquad (3.1)$$

where (ϕ) represents the empty set (no events in it).

We define a set of m events to be *exhaustive* if between them they contain all possible outcomes of the experiment. By definition the set of n simple events is exhaustive. In experiment 3 the two (compound) events A_2 and A_3 are exhaustive, as are the events A_1 and A_2.

Finally we define the *union* of two events to be the set of simple events that are contained in *either* one *or* the other *or both* of the two events. Denoting the union of A_1 and A_2 by (A_1 or A_2), then if

$$A_1 = (E_1, E_2, \ldots, E_h)$$

and

$$A_2 = (E_g, E_{g+1}, \ldots, E_{g+k})$$

then

$$(A_1 \text{ or } A_2) = (E_1, E_2, \ldots, E_h, E_g, E_{g+1}, \ldots, E_{g+k}).$$

Note that if $(A_1$ and $A_2)$ is non-empty the above list of simple events in $(A_1$ or $A_2)$ has the simple events of $(A_1$ and $A_2)$ repeated twice. The event $(A_1$ and $A_2)$ occurs if either A_1 *or* A_2 *or both* occur. In figure 3.3 we illustrate the intersection and union of two events in a general sample space. (The actual points representing the simple events are omitted; each compound event is understood to consist of all points contained within it.)

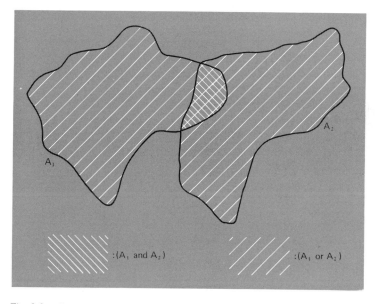

Fig. 3.3   Intersection and union of two events

Having defined some terminology with respect to the outcomes of experiments we must now turn our attention to attaching probabilities to the various outcomes. As we will see, we can concentrate on attaching probabilities to all the simple events, since once we have done this we can easily derive the probabilities of any compound events.

Unfortunately, deciding on the probabilities of each simple event is not as easy as might appear, even for our simple experiments. Consider the two simple events $(E_1:$ head, $E_2:$ tail) of experiment 1 (tossing a coin once). We could argue that, if the coin was fair, then E_1 and E_2 both have a 50:50 chance of happening and so we can attach a probability of ½ to each. This argument is, however, circular since it starts from the supposition of fairness, and if pressed, we would have to define 'fairness' as both events being equally probable.

If we had never seen the coin before, we could argue (using the famous *principle of insufficient reason*) that we have no reason or evidence to suppose that the coin is biased one way or the other, and thus, given our ignorance, we would have to impute equal probabilities to both outcomes. Using this argument we would be forced by logical necessity, as we gained evidence about the coin's behaviour (through repetition of the experiment), to change the values of the probabilities we attach to the simple events. We can no longer argue from ignorance when we have evidence. Now the idea of the probability of a particular coin landing 'heads' changing through time is rather disturbing. Do we really wish to assert that this probability is changing? It *may* change through time (as the coin becomes distorted or worn by repeated tossing), but what we are really measuring is our subjective assessment of the probability, and it is our subjective assessment that is changing through time, not the probability itself. Another unfortunate consequence of using such subjective probabilities is that they may differ from person to person, as different people observe different evidence or interpret the same evidence differently. For example, after observing a coin come down 'heads' 20 times in succession, one person may argue that the coin is obviously biased in favour of heads and thus attach a probability greater than ½ to the chance of a head on the 21st toss; another person may argue that 'after all these heads, a tail is virtually certain to come up' and thus attach a probability greater than ½ to the chance of a tail on the 21st toss. This last argument, although very common, must surely be fallacious as it implies that the coin has a memory, and that the outcome on any one toss depends on the outcomes of previous tosses. For this argument to be valid the person using it must demonstrate that past outcomes influence future outcomes. The fallacy originates from a misunderstanding of the idea of a 'fair coin' and of the 'law of averages'. Asked to define a fair coin most people would say it was one for which *on average* we would observe equal numbers of tails and heads. If pressed to give the number of tosses to which the 'on average' refers we would find ourselves in difficulties: if we tossed it 20 times, but did not get 10 heads and 10 tails, is the coin (necessarily) unfair? If we tossed it 100 times, but did not get exactly 50 heads and 50 tails, is the coin unfair? Forced into such a situation we would probably say the 'on average' refers to a 'large' number of tosses. But how large is 'large'? Is it 1,000 or 10,000? If we tossed the coin 10,000 times and observed 5,001 heads and 4,999 tails, we might still be unhappy at concluding that the coin was unfair. We are thus forced to the conclusion that, on this basis, we must define a coin as fair if the ratio of heads to tosses approaches one-half as the number of tosses increases without limit.

This leads us to a general definition of probability based on the evidence of successive repetitions of the experiment. Suppose an experiment has n simple events E_1, E_2, ..., E_n. Suppose that after N repetitions of the experiment, event E_1 has happened N_1 times, E_2 has happened N_2 times, ..., E_n has happened N_n times. Of course we must have:

$$\sum_1^n N_i = N_1 + N_2 + \ldots + N_n = N \tag{3.2}$$

since on each repetition one and only one of the n simple events may occur.

Consider the ratio N_i/N: this is the proportion of the repetitions for which outcome E_i resulted. By definition it cannot be greater than one, or negative:

$$0 \leqslant \frac{N_i}{N} \leqslant 1 \tag{3.3}$$

and it takes the extreme value zero only when E_i has never happened and the extreme value one only when E_i has always resulted. We note that the ratio N_i/N must *inevitably* change as N changes. Consider the $(N + 1)$th repetition: the outcome is either E_i or it is not E_i (i.e. it is E_j for some $j \neq i$, $1 \leqslant j \leqslant n$). If the outcome of the $(N + 1)$th repetition is E_i then the proportion of the $(N + 1)$ repetitions for which E_i resulted is $(N_i + 1)/(N + 1)$ and this is *necessarily* greater than the proportion N_i/N after N repetitions (unless $N_i = N$). Similarly if the outcome of the $(N + 1)$th repetition is not E_i then the proportion of the $(N + 1)$ repetitions for which E_i resulted is $(N_i)/(N + 1)$, and this is *necessarily* less than the proportion N_i/N after N repetitions (unless $N_i = 0$).

However, although the ratio N_i/N must change as N changes, it is intuitively reasonable to suppose that the ratio fluctuates about some average value, and that the ratio approaches this average as N gets larger and larger. We therefore define the probability of the simple event E_i as the limit of the ratio N_i/N as N grows without limit. Formally, denoting the probability of E_i by $P[E_i]$, we define

$$P[E_i] = \underset{N \to \infty}{\text{Lt}} \frac{N_i}{N} \tag{3.4}$$

where '$\underset{N \to \infty}{\text{Lt}} \dfrac{N_i}{N}$' means 'the limit of the ratio N_i/N as N approaches infinity'.

We note that N_i/N is just the relative frequency of E_i after N repetitions, and we are defining the probability of the ith event as the relative frequency of the ith event after an infinite number of repetitions.

This definition of probability appears to be a fruitful one, and it also appears to have the advantage of being objective. However in attempting to apply this definition in practice we immediately run into problems. The most obvious problem is that if each repetition takes some time, however little, then to repeat the experiment an infinite number of times we need an infinite amount of time! A second problem is that the experiment may change through time: the coin we are tossing may become distorted or worn. A third problem is that experimentation may be costly, not only in time but also in money, and we do not have infinite resources (consider the problem of determining the probability of caviare on toast landing caviare side up when we drop it from a height of five feet on to a carpeted floor!).

All these problems could be avoided if we are prepared to assert that our definition of probability as given by (3.4) should be interpreted as the *expected* limit of the relative frequency. For a large set of experiments this may provide a satisfactory definition, but for others we immediately run into the problem of differing expectations, once again giving us a subjective probability measure.

Further, this definition may not be applicable to certain events to which we would like to attach probabilities. Consider for example the probability of Edward Kennedy being the next U.S. President. Definition (3.4) is hardly applicable in this case! Consider also the probability of there being a monster in Loch Ness. As a straight scientific fact there either is or there is not such a monster. We ought therefore (on an objective basis) to attach a probability of either one or zero to this event — whether it should be one or zero we do not know. However, probabilities other than zero or one *are* attached to this event, and these probabilities are necessarily subjective, reflecting individual evaluations and also indicating the odds at which individuals would be prepared to lay bets.

The discussion above shows some of the many approaches to defining probability; it also shows that none of the approaches is entirely satisfactory. Each approach has particular merits which make it useful for certain applications. For instance, a subjective probability approach may well be fruitful if we wish to explain individual decision-making under uncertainty. The (expected) relative frequency approach (as defined by (3.4)), despite its drawbacks, is best suited to the type of applications we will be discussing in the remainder of the book. It is for this reason, rather than any inherent superiority, that we will base our subsequent analysis on the relative frequency definition of probability. One point to note however is that, although there may be disagreement about how to attach probabilities to the simple events, once this is done there is universal agreement about the derivation of the probabilities of compound events.

We re-state our definition of probability. Suppose that an experiment has n simple events E_i $(i = 1, 2, \ldots, n)$ and that after N repetitions event E_i has been observed N_i times $(i = 1, 2, \ldots, n)$. Then the probability of the event E_i, denoted by $P[E_i]$, is defined by:

$$P[E_i] = \underset{N \to \infty}{\text{Lt}} \frac{N_i}{N} \tag{3.4}$$

(where we interpret the limit to be the expected limit; that is, the limit which we expect the ratio N_i/N to approach). The probability of an event is its expected relative frequency. Since, as we have already shown, $0 \leqslant N_i/N \leqslant 1$ it follows that*

$$0 \leqslant P[E_i] \leqslant 1 \quad \text{(for all } i) \tag{3.5}$$

Now from (3.2),

$$\overset{n}{\underset{1}{\Sigma}} N_i = N_1 + N_2 + \ldots + N_n = N$$

and so

$$\overset{n}{\underset{1}{\Sigma}} \left(\frac{N_i}{N} \right) = \frac{N_1}{N} + \frac{N_2}{N} + \ldots + \frac{N_n}{N} = 1. \tag{3.6}$$

(3.6) of course just says that the relative frequencies sum to one (cf. chapter 2). From (3.4) and (3.6) it immediately follows that:

$$\overset{n}{\underset{1}{\Sigma}} P[E_i] = P[E_1] + P[E_2] + \ldots + P[E_n] = 1 \tag{3.7}$$

that is, the probabilities of the n simple events sum to one. This is a very important result.

We now proceed to derive the probability of a compound event from the probabilities of the simple events. Consider the compound event $A_1 = (E_1, E_2, \ldots, E_h)$. A_1 occurs if any of $E_1, E_2, \ldots, E_h$ occurs. After N

*A little care needs to be exercised over the extreme cases $P[E_i] = 0$ and $P[E_i] = 1$. For the former case we can loosely say that event E_i is impossible, but this is not strictly true as E_i may have happened a finite number of times out of the infinite number of repetitions (any finite number divided by infinity is zero). Similarly if $P[E_i] = 1$ it effectively means that E_i is always the outcome of the experiments, but here again some of the other events may have happened a finite number of times in the infinite number of repetitions. We will, however, refer to an event with probability zero as impossible, and an event with probability one as certain.

repetitions, E_1 has occurred N_1 times, E_2 has occurred N_2 times, . . . , E_h has occurred N_h times. Thus after N repetitions A_1 has occurred $(N_1 + N_2 + . . . + N_h)$ times, giving a relative frequency of occurrence of A_1 of $(N_1 + N_2 + . . . + N_h)/N$. Thus the probability of A_1 is:

$$P[A_1] = \operatorname*{Lt}_{N \to \infty} \left(\frac{N_1 + N_2 + . . . + N_h}{N} \right)$$

$$= \operatorname*{Lt}_{N \to \infty} \frac{N_1}{N} + \operatorname*{Lt}_{N \to \infty} \frac{N_2}{N} + . . . + \operatorname*{Lt}_{N \to \infty} \frac{N_h}{N}$$

Thus

$$P[A_1] = P[E_1] + P[E_2] + . . . + P[E_h] = \sum_1^h P[E_i] \tag{3.8}$$

and so the probability of a compound event is the sum of the probabilities of the simple events from which it is compounded.

For illustration, consider experiment 3 (tossing a coin twice). We should be able to agree (if the coin is fair) that the probability of each of the four simple events is ¼, that is $P[E_i] = ¼$ $(i = 1, 2, 3, 4)$. Thus, the probability of $A_1 = (E_1, E_2, E_3)$ is

$$P[A_1] = P[E_1] + P[E_2] + P[E_3] = ¾$$

That is, the probability of getting at least one head in the two tosses is 3/4. Similarly, for $A_2 = (E_3, E_4)$,

$$P[A_2] = P[E_3] + P[E_4] = ½$$

a result which should be reassuring!

We now consider how to derive the probability of the intersection and union of two events. Call them A_1 and A_2. Suppose first that they are mutually exclusive. Then $(A_1 \text{ and } A_2) = (\phi)$ from (3.1). In other words, the intersection $(A_1 \text{ and } A_2)$ of mutually exclusive events is empty, and so $(A_1 \text{ and } A_2)$ can never happen. Thus if A_1 and A_2 are mutually exclusive events then

$$P[A_1 \text{ and } A_2] = 0. \tag{3.9}$$

Further, suppose

$$A_1 = (E_1, E_2, . . . , E_h)$$

and

$$A_2 = (E_g, E_{g+1}, . . . , E_{g+k})$$

where $h < g$ if A_1 and A_2 are mutually exclusive. Then

$$(A_1 \text{ or } A_2) = (E_1, E_2, \ldots, E_h, E_g, E_{g+1}, \ldots, E_{g+k}) \qquad (3.10)$$

where no E_i is listed twice in (3.10). Thus, using (3.8),

$$P[A_1 \text{ or } A_2] = P[E_1] + P[E_2] + \ldots + P[E_h] + P[E_g] + P[E_{g+1}]$$
$$+ \ldots + P[E_{g+k}]$$

Now

$$P[A_1] = P[E_1] + P[E_2] + \ldots + P[E_h]$$

from (3.8) and

$$P[A_2] = P[E_g] + P[E_{g+1}] + \ldots + P[E_{g+k}]$$

$$\left. \right\} \qquad (3.11)$$

also from (3.8). It immediately follows that, for mutually exclusive events A_1 and A_2,

$$P[A_1 \text{ or } A_2] = P[A_1] + P[A_2]. \qquad (3.12)$$

This is known as the *addition rule* for mutually exclusive events.

Suppose now that A_1 and A_2 are not mutually exclusive; that is, for A_1 and A_2 as defined above, $g \leqslant h$ and simple events $E_g, \ldots, E_h$ are common to both A_1 and A_2. Then

$$(A_1 \text{ and } A_2) = (E_g, \ldots, E_h). \qquad (3.13)$$

Consider now the union $(A_1 \text{ or } A_2)$ where we are careful not to list any simple event twice. Then we can write either:

$$(A_1 \text{ or } A_2) = (E_1, E_2, \ldots, E_h, E_{h+1}, \ldots, E_{g+k})$$

or:

$$(A_1 \text{ or } A_2) = (E_1, E_2, \ldots, E_g, E_{g+1}, \ldots, E_{g+k}).$$

Both lists are of course identical and include all simple events E_i for $1 \leqslant i \leqslant g + k$. More simply, we can write:

$$(A_1 \text{ or } A_2) = (E_1, E_2, \ldots, E_{g+k}).$$

Thus, using (3.8)

$$P[A_1 \text{ or } A_2] = P[E_1] + P[E_2] + \ldots + P[E_{g+k}]. \qquad (3.14)$$

Now we can see, from (3.11), that $P[A_1] + P[A_2]$ is not the same as $P[A_1 \text{ or } A_2]$, since, for $g < h$, we have $P[A_1] + P[A_2] = P[E_1] + \ldots + P[E_{g-1}] + 2P[E_g] + 2P[E_{g+1}] + \ldots + 2P[E_h] + P[E_{h+1}] + \ldots + P[E_{g+k}]$; that is, the probabilities of the simple events common to both A_1 and A_2 are counted twice in the sum of $P[A_1]$ and $P[A_2]$. We can therefore see that

$$P[A_1 \text{ or } A_2] = P[A_1] + P[A_2] - \{P[E_g] + P[E_{g+1}] + \ldots + P[E_h]\}. (3.15)$$

The term in square brackets on the right hand side of (3.15) is just $P[A_1 \text{ and } A_2]$ (cf. (3.13)). Thus:

$$P[A_1 \text{ or } A_2] = P[A_1] + P[A_2] - P[A_1 \text{ and } A_2]. \tag{3.16}$$

This is the general addition rule and holds true whether A_1 and A_2 are mutually exclusive or not. (Note that (3.16) reduces to (3.12) when (3.9) holds.)

Let us illustrate the use of the addition rule using experiment 3. Remember that the simple events were: E_1: both tosses heads; E_2: head first then tail; E_3: tail first then head; E_4: both tails. We have already shown that $A_2 = (E_3, E_4)$ and $A_3 = (E_1, E_2)$ are mutually exclusive, thus $(A_2 \text{ and } A_3) = (\phi)$ and so $P[A_2 \text{ and } A_3] = 0$, which simply means that it is impossible to get both a head on the first toss and a tail on the first toss (at the same time). Now $P[A_2] = P[A_3] = \frac{1}{2}$, and so, using (3.16), $P[A_2$ or $A_3] = \frac{1}{2} + \frac{1}{2} - 0 = 1$; in words, we are certain to get *either* a head on the first toss *or* a tail on the first toss. Consider also the event (A_1 or A_3) where $A_1 = (E_1, E_2, E_3)$. A_1 and A_3 are not mutually exclusive, and $(A_1 \text{ and } A_3) = (E_1, E_2)$. Using (3.13) we get

$$P[A_1 \text{ or } A_3] = P[A_1] + P[A_3] - P[A_1 \text{ and } A_3]$$

$$= \frac{3}{4} + \frac{1}{2} - \frac{1}{2} = \frac{3}{4}.$$

You should interpret this result yourself.

We now introduce the idea of *conditional probability*. We denote by $P[A_1 | A_2]$ the conditional probability of A_1 occurring *given* that A_2 occurs. For simple events we get some particularly simple expressions for conditional probabilities. Consider one repetition of an experiment, and suppose we are told that E_j was the outcome. What can we deduce about the probability of E_i having occurred given the information that E_j has occurred? Since we know that, in any single repetition, one and only one of the n simple events can occur, we get the trivial result:

$$P[E_i | E_j] = \begin{cases} 1 \text{ if } i = j \\ 0 \text{ if } i \neq j. \end{cases} \tag{3.17}$$

The conditional probabilities of compound events given simple events are similarly easy to evaluate. If E_j has occurred and E_j is one of the simple events contained in the compound event A_1, then A_1 has occurred. If E_j is not in A_1 then A_1 can not have occurred. We thus get

$$P[A_1 | E_j] = \begin{cases} 1 \text{ if } E_j \text{ is contained in } A_1 \\ 0 \text{ if } E_j \text{ is not contained in } A_1. \end{cases} \tag{3.18}$$

Notice that we could have deduced this from (3.17) and our rule for probabilities of compound events: suppose $A_1 = (E_1, E_2, \ldots, E_h)$, then

$$P[A_1 \mid E_j] = P[E_1 \mid E_j] + P[E_2 \mid E_j] + \ldots + P[E_h \mid E_j]$$

and each term $P[E_i \mid E_j]$ $(i = 1, 2, \ldots, h)$ on the right-hand side of this expression is zero unless $i = j$ when it is one (from (3.17)), and of course i can equal j (for $1 \leqslant i \leqslant h$) only if $1 \leqslant j \leqslant h$, that is if E_j is contained in A_1.

Conditional probabilities for simple events given a compound event are somewhat more difficult to evaluate. Consider $P[E_i \mid A_2]$ where $A_2 = (E_g, E_{g+1}, \ldots, E_{g+k})$. If we know that A_2 has occurred then we know that one (and only one) of the simple events $E_g, E_{g+1}, \ldots, E_{g+k}$ must have occurred. If E_i is not contained in A_2 (that is $1 \leqslant i < g$ or $g + k < i \leqslant n$) then we can conclude that E_i cannot have occurred if A_2 has occurred. Thus, if E_i is not contained in A_2,

$$P[E_i \mid A_2] = 0. \tag{3.19}$$

However if E_i *is* contained in A_2 (that is $g \leqslant i \leqslant g + k$), then we know that E_i *may* have occurred given that A_2 has occurred. Now, suppose that in N repetitions of the experiment event E_j occurs N_j times $(j = 1, 2, \ldots, n)$. Then A_2 occurs $(N_g + N_{g+1} + \ldots + N_{g+k})$ times and E_i occurs N_i times. Thus, for E_i contained in A_2, the proportion of times that E_i occurs when A_2 occurs is

$$\frac{N_i}{N_g + N_{g+1} + \ldots + N_{g+k}}.$$

This is the relative frequency of E_i given A_2. Thus, using our definition of probability (3.4), the conditional probability of E_i given A_2 (for E_i contained in A_2) is given by the limit of the above expression as the number of occurrences of A_2 approaches infinity. Now, if the number of occurrences of A_2 approaches infinity, then so must N, the total number of repetitions of the experiment (otherwise $P[A_2]$ would be greater than 1). Thus

$$P[E_i \mid A_2] = \underset{N \to \infty}{\mathrm{Lt}} \left\{ \frac{N_i}{N_g + N_{g+1} + \ldots + N_{g+k}} \right\}.$$

To evaluate this, we note that

$$\frac{N_i}{N_g + N_{g+1} + \ldots + N_{g+k}} = \left(\frac{N_i}{N} \right) \Big/ \left(\frac{N_g + N_{g+1} + \ldots + N_{g+k}}{N} \right)$$

and that

$$\underset{N \to \infty}{\mathrm{Lt}} \left(\frac{N_i}{N} \right) = P[E_i] \quad \text{and} \quad \underset{N \to \infty}{\mathrm{Lt}} \left(\frac{N_g + N_{g+1} + \ldots + N_{g+k}}{N} \right) = P[A_2]$$

Thus*

$$P[E_i \mid A_2] = P[E_i]/P[A_2]$$

Combining this with (3.19) we get:

$$P[E_i \mid A_2] = \begin{cases} \dfrac{P[E_i]}{P[A_2]} & \text{if } E_i \text{ is contained in } A_2 \\ 0 & \text{if } E_i \text{ is not contained in } A_2. \end{cases} \tag{3.20}$$

We are now in a position to express the most general conditional probability $P[A_1 \mid A_2]$. Suppose $A_1 = (E_1, E_2, \ldots, E_h)$. Then:

$$P[A_1 \mid A_2] = P[E_1 \mid A_2] + P[E_2 \mid A_2] + \ldots + P[E_h \mid A_2]. \tag{3.21}$$

If A_1 and A_2 are mutually exclusive, then none of $E_1, E_2, \ldots, E_h$ are contained in A_2, and so, using (3.20), $P[A_1 \mid A_2] = 0$. However if A_1 and A_2 are not mutually exclusive, so that $(A_1 \text{ and } A_2) = (E_g, \ldots, E_h)$ is not empty, then (using (3.20))

$$P[E_i \mid A_2] = \frac{P[E_i]}{P[A_2]} \quad \text{for } E_i \text{ contained in } (A_1 \text{ and } A_2)$$

$$= 0 \quad \text{for } E_i \text{ not in } (A_1 \text{ and } A_2)$$

Thus, from (3.21),

$$P[A_1 \mid A_2] = \frac{P[E_g] + \ldots + P[E_h]}{P[A_2]}.$$

The numerator of this expression is $P[A_1 \text{ and } A_2]$ and so we get

$$P[A_1 \mid A_2] = \frac{P[A_1 \text{ and } A_2]}{P[A_2]}. \tag{3.22}$$

This, of course, gives $P[A_1 \mid A_2] = 0$ for A_1 and A_2 mutually exclusive. By symmetry it follows from (3.22) that

$$P[A_2 \mid A_1] = \frac{P[A_1 \text{ and } A_2]}{P[A_1]}$$

Combining this with (3.22) we get the *multiplication rule*:

$$P[A_1 \text{ and } A_2] = P[A_2 \mid A_1]P[A_1] = P[A_1 \mid A_2]P[A_2]. \tag{3.23}$$

To illustrate the results on conditional probabilities consider experiment 3. Suppose we were told that on the first toss a tail was observed; that is, event $A_2 = (E_3, E_4)$ has occurred. Given this information, what is

---

*This uses the important result that the limit of the ratio of two variables equals the ratio of the limits of the two variables (see (14) page 273).

the probability that (on the two tosses) event $A_1 = (E_1, E_2, E_3)$, 'getting at least one head', occurs? Now, we know that $(A_1$ and $A_2) = (E_3)$, and so

$$P[A_1 \text{ and } A_2] = P[E_3] = \tfrac{1}{4}$$

Further

$$P[A_2] = P[E_3] + P[E_4] = \tfrac{1}{2}$$

Thus, using (3.22)

$$P[A_1 \mid A_2] = \frac{P[A_1 \text{ and } A_2]}{P[A_2]} = \frac{\tfrac{1}{4}}{\tfrac{1}{2}} = \frac{1}{2}$$

and so, the probability of observing at least one head on the two tosses, given that a tail was observed on the first, is ½. This result accords with intuition.

We note that the unconditional probability $P[A_1]$ $(= \tfrac{3}{4})$ is different from the conditional probability $P[A_1 \mid A_2]$ $(= \tfrac{1}{2})$. In other words, the occurrence of A_2 changes the probability of A_1 occurring. However consider the two events $A_2 = (E_3, E_4)$ ('getting a tail on the first toss') and $A_4 = (E_2, E_4)$ ('getting a tail on the second toss'). We have $(A_2$ and $A_4) = (E_4)$ and so $P[A_2 \text{ and } A_4] = P[E_4] = \tfrac{1}{4}$. Also $P[A_2] = P[A_4] = \tfrac{1}{2}$. Thus, using (3.22),

$$P[A_4 \mid A_2] = \frac{P[A_2 \text{ and } A_4]}{P[A_2]} = \frac{\tfrac{1}{4}}{\tfrac{1}{2}} = \frac{1}{2}.$$

Thus $P[A_4 \mid A_2] = \tfrac{1}{2} = P[A_4]$, and so the occurrence of A_2 does not affect the probability of the occurrence of A_4; getting a tail on the first toss does not change the probability of a tail on the second toss.

Whether the occurrence of one event affects the probability of another event occurring depends on the relation between the two events. We call two events *independent* if the occurrence of one does not affect the probability of the other occurring. Formally, events A_1 and A_2 are independent if

$$P[A_1 \mid A_2] = P[A_1] \text{ and } P[A_2 \mid A_1] = P[A_2]. \tag{3.24}$$

Either one of these conditions implies the other (see exercise 3.3). For independent events the multiplication rule (3.23) reduces to

$$P[A_1 \text{ and } A_2] = P[A_1] \, P[A_2].$$

Thus, in experiment 3, events A_2 and A_4 are independent but A_1 and A_2 are not independent.

The main results proved so far in this section are:

(a) the addition rule: $P[A_1 \text{ or } A_2] = P[A_1] + P[A_2] - P[A_1 \text{ and } A_2]$;
(b) the multiplication rule: $P[A_1 \text{ and } A_2] = P[A_2 | A_1]\ P[A_1] = P[A_1 | A_2]\ P[A_2]$;
(c) for mutually exclusive events A_1, A_2: $P[A_1 \text{ and } A_2] = 0$;
(d) for independent events A_1, A_2: $P[A_1 | A_2] = P[A_1]$; $P[A_2 | A_1] = P[A_2]$.

To complete this section we give two illustrations of the main results. Suppose in a certain area we can divide all households into two classes: an upper class and a lower class. Suppose that one-quarter of the households are in the upper class. Suppose further that 80 per cent of the upper-class households own colour television sets while only 30 per cent of the lower-class households own colour sets. Consider then the 'experiment' of picking one household at random from all the households in the area, and suppose we wish to calculate the following probabilities:

(a) the probability that the household selected at random owns a colour television set and belongs to the upper class;
(b) the probability that the randomly selected household owns a colour television set;
(c) the probability that the randomly selected household belongs to the upper class given that it owns a colour television set.

There are many ways to evaluate these probabilities, and we are not suggesting that this particular way is best; you may prefer some alternative method of solution. We first express the given information in terms of probability statements. Define, with reference to the experiment of picking one household at random, the following events (outcomes):

A$_1$: household selected is upper-class
A$_2$: household selected is lower-class
A$_3$: household selected owns a colour television set
A$_4$: household selected does not own a colour television set

We note that A_1 and A_2 are mutually exclusive and exhaustive. A_3 and A_4 are also mutually exclusive and exhaustive. Our given information can be expressed by:

$P[A_1] = 1/4$ $\qquad P[A_2] = 3/4$ $\qquad$ (one-quarter are upper-class)

$P[A_3 | A_1] = 4/5$ $\qquad P[A_3 | A_2] = 3/10$

(80 per cent (i.e. $4/5$) of upper class own colour; 30 per cent (i.e. $3/10$) of lower class own colour).

The probabilities ((a), (b) and (c) above) that we wish to evaluate are $P[A_1 \text{ and } A_3]$, $P[A_3]$ and $P[A_1 | A_3]$ respectively. Consider the first. By the multiplication rule (3.23) $P[A_1$ and $A_3] = P[A_3 | A_1]P[A_1] = (\frac{4}{5})$ $(\frac{1}{4}) = \frac{1}{5}$. The interpretation of this result is that one-fifth of all households are colour television-owning upper-class households. Evaluation of $P[A_3]$ is a little trickier. However we can write $A_3 = ((A_1$ and $A_3)$ or $(A_2$ and $A_3))$ which reads 'a colour television-owning household is either an upper-class colour television-owning household or a lower-class colour television-owning household'. This is manifestly correct. Now we note that the events $(A_1$ and $A_3)$ and $(A_2$ and $A_3)$ are mutually exclusive (even if you own a colour television you cannot be in both classes at once). Thus, using the addition rule for mutually exclusive events (3.12), we have

$$P[A_3] = P[(A_1 \text{ and } A_3) \text{ or } (A_2 \text{ and } A_3)]$$
$$= P[A_1 \text{ and } A_3] + P[A_2 \text{ and } A_3].$$

We have already evaluated the first probability on the right-hand side. By similar methods we can evaluate the second probability:

$$P[A_2 \text{ and } A_3] = P[A_3 | A_2]P[A_2] = (3/10)(3/4) = 9/40$$

Thus,

$$P[A_3] = 1/5 + 9/40 = 17/40$$

and so 17/40ths of the households own colour television. Finally,

$$P[A_1 | A_3] = \frac{P[A_1 \text{ and } A_3]}{P[A_3]} \quad \text{(using (3.22))}$$

$$= \frac{1/5}{17/40} = \frac{8}{17},$$

that is, 8/17ths of colour television-owning households are upper-class.

Finally, to illustrate the concept of independence, consider the experiment of drawing one card at random from a well-shuffled complete pack of 52 cards. Define A_1 as the event of drawing a king; A_2 as drawing a heart. Obviously $P[A_1] = 1/13$; $P[A_2] = 1/4$. The event $(A_1$ and $A_2)$ contains one element, the king of hearts, and so $P[A_1$ and $A_2] = 1/52$. Hence:

$$P[A_1 | A_2] = \frac{P[A_1 \text{ and } A_2]}{P[A_2]} \quad \text{(using (3.22))}$$

$$= \frac{1/52}{1/4} = 1/13 = P[A_1].$$

Similarly

$$P[A_2 \mid A_1] = \frac{P[A_1 \text{ and } A_2]}{P[A_1]} = \frac{1/52}{1/13} = 1/4 = P[A_2].$$

Thus A_1 and A_2 are independent; this should be intuitively reasonable, for knowing that a particular card is a king does not give us any information about its suit, nor does knowing that a card is a heart give us any information about its face value.

However suppose the pack we were using had the king of hearts missing. In this case

$$P[A_1 \mid A_2] = 0 \neq 3/51 = P[A_1]$$

and

$$P[A_2 \mid A_1] = 0 \neq 12/51 = P[A_2]$$

and so A_1 and A_2 are *not* independent: knowledge of the suit does give us some information about the face value.

3.3 Random Variables

The previous section was concerned solely with the application of probability to events that were the outcomes of experiments. In economics we are more concerned with application of probability to the values that a variable may take. We stay within the framework of idealised experiments for the moment, and introduce the idea of a *random variable*, which is a variable whose value is defined at each point in the sample space, or equivalently a variable with a given value for each simple event of the experiment. We can, of course, define several random variables on the same experiment. The word 'random' refers to the fact that, before the experiment is performed, we do not know which event will result, and thus do not know which value the random variable will take. However, once the experiment has been performed and we know which event resulted, we do know the value the random variable takes.

Consider, for example, experiment 2 of section 3.2, that of rolling a six-sided die once. For this experiment there were six simple events: E_1: '1'; E_2: '2', ... ; E_6: '6', each with probability 1/6 (for a fair die). On this experiment we could define a random variable X by:

at E_1, E_3 and E_5, $X = 0$

at E_2, E_4 and E_6, $X = 1$

(i.e., X takes the value zero for odd numbers showing on the die and the value 1 for even numbers). Given the probability of each simple event we can calculate the probability of X taking its two values. Thus:

$$P[X = 0] = P[E_1] + P[E_3] + P[E_5] = \tfrac{1}{2}$$
$$P[X = 1] = P[E_2] + P[E_4] + P[E_6] = \tfrac{1}{2}$$

where of course $P[X = 1]$ is to be read 'the probability that the random variable X takes the value 1'.

On the same experiment we could define a second random variable Y by:

at E_1, $Y = 1$ at E_2, $Y = 2$ at E_3, $Y = 3$

at E_4, $Y = 4$ at E_5, $Y = 5$ at E_6, $Y = 6$

(i.e., the value of Y is the number showing on the die). Thus $P[Y = 1] = P[Y = 2] = \ldots = P[Y = 6] = 1/6$. Obviously, we could define a whole variety of random variables on the same experiment.

Table 3.1   Probability distribution of two random variables

i	X_i	$P[X_i]$	i	Y_i	$P[Y_i]$
1	0	1/2	1	1	1/6
2	1	1/2	2	2	1/6
Σ		1	3	3	1/6
			4	4	1/6
			5	5	1/6
			6	6	1/6
			Σ		1

When the random variable to which the probability statement refers is unambiguous, instead of writing $P[X = 1]$ we can simply write $P[1]$; and, in general instead of writing $P[X = X_i]$ we can write $P[X_i]$. The values taken by a random variable together with their respective probabilities are known as the *probability distribution* of the random variable. It is often useful to express the probability distribution in tabular form. Table 3.1 gives the probability distribution of the random variables X and Y as defined above on experiment 2. Notice that the sum of the probabilities over all values of the random variable is one in each case. This is a consequence of the fact that the sum of the probabilities over all the simple events is one (and for each simple event the random variable takes one and only one value).

In general, if a random variable X takes the n values $X_1, X_2, \ldots, X_n$, then it must follow that

$$\sum_1^n P[X_i] = P[X_1] + P[X_2] + \ldots + P[X_n] = 1 \qquad (3.25)$$

and of course for each i

$$0 \leqslant P[X_i] \leqslant 1. \tag{3.26}$$

Obviously there are very close connections between frequency distributions and probability distributions, particularly as we have defined probability as being (expected) relative frequency. It follows therefore that we can carry over the ideas of chapter 2 on visual presentation of frequency distributions to the visual presentation of probability distributions. We can thus present a probability distribution in diagrammatical form by means of a probability histogram or, where this may give a distorted picture, by means of a probability density histogram. The probability histogram is directly analogous to the relative frequency histogram and the probability density histogram is directly analogous to the relative frequency density histogram. (We define the probability density more formally shortly, but you should try and work out its meaning yourself now.) The probability histogram of the random variables X and Y whose distributions are given in table 3.1 are pictured in figure 3.4.

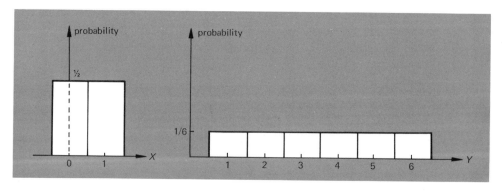

Fig. 3.4   Probability histograms of two random variables

We now consider a slightly more interesting random variable, defined on the experiment (number 4 in section 3.2) of rolling two distinguishable dice. On this experiment we define the random variable X as the sum of the two numbers showing on the dice. Figure 3.5 shows the sample space of this experiment and the values of X associated with each of the 36 simple events. X takes the 11 values 2 to 12. If the two dice are fair, then each of the 36 simple events has probability 1/36. Now, for example, X takes the value 4 for three of the simple events (first die 3, second 1; first 2, second 2; first 1, second 3) and so the probability that X is 4 is 3/36. Similarly $P[X = 6] = 5/36$ and $P[X = 12] = 1/36$. Continuing in this way

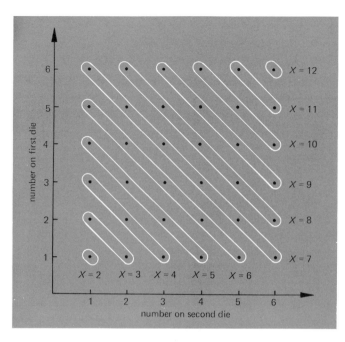

Table 3.2   Probability distribution of the sum of the numbers uppermost on two fair dice

i	X_i	$P[X_i]$
1	2	1/36
2	3	2/36
3	4	3/36
4	5	4/36
5	6	5/36
6	7	6/36
7	8	5/36
8	9	4/36
9	10	3/36
10	11	2/36
11	12	1/36
Σ		1

Fig. 3.5   Sample space for experiment 4

we get the probability distribution of X that is given in tabular form in table 3.2 and in visual form in figure 3.6. We note that the distribution of X is symmetrical and is centred on the value 7. On the same experiment we could also define a random variable Y as the difference between the number showing on the first die and the number showing on the second

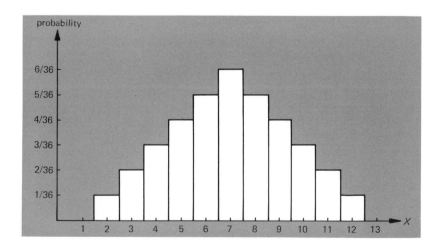

Fig. 3.6   Probability histogram of the sum on two dice

die. Y takes the 11 values -5 to 5. You should verify that the shape of the probability distribution of Y is exactly the same as the shape of the probability distribution of X, except that it is centred on zero instead of on 7.

As a final illustration, consider the experiment of tossing a fair coin three times, and define the random variable X as being the number of heads obtained in the three tosses. As we can either get 0, 1, 2 or 3 heads, these are the values X takes. Now X takes the value zero if we get three tails in a row, thus $P[X = 0] = P[$(tail on first toss) and (tail on second toss) and (tail on third toss)$]$. The three events 'tail on first toss', 'tail on second toss' and 'tail on third toss' are independent since the outcome of any one toss cannot affect the outcome on any other (the coin has no memory). Thus, using the multiplication rule (3.23) for independent events, we have $P[X = 0] = P$ [tail on first toss] P[tail on second toss] P[tail on third toss]. Each of these three probabilities is ½ (since the coin is fair) and so $P[X = 0] = \frac{1}{8}$. Now X takes the value one if any of the following three events occur: HTT or THT or TTH (where, of course, HTT means head on first toss, tail on second, tail on third). Thus

$$P[X = 1] = P[\text{HTT or THT or TTH}].$$

But the three events HTT, THT and TTH are mutually exclusive; thus using the addition rule for mutually exclusive events (3.12):

$$P[X = 1] = P[\text{HTT}] + P[\text{THT}] + P[\text{TTH}]$$

$$= P[\text{H}]P[\text{T}]P[\text{T}] + P[\text{T}]P[\text{H}]P[\text{T}] + P[\text{T}]P[\text{T}]P[\text{H}]$$

Table 3.3 Probability distribution of the number of heads in three tosses of a fair coin

i	X_i	$P[X_i]$
1	0	1/8
2	1	3/8
3	2	3/8
4	3	1/8
Σ		1

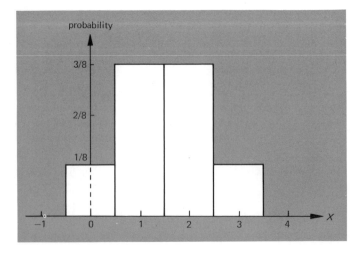

Fig. 3.7   Probability histogram of the number of heads in three tosses of a fair coin

since successive outcomes are independent. Thus

$$P[X = 1] = 3P[\text{H}]P[\text{T}]P[\text{T}] = 3 \cdot \tfrac{1}{2} \cdot \tfrac{1}{2} \cdot \tfrac{1}{2} = \tfrac{3}{8}$$

Continuing in this way we get the probability distribution of X, given in tabular form in table 3.3 and in histogram form in figure 3.7.

You should make sure that you fully understand this example as it is a particular case of a more general distribution which we will be discussing in the next chapter and which forms the basis for several applications of statistics in economics.

We have already noted that several random variables can be defined on the same experiment. Similarly, random variables with exactly the same distribution can be defined independently on different experiments. Consider, for example, defining X as the number of heads when a fair coin is tossed once, and defining Y as taking the value zero when a fair die shows a number three or less when rolled once and the value one when a number four or greater is obtained. A random variable is not unique to a particular experiment.

We continue the analogy between frequency distributions and probability distributions by considering how we might define summary measures for random variables. First we discuss measures of location for random variables, that is indicators of the 'average' value that a random variable takes. The notation and terminology differ slightly, but the ideas are the same.

The direct analogy of the (arithmetic) mean is the *mean* or the *expected value* of the random variable. For a random variable X which takes values X_i with probability $P[X_i]$ $(i = 1, 2, \ldots, n)$, the expected value is denoted by EX and is defined by:

$$EX = \sum_{1}^{n} X_i \, P[X_i]. \tag{3.27}$$

EX is just the weighted average of the X_i where the weights are their respective probabilities. Remembering that these weights sum to one (see (3.25)), and that probability is just (expected) relative frequency, examination of (2.6) and (2.7) in chapter 2 should convince you that the expected value (or mean) of a probability distribution is defined in exactly the same way as the mean of a frequency distribution. The expected values of the two random variables whose distributions are given in tables 3.2 and 3.3 are calculated in table 3.4. These show that the expected value of the sum of the numbers uppermost on two fair dice is 7, and the expected value of the number of heads in three tosses of a fair coin is 1½. These results show the appropriateness of the term 'expected

Table 3.4   Expected values of random variables of tables
3.2 and 3.3

X_i	$P[X_i]$	$X_iP[X_i]$	X_i	$P[X_i]$	$X_iP[X_i]$
2	1/36	2/36	0	1/8	0
3	2/36	6/36	1	3/8	3/8
4	3/36	12/36	2	3/8	6/8
5	4/36	20/36	3	1/8	3/8
6	5/36	30/36			
7	6/36	42/36		$EX = 12/8$	
8	5/36	40/36		$= 3/2$	
9	4/36	36/36			
10	3/36	30/36			
11	2/36	22/36			
12	1/36	12/36			
Σ	1	$EX = 252/36$ $= 7$			

value', as it is the *value* which the mean of the random variable (over several repetitions) is *expected* to approach (as the number of repetitions approaches infinity).

We could also define the median and the mode of a random variable: the median as the value above/below which the variable lies half the time; and the mode as the value with the highest probability (the one that occurs most frequently). Study of figure 3.6 will show that for this random variable, the mean (expected value), median and mode all have the same value (7). This is due to the symmetry of the distribution.

As a measure of dispersion of a random variable we define the variance:

$$\text{var } X = \sum_{1}^{n} (X_i - EX)^2 P[X_i]. \tag{3.28}$$

Thus, the variance is the weighted average of the squared deviations from the mean, the weights being their respective probabilities. You should convince yourself that this is entirely analagous to the definition of variance used in chapter 2. The standard deviation is again defined as the (positive) square root of the variance.

In general suppose $Y \equiv g(X)$ is some function of the random variable X. When $X = X_i$, $Y = Y_i = g(X_i)$ $(i = 1, 2, \ldots, n)$; thus $P[Y = Y_i] = P[X = X_i]$, for all i. Thus the expected value of the random variable Y is (from (3.27))

$$EY = \sum_{1}^{n} Y_iP[Y_i] = \sum_{1}^{n} g(X_i)P[X_i]$$

and so

$$Eg(X) = \sum_{1}^{n} g(X_i)P[X_i]. \tag{3.29}$$

The expected value of any function of X is the weighted average of that function, the weights being their respective probabilities.

Consider the particular case: $g(X) = (X - EX)^2$. Then, from (3.29)

$$E(X - EX)^2 = \sum_{1}^{n} (X_i - EX)^2 P[X_i].$$

Comparing this with (3.28), we see that the variance can be expressed as:

$$\text{var } X = E(X - EX)^2. \tag{3.30}$$

The variance is the expected value of the squared deviations from the expected value.

Table 3.5   Calculation of variance

X_i	$P[X_i]$ (x 36)	$X_i P[X_i]$ (x 36)	$(X_i - EX)$	$(X_i - EX)^2$	$(X_i - EX)^2 P[X_i]$ (x 36)
2	1	2	−5	25	25
3	2	6	−4	16	32
4	3	12	−3	9	27
5	4	20	−2	4	16
6	5	30	−1	1	5
7	6	42	0	0	0
8	5	40	1	1	5
9	4	36	2	4	16
10	3	30	3	9	27
11	2	22	4	16	32
12	1	12	5	25	25
Σ	36	252			210

$$EX = \frac{252}{36} = 7 \qquad \text{Var } X = \frac{210}{36} = \frac{35}{6}$$

$$\text{s.d. } (X) = 2.4152$$

In table 3.5 we calculate the variance of the sum of the numbers uppermost on two fair dice. Notice that the probability of X lying within two standard deviations of its mean (i.e. between $7 \pm 2 \times 2.4152$; i.e. between 2.1696 and 11.8304) is 34/36 (= 0.944); and that the probability of X lying within one standard deviation of its mean (i.e. between 4.5848 and 9.4152) is 24/36 (= 0.666).

So far, our illustrations and discussion have been confined to discrete random variables. However we are also interested in random variables that are continuous. Consider an experiment that involves spinning a pointer, as illustrated in figure 3.8, where the circumference of the circle is calibrated continuously from 0 to 1 as shown. On this experiment we define the random variable X as the number on the circumference to

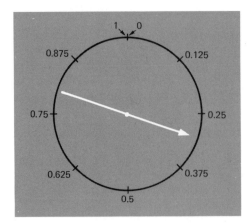

Fig. 3.8   A continuous random variable

which the pointer points when it comes to rest. If the pointer is not biased in any way, so that it is equally likely to come to rest at any point, we can deduce, for example, the following probabilities about X:

$P[0 \leqslant X \leqslant 0.5] = \tfrac{1}{2}$

$P[0.25 \leqslant X \leqslant 0.5] = \tfrac{1}{4}$

$P[0.625 \leqslant X \leqslant 0.875] = \tfrac{1}{4}$ etc.

How might we represent the probability distribution of X in a visual form? Can we plot the probability of X taking certain values against X? Consider the following probabilities:

$P[0.5 \leqslant X \leqslant 1] = 0.5$ $P[0.2 \leqslant X \leqslant 0.7] = 0.5$

$P[0.5 \leqslant X \leqslant 0.75] = 0.25$ $P[0.2 \leqslant X \leqslant 0.45] = 0.25$

$P[0.5 \leqslant X \leqslant 0.625] = 0.125$ $P[0.2 \leqslant X \leqslant 0.325] = 0.125$

$P[0.5 \leqslant X \leqslant 0.6] = 0.1$ $P[0.2 \leqslant X \leqslant 0.3] = 0.1$

We can see that, in general

$P[X_0 \leqslant X \leqslant X_0 + h] = h$ (where $0 \leqslant X_0 \leqslant X_0 + h \leqslant 1$)

Consider what happens to both sides of this result as h approaches zero. The left-hand side approaches $P[X_0 \leqslant X \leqslant X_0]$ i.e. $P[X = X_0]$; the right-hand side approaches zero. We thus get the result

$P[X = X_0] = 0$

that is, the probability of X *exactly equalling* a particular number is zero. This result may appear surprising initially, but not when we remember that between 0 and 1 there are an infinite number of possible values that

X may take. Plotting probability against X is not going to be a very useful visual presentation! However, instead of drawing the probability histogram, we can draw the probability density histogram. The probability density of a random variable over the range $X_0 \leqslant X \leqslant X_0 + h$ is defined as the probability of X lying in this range divided by the width of the range (which is of course h):

$$\frac{P[X_0 \leqslant X \leqslant X_0 + h]}{h}$$

(compare this with the definition of relative frequency density). The probability density of a random variable at the value X_0 is defined as the limit of the probability density over the range $X_0 \leqslant X \leqslant X_0 + h$ as the width of the range (h) approaches zero:

$$\underset{h \to 0}{\mathrm{Lt}} \frac{P[X_0 \leqslant X \leqslant X_0 + h]}{h}.$$

For the random variable defined as in figure 3.8, we have already shown that $P[X_0 \leqslant X \leqslant X_0 + h] = h$, thus the probability density over the range $X_0 \leqslant X \leqslant X_0 + h$ is

$$\frac{P[X_0 \leqslant X \leqslant X_0 + h]}{h} = \frac{h}{h} = 1.$$

Since this density does not depend on X_0, it follows that the probability density of X at any value (between 0 and 1) is 1. Figure 3.9 illustrates this result.

We denote the probability density by $f(X)$. Thus

$$f(X_0) = \underset{h \to 0}{\mathrm{Lt}} \frac{P[X_0 \leqslant X \leqslant X_0 + h]}{h}. \tag{3.31}$$

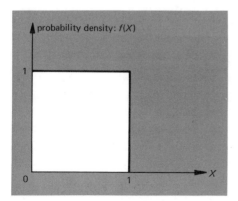

probability density: $f(X)$

Fig. 3.9   Probability density of variable defined in figure 3.8

For the random variable of figure 3.8, we have

$$f(X) = \begin{cases} 1 & 0 \leqslant X \leqslant 1 \\ 0 & \text{otherwise.} \end{cases}$$

As probability and (expected) relative frequency are synonymous, and as we have already seen that the relative frequency density histogram has some interesting area properties, it would be reasonable to expect that the probability density histogram also has some interesting area properties. Indeed it has. Consider figure 3.9; the total area under the histogram is unity, and the area between any two values of X is the probability of X lying between those two values. (For example, the area under the histogram between 0.5 and 0.75 is ¼ (¼ of the total area which is 1); $P[0.5 \leqslant X \leqslant 0.75] = $ ¼.) These two area properties hold true for all probability density histograms or curves. Conversely, if a curve does not satisfy these properties it cannot be a probability density function.

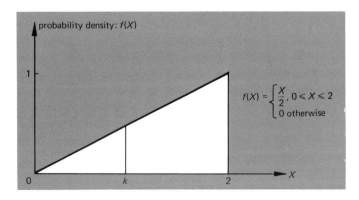

Fig. 3.10   'Triangular' probability distribution

Consider the probability density function as drawn in figure 3.10. What kind of experiment might give rise to such a probability distribution? First we check that figure 3.10 can be a probability density function by checking that it satisfies the first area property — that the total area under it is unity (area of a triangle is half base times height). Suppose we use the pointer experiment of figure 3.8, but calibrate the circumference differently so that the experiment generates the probability density function of figure 3.10. Obviously the circumference must be numbered from 0 right round to 2 since this is the range of X. From figure 3.10, and using the second of the area properties, we see that for any $k(0 \leqslant k \leqslant 2)$, the probability of X lying between 0 and k is the area under the density function between 0 and k. Now $f(X) = X/2$ (from figure 3.10), thus the density at k is $k/2$. Thus the area of the triangle to the left of k is

$(\frac{1}{2}k)k/2 = k^2/4$ and so

$$P[0 \leqslant X \leqslant k] = \frac{k^2}{4}. \tag{3.32}$$

Putting $k = 1$ in (3.32) we have $P[0 \leqslant X \leqslant 1] = \frac{1}{4}$. It also follows from (3.32) that $P[0 \leqslant X \leqslant \sqrt{2}] = \frac{1}{2}$; $P[0 \leqslant X \leqslant \sqrt{3}] = \frac{3}{4}$ etc.

This information enables us to start the calibration of the circumference; the above values are marked in figure 3.11. You should try to calibrate it further yourself.

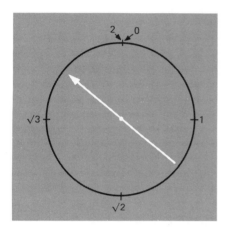

Fig. 3.11   'Pointer' experiment for distribution of figure 3.10

If we know the probability density function of a random variable, then we have all the information we need to make probability statements about it, since probabilities are given by the area under the function. Consider for example the distribution as shown in figure 3.12. Although the prob-

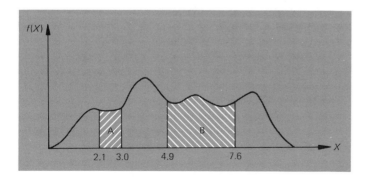

Fig. 3.12   A general probability distribution

ability density is somewhat weird (and we may well have difficulty in finding an experiment that gives rise to such a function), we can calculate any desired probability by measuring the appropriate area. Thus, for example, $P[2.1 \leqslant X \leqslant 3.0]$ = area A and $P[4.9 \leqslant X \leqslant 7.6]$ = area B.

Reverting to the discrete random variables whose distributions are drawn in figures 3.4, 3.6 and 3.7, we note that in fact the same area properties apply to them: the total area under the histogram is one, and the area in any block is the probability of the random variable taking the value on which the block is centred. This happens because, in each case, the width of the blocks is one, and so the probability densities are the same as the corresponding probabilities. Thus, figures 3.4, 3.6 and 3.7 are also probability density histograms. Suppose, though, that in figure 3.7 the values that X took were 0, 2, 4 and 6 (instead of 0, 1, 2 and 3), then the width of each block would be 2. The probability densities would thus be found by dividing all probabilities by 2. We would then be in a position to represent diagrammatically the probability distribution in a way that possesses the useful area properties. From now on we will represent all probability distributions in probability density form, so that we may calculate probabilities by calculating the corresponding area under the density function.

For continuous random variables the mean and variance are defined in exactly the same way as for discrete variables; that is, the mean is the weighted average of the random variable, the weights being the probabilities, and the variance is the weighted average of the squared deviations of the variable from its mean, the weights, once again, being the probabilities. However, for continuous random variables, a slight difficulty arises, since, as we have already seen, a continuous random variable takes an infinite number of values, and the probability of it *exactly equalling* a particular value is zero. Thus, if we tried to apply the definition of the mean as given by (3.27), it would appear that the mean was the sum of an infinite number of zeros, which is, in general, indeterminate!* It would also appear difficult to show that condition (3.25) was satisfied, since

---

*This remark may well be difficult to understand. However, the following simple example should aid comprehension. Consider the sum (for $b > 0$):

$$\sum_{1}^{n} \frac{a}{n^b} = n\frac{a}{n^b} = \frac{a}{n^{b-1}}.$$

Now let n approach infinity; the left hand side of the above equation approaches '$\sum_{1}^{\infty} 0$', that is, 'the sum of an infinite number of zeros', while the right hand side approaches either ∞, a, or 0 depending on whether b is less than 1, equal to 1, or greater than 1. Thus 'the sum of an infinite number of zeros' may take any value.

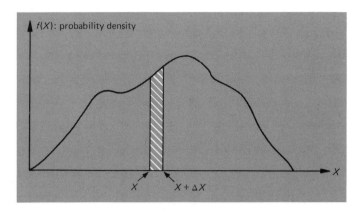

Fig. 3.13   Calculation of mean of continuous random variable

the left hand side is also an infinite sum of zeros. However the integral calculus comes to our rescue. (Readers unfamiliar with the integral calculus are reassured that it is not used in the remainder of the book, and is referred to in these closing paragraphs of this section only for purposes of pedagogical completeness.) Consider figure 3.13, which represents the density function of a continuous random variable. The shaded area is the probability that the random variable lies between X and $X + \Delta X$. The shaded area is approximately rectangular, of height $f(X)$ and width ΔX. Thus approximately the probability that the random variable lies between X and $X + \Delta X$ is $f(X)\Delta X$, and the approximation improves as ΔX gets smaller. Now we know that the total area under the curve is unity, so the sum $\Sigma f(X)\Delta X$ is approximately unity, where the sum is over all possible such shaded areas, the approximation improving as ΔX gets smaller. However as ΔX gets smaller, the number of terms in the above sum gets bigger. Now, from the integral calculus

$$\underset{\Delta X \to 0}{\text{Lt}} \ \Sigma f(X)\Delta X \equiv \int_R f(X)dX$$

where the integration is over R, the range of X; thus

$$\int_R f(X)dX = 1.$$

If you are unfamiliar with the integral calculus, there is no need to worry, as all this result says is that the area under the density function is one, and this we knew already. The analogous statement for discrete random variables is (3.25), that is

$$\sum_i P[X_i] = 1.$$

By reasoning as above, it can be shown that the mean of a continuous random variable is

$$EX = \int_R Xf(X)dX.$$

For a discrete random variable the mean is (see (3.27))

$$EX = \sum_i X_i P[X_i]$$

The variances for continuous and discrete random variables are respectively

$$\text{var } X = \int_R (X - EX)^2 f(X)dX$$

and

$$\text{var } X = \sum_i (X_i - EX)^2 P[X_i].$$

Thus the only differences are that wherever a Σ (summation sign) appears for discrete random variables it is replaced by $\int$ (integral sign) for continuous random variables, and $P[X]$ is replaced by $f(X)\ dX$.

If you are familiar with the integral calculus you may enjoy applying this knowledge to the solution of the exercises at the end of this chapter and to a few unproved results in other chapters. However the exercises at the end of this chapter, and the material covered in the remainder of the book, can be tackled without knowledge of the integral calculus; all that is required is the ability to calculate certain simple areas.

In the next chapter we consider particular random variables whose distributions are commonly found in economic applications. First, we briefly consider the generalisation of this section to jointly distributed random variables, and generalise the concept of independence between events to independence between random variables.

3.4 Joint Probability Distributions

By now you should appreciate the strong connections between the material in chapter 2 on frequency distributions and the material in this chapter on probability distributions. This section is similarly connected with section 2.4 on joint frequency distributions. We thus proceed fairly quickly in this section to avoid repetition of concepts and discussion common to both sections.

Suppose that on a particular experiment we define two random variables X and Y, where X takes the values $X_i (i = 1, 2, \ldots, n)$ and Y the values $Y_j (j = 1, 2, \ldots, m)$. By $P[X = X_i, Y = Y_j]$, or more simply $P[X_i, Y_j]$, we denote the probability that for a particular repetition of the experiment X takes the value X_i and Y the value Y_j.

Consider, for example, the experiment of drawing one card at random from a (complete) well-shuffled pack of cards. Define the random variables X and Y by:

$$X = \begin{cases} 1 \text{ if ace drawn} \\ 2 \text{ if king drawn} \\ 3 \text{ if queen or jack drawn} \\ 4 \text{ otherwise} \end{cases} \qquad Y = \begin{cases} 1 \text{ if card is red} \\ 2 \text{ if card is a club} \\ 3 \text{ if card is a spade} \end{cases}$$

Thus, $X = 1$ and $Y = 2$ only if we draw the ace of clubs, and so

$$P[X = 1, Y = 2] \equiv P[1, 2] = 1/52.$$

Similarly

$$P[X = 2, Y = 1] \equiv P[2, 1] = 2/52 \text{ (two red kings)}.$$

Continuing in this way we find the *joint probability distribution* of X and Y, which we present in tabular form in table 3.6. The last column is just the probability distribution of X (by itself), and the last row is just the probability distribution of Y (by itself).

Table 3.6   Joint probability distribution

The entries in the table show the probabilities that $X = X_i$ and $Y = Y_j$ ($i = 1, 2, 3, 4: j = 1, 2, 3$).

X_i \ Y_j	1	2	3	Totals
1	2/52	1/52	1/52	4/52
2	2/52	1/52	1/52	4/52
3	4/52	2/52	2/52	8/52
4	18/52	9/52	9/52	36/52
Totals	26/52	13/52	13/52	52/52

This last result must hold generally; that is

$$P[X_i] = \sum_{j=1}^{m} P[X_i, Y_j] \qquad \text{(for all } i\text{)} \tag{3.33}$$

$$P[Y_j] = \sum_{i=1}^{n} P[X_i, Y_j] \qquad \text{(for all } j\text{)} \tag{3.34}$$

and since

$$\sum_{1}^{n} P[X_i] = \sum_{1}^{m} P[Y_j] = 1$$

we have

$$\sum_{i=1}^{n} \sum_{j=1}^{m} P[X_i, Y_j] = 1. \tag{3.35}$$

In section 3.2 we showed that the conditional probability of event A_1 given event A_2 is (see (3.22)):

$$P[A_1 | A_2] = \frac{P[A_1 \text{ and } A_2]}{P[A_2]}.$$

Similarly the conditional probability that X takes the value X_i given that Y takes the value Y_j is:

$$P[X_i | Y_j] = \frac{P[X_i, Y_j]}{P[Y_j]}. \tag{3.36}$$

In section 3.2 we defined two events A_1, A_2 as being independent if $P[A_1 | A_2] = P[A_1]$ and $P[A_2 | A_1] = P[A_2]$.

Similarly we define two random variables X, Y as being independent if:

$$P[X_i | Y_j] = P[X_i] \text{ and } P[Y_j | X_i] = P[Y_j] \qquad \text{(for all } i, j) \tag{3.37}$$

(note that either condition implies the other). Combining (3.36) and (3.37), we see that X and Y are independent random variables if:

$$P[X_i, Y_j] = P[X_i] P[Y_j] \qquad \text{(for all } i, j) \tag{3.38}$$

(compare this with (3.23) as applicable to independent events).

Finally, we define the *covariance* between two random variables (a measure of their association) by

$$\text{cov}(X, Y) = \sum_{i=1}^{n} \sum_{j=1}^{m} (X_i - EX)(Y_j - EY)P[X_i, Y_j] \tag{3.39}$$

If X and Y are independent we can substitute from (3.38) into (3.39) giving

$$\text{cov}(X, Y) = \sum_{i=1}^{n} \sum_{j=1}^{m} (X_i - EX)(Y_j - EY)P[X_i]P[Y_j]$$

$$\text{cov}(X, Y) = \left\{ \sum_{i=1}^{n} (X_i - EX)P[X_i] \right\} \left\{ \sum_{j=1}^{m} (Y_j - EY)P[Y_j] \right\}. \tag{3.40}$$

Consider the first bracketed term on the right-hand side of (3.40):

$$\sum_{1}^{n} (X_i - EX)P[X_i] = \sum_{1}^{n} X_i P[X_i] - EX \sum_{1}^{n} P[X_i]$$

$$= EX - EX = 0, \quad \text{using (3.27) and (3.25)}.$$

Similarly, the second bracketed term is zero. Thus, if two random variables are independent, their covariance is zero. Note, however, that the converse is not true (cf. section 2.4).

To illustrate these results we consider the jointly distributed random variables defined earlier in this section, and for which table 3.6 gives the joint distribution. Using (3.36) and the entries in table 3.6 we can obtain the following results:

$$P[X = 1 \mid Y = 2] = \frac{P[X = 1, Y = 2]}{P[Y = 2]} = \frac{1/52}{13/52} = \frac{1}{13} = P[X = 1]$$

and

$$P[Y = 1 \mid X = 4] = \frac{P[X = 4, Y = 1]}{P[X = 4]} = \frac{18/52}{36/52} = \frac{1}{2} = P[Y = 1].$$

By continuing in this way we can verify that all conditional probabilities equal the corresponding unconditional probabilities and so X and Y are independent. We have already proved the general result that independence implies zero covariance; the calculation of the covariance between X and Y, as shown in table 3.7, verifies this general result.

Table 3.7   Calculation of covariance

X_i	$P[X_i]$ (× 13)	$X_i P[X_i]$ (× 13)	Y_j	$P[Y_j]$ (× 4)	$Y_j P[Y_j]$ (× 4)
1	1	1	1	2	2
2	1	2	2	1	2
3	2	6	3	1	3
4	9	36			
Σ	13	45	Σ	4	7
$EX = \dfrac{45}{13}$			$EY = \dfrac{7}{4}$		

X_i	Y_i	$P[X_i, Y_j]$ (× 52)	$X_i - EX$ (× 13)	$Y_j - EY$ (× 4)	$(X_i - EX)(Y_j - EY)P[X_i, Y_j]$ (× 2704)
1	1	2	−32	−3	192
1	2	1	−32	1	−32
1	3	1	−32	5	−160
2	1	2	−19	−3	114
2	2	1	−19	1	−19
2	3	1	−19	5	−95
3	1	4	−6	−3	72
3	2	2	−6	1	−12
3	3	2	−6	5	−60
4	1	18	7	−3	−378
4	2	9	7	1	63
4	3	9	7	5	315
Σ		52			0
				cov$(X, Y) =$	0

If the pack from which we were drawing cards had one card missing (say the king of hearts) then X and Y would not be independent. The demonstration of this is left as an exercise for the reader.

The correlation coefficient r between two random variables is defined, as before, by:

$$r = \frac{\text{cov}(X, Y)}{\sqrt{(\text{var } X \cdot \text{var } Y)}}. \tag{3.41}$$

Finally we use the ideas developed in this section to derive some important results about functions of jointly distributed random variables. Suppose X and Y are jointly distributed random variables with probabilities $P[X_i, Y_j]$ $(i = 1, 2, \ldots, n; j = 1, 2, \ldots, m)$.

Define a random variable Z in terms of X and Y by $Z = X + Y$; that is,

$$Z_{ij} = X_i + Y_j, \quad (i = 1, 2, \ldots, n; \ j = 1, 2, \ldots, m). \tag{3.42}$$

Thus, when $X = X_i$ and $Y = Y_j$ then $Z = Z_{ij}$, and so

$$P[Z = Z_{ij}] \equiv P[Z_{ij}] = P[X_i, Y_j] \quad \text{(all } i, j\text{)}. \tag{3.43}$$

The expected value of Z is thus given by:

$$EZ = \sum_{i=1}^{n} \sum_{j=1}^{m} Z_{ij} P[Z_{ij}]$$

$$= \sum_{i=1}^{n} \sum_{j=1}^{m} (X_i + Y_j) P[X_i, Y_j] \quad \text{from (3.42) and (3.43)}$$

$$= \sum_{i=1}^{n} X_i \sum_{j=1}^{m} P[X_i, Y_j] + \sum_{j=1}^{m} Y_j \sum_{i=1}^{n} P[X_i, Y_j]$$

$$= \sum_{1}^{n} X_i P[X_i] + \sum_{1}^{m} Y_j P[Y_j] \quad \text{from (3.33) and (3.34)}.$$

Thus,

$$EZ = EX + EY$$

that is,

$$E(X + Y) = EX + EY. \tag{3.44}$$

The expected value of the sum of two random variables is the sum of their expected values.

Consider also the variance of Z:

$$\text{var } Z = \sum_{i=1}^{n} \sum_{j=1}^{m} (Z_{ij} - EZ)^2 P[Z_{ij}]$$

$$= \sum_{i=1}^{n} \sum_{j=1}^{m} \{(X_i - EX) + (Y_j - EY)\}^2 P[X_i, Y_j]$$

$$\text{(using (3.42), (3.43), (3.44))}$$

$$= \sum_{i=1}^{n} \sum_{j=1}^{m} \{(X_i - EX)^2 + 2(X_i - EX)(Y_j - EY) + (Y_j - EY)^2\} P[X_i, Y_j]$$

$$= \sum_{i=1}^{n} (X_i - EX)^2 \sum_{j=1}^{m} P[X_i, Y_j]$$

$$+ 2 \sum_{i=1}^{n} \sum_{j=1}^{m} (X_i - EX)(Y_j - EY) P[X_i, Y_j]$$

$$+ \sum_{j=1}^{m} (Y_j - EY)^2 \sum_{i=1}^{n} P[X_i, Y_j]$$

Thus,

$$\text{var } Z = \text{var } X + 2 \text{ cov } (X, Y) + \text{var } Y \tag{3.45}$$

(using (3.28), (3.33), (3.34) and (3.39)).

Notice that, for independent random variables, $\text{cov}(X, Y)$ is zero and so (3.45) reduces to

$$\text{var } Z = \text{var } X + \text{var } Y$$

that is

$$\text{var } (X + Y) = \text{var } X + \text{var } Y. \tag{3.46}$$

The variance of the sum of two independent random variables is the sum of their variances.

It can also be shown (see exercise 3.24) that, for any random variables, $E(X - Y) = EX - EY$ and $\text{var } (X - Y) = \text{var } X - 2 \text{ cov } (X, Y) + \text{var } Y$, which reduces to $\text{var } (X - Y) = \text{var } X + \text{var } Y$ when X and Y are independent.

Consider now the variable Z defined by

$$Z = (X - EX)(Y - EY)$$

that is,

$$Z_{ij} = (X_i - EX)(Y_j - EY) \quad (i = 1, 2, \ldots, n; \; j = 1, 2, \ldots, m) \tag{3.47}$$

By exactly the same reasoning as above,

$$P[Z_{ij}] = P[X_i, Y_j] \quad \text{(all } i, j) \tag{3.48}$$

and so

$$EZ = \sum_{i=1}^{n} \sum_{j=1}^{m} Z_{ij} P[Z_{ij}]$$

$$= \sum_{i=1}^{n} \sum_{j=1}^{m} (X_i - EX)(Y_j - EY) P[X_i, Y_j]$$

(using (3.47) and (3.48)). Thus

$$EZ = \text{cov } (X, Y)$$

by definition (3.39); that is

$$E(X - EX)(Y - EY) = \text{cov}(X, Y). \tag{3.49}$$

Expanding the term on the left-hand side of (3.49), we have

$$\begin{aligned}
E(X - EX)(Y - EY) &= E\{XY - (EX)Y - X(EY) + (EX)(EY)\} \\
&= E(XY) - (EX)(EY) - (EX)(EY) + (EX)(EY) \\
&= E(XY) - (EX)(EY)
\end{aligned}$$

and so

$$E(XY) = (EX)(EY) + \text{cov}(X, Y). \tag{3.50}$$

If X and Y are independent this reduces to

$$E(XY) = (EX)(EY). \tag{3.51}$$

The expected value of the product of two independent random variables is the product of their expected values.

For our illustrative example, X and Y are independent, so results (3.46) and (3.51) should hold. Table 3.8 verifies that (3.51) holds. The verification of (3.46) is left as an exercise for the reader.

Table 3.8   Verification of $E(XY) = (EX)(EY)$ for X, Y independent

X_i	Y_j	$P[X_i, Y_j]$ (× 52)	$Z_{ij} = X_i Y_j$	$P[Z_{ij}]$ (× 52)	$Z_{ij}P[Z_{ij}]$ (× 52)
1	1	2	1	2	2
1	2	1	2	1	2
1	3	1	3	1	3
2	1	2	2	2	4
2	2	1	4	1	4
2	3	1	6	1	6
3	1	4	3	4	12
3	2	2	6	2	12
3	3	2	9	2	18
4	1	18	4	18	72
4	2	9	8	9	72
4	3	9	12	9	108
Σ		52		52	315

$$EZ = \frac{315}{52} = \left(\frac{45}{13}\right)\left(\frac{7}{4}\right) = (EX)(EY)$$

We could have proved (3.45) by use of (3.49) as follows:

$$\begin{aligned}
\text{var}(X + Y) &= E\{(X + Y) - E(X + Y)\}^2 \quad \text{from (3.30)} \\
&= E\{(X - EX) + (Y - EY)\}^2 \\
&= E(X - EX)^2 + 2E(X - EX)(Y - EY) + E(Y - EY)^2
\end{aligned}$$

that is

$$\text{var}\,(X + Y) = \text{var}\,X + 2\,\text{cov}\,(X, Y) + \text{var}\,Y.$$

The results of this section, although derived in terms of discrete random variables, are also true for continuous random variables.

3.5 Transformations

The results proved in this section are entirely analogous to those in section 2.8. Suppose the random variable X takes the values X_i with probability $P[X_i]$ ($i = 1, 2, \ldots, n$). Define the variable Y in terms of X by

$$Y_i = a + bX_i \tag{3.52}$$

where a and b are constants. Then $Y = Y_i$ when $X = X_i$ (for all i); thus $P[Y = Y_i] = P[X = X_i]$, all i, or more succinctly

$$P[Y_i] = P[X_i]. \tag{3.53}$$

Thus

$$EY = \sum_1^n Y_i P[Y_i] = \sum_1^n (a + bX_i)P[X_i], \text{ using (3.52), (3.53)}$$

$$= a \sum_1^n P[X_i] + b \sum_1^n X_i P[X_i].$$

But

$$\sum_1^n P[X_i] = 1$$

(from (3.25)), since the probabilities must sum to one, and

$$\sum_1^n X_i P[X_i] = EX$$

from (3.27). Thus

$$EY = a + bEX. \tag{3.54}$$

Further, from (3.52) and (3.54):

$$Y_i - EY = b(X_i - EX)$$

Thus

$$\text{var}\,Y = \sum_1^n (Y_i - EY)^2 P[Y_i] = \sum_1^n b^2 (X_i - EX)^2 P[X_i]$$

and so

$$\text{var}\,Y = b^2\,\text{var}\,X. \tag{3.55}$$

Results (3.54) and (3.55) enable us, as before, to standardise any random variable: define Z in terms of X by

$$Z = \frac{X - EX}{\sqrt{\text{var } X}} \tag{3.56}$$

i.e.

$$Z = a + bX \quad \text{where} \quad a = \frac{-EX}{\sqrt{\text{var } X}} ; \ b = \frac{1}{\sqrt{\text{var } X}}$$

then

$$EZ = 0$$

and

$$\left.\begin{array}{c}\\ \\ \\ \text{var } Z = 1. \end{array}\right\} \tag{3.57}$$

As a very simple illustration of this standardisation, consider the problem of standardising the random variable X, which is defined as the number of heads obtained when we toss a fair coin once. Obviously X takes the values 0 and 1 each with probability ½. The mean of X is ½ and the variance, as table 3.9 shows, is ¼. Table 3.9 also shows the values of the standardised variable Z corresponding to each X value. You should confirm that $EZ = 0$ and var $Z = 1$.

Table 3.9   Standardisation of a random variable

X_i	$P[X_i]$ (× 2)	$X_i P[X_i]$ (× 2)	$(X_i - EX)$ (× 2)	$(X_i - EX)^2 P[X_i]$ (× 8)	$Z = \dfrac{X_i - EX}{\sqrt{\text{var } X}}$
0	1	0	−1	1	−1
1	1	1	1	1	1
Σ	2	1		2	

$$EX = \frac{1}{2} \qquad \text{var } X = \frac{2}{8} = \frac{1}{4}$$

3.6 Summary

In this chapter we introduced the concepts of probability theory with reference to simple experiments. We saw that there is no entirely satisfactory definition of probability, but asserted that the definition of probability as (expected) relative frequency is the most suitable for our particular needs. (Hopefully the validity of this assertion will become apparent in subsequent chapters.) As a consequence of using this definition of probability we were able to emphasise the strong connections between frequency distributions and probability distributions. We were thus able to apply the concepts of chapter 2 to probability distributions of random variables, in particular the methods of visual presentation and the idea of summary measures. We showed that probability density histograms or curves have important area properties. The concepts of joint frequency distributions were carried over to joint probability distributions. Finally, we showed that any random variable can be transformed into a standardised random variable, that is a variable with zero mean and unit variance.

3.7 Exercises

3.1. Discuss alternative methods of defining probability, giving examples to illustrate the advantages and disadvantages of each method.

3.2. Some statisticians avoid the problem of defining probability by using an axiomatic approach to probability theory. A typical set of axioms used is the following:

(a) Every event has an associated probability: $P[A] \geqslant 0$ for all A.
(b) If A_1 and A_2 are mutually exclusive events then $P[A_1 \text{ or } A_2] = P[A_1] + P[A_2]$.
(c) If an event is certain to occur its probability is unity.

Given these axioms can you *deduce* the following results:

(d) $P[A_1 \text{ or } A_2] = P[A_1] + P[A_2] - P[A_1 \text{ and } A_2]$ for any A_1 , A_2 ;
(e) $P[A_1 \mid A_2] = P[A_1 \text{ and } A_2]/P[A_2]$?

3.3. Show that $P[A_1 \mid A_2] = P[A_1]$ if and only if $P[A_2 \mid A_1] = P[A_2]$.

3.4. Consider the experiment of rolling two distinguishable fair dice. Find the probabilities of the following events:

(a) the total of the two dice is 12;
(b) at least one of the two dice shows a 1;
(c) the total of the two dice is less than 6;
(d) one of the dice shows a 6 and the total is greater than 9;
(e) one of the dice shows a 5 and the total is less than 6.

Find the following conditional probabilities:

(f) the total is 6 given that at least one die shows a 2;
(g) at least one die shows a 2 given that the total is 6.

Recalculate probabilities (a) to (g) for the case where one of the dice has its '6' replaced by a second '1'. Are the events 'total is 6' and 'both dice are greater than 3' mutually exclusive? Are they independent? Are

the events 'first die is less than 3' and 'second die is greater than 3'
mutually exclusive? Are they independent?

3.5. A box contains three red and two black balls. Two balls are to be
drawn from the box, without replacing the first ball before the
second is drawn. Construct the sample space on the assumption that balls
of the same colour are distinguishable, and assign probabilities to all
points in the sample space. Construct the sample space on the assumption
that balls of the same colour are not distinguishable, and assign
probabilities to all points in the sample space. How would the sample
spaces drawn above differ if the first ball was replaced before the second
drawing?

3.6. Show that if two events are mutually exclusive they cannot be
independent.

3.7. If two events are independent can you say whether they are mutually
exclusive or not?

3.8. Suppose there are equal numbers of male and female students in a
university, and that 1/5 of the male students and 1/20 of the female
students are economists. Calculate the following probabilities:

(a) that a student selected at random will be a male economist;
(b) that a student selected at random will be an economist;
(c) that an economist selected at random will be male.

3.9. In the single-handed transatlantic race the probability that competitor A will beat the existing record time is ½, that B will beat the
record is ⅓ and that C will beat the record is ¼. Calculate the following
probabilities:

(a) that all three will beat the record;
(b) that none will beat the record;
(c) that only one will beat the record.

Discuss critically and at length the validity of the assumptions you have
used to derive these probabilities. Discuss also the concept of probability
used in this question, and compare it with that used in question 3.8.

3.10. A fair coin is tossed; if it comes up heads, a die is rolled and you are
paid the amount showing in dollars; if it comes up tails, two dice are
rolled, and you are paid in dollars the sum of the numbers showing. What
is the probability that you will be paid at most $4? Let X be the number
of dollars paid to you for any one outcome. Derive the probability

distribution of X, and draw the probability density histogram. Calculate the expected value of X. What interpretation do you attach to this expected value?

3.11. A box contains four cards: the two, three, four and five of hearts. An experiment is performed consisting of drawing two cards from the box, without replacing the first card before the second is drawn. Let X be the random variable defined as the sum of the numbers on the two cards drawn. Derive, and depict, the probability distribution of X, and find its mean (expected value) and variance. Would you expect the mean and variance to have different values if the first card had been replaced before the second was drawn?

3.12. Suppose a sample of ten is taken from a day's output of a machine that normally produces 5 per cent defective parts. If the whole day's production is thrown away whenever the sample of ten contains two or more defectives, what is the probability that the whole day's production will be thrown away even though the machine is functioning normally?

Exercises 3.13 and 3.14 do not have objectively 'correct' answers; they are both concerned with what *you* would do in particular circumstances. The methods of this chapter should be useful in arriving at your decisions.

3.13. Suppose you are offered, for a positive price, the chance to play the following game once: a fair coin is tossed repeatedly until a tail appears, and the total number of tosses made (including the one first showing a tail) will be called n. Then you will get a prize of 2^n dollars. What is the highest price you would be prepared to pay to play this game?

3.14. (a) Suppose that you had your choice of either of the following two gambles:
In gamble 1 the reward would be \$5,000 with certainty.
In gamble 2 the possible rewards would be either \$25,000 with probability 0.10, \$5,000 with probability 0.89, or \$0 with probability 0.01.

Which gamble would you prefer? You cannot lose under any circumstances, and it should be assumed that all rewards are tax-free.

(b) Now suppose that instead of the above choice, you had your choice of either of the following two gambles:
In gamble 3 the possible rewards would be either \$5,000 with probability 0.11, or \$0 with probability 0.89.

In gamble 4 the possible rewards would be either \$25,000 with probability 0.10, or \$0 with probability 0.90.
Which of these gambles would you prefer?

3.15. Consider the distribution given by the probability density function

$$f(X) = \frac{X}{8} \qquad 0 \leqslant X \leqslant 4.$$

Calculate the following probabilities

$$P[0 \leqslant X \leqslant 1] \, ; P[1 \leqslant X \leqslant 2] \, ; P[2 \leqslant X \leqslant 3] \, ; P[3 \leqslant X \leqslant 4] \, ; P[X = 2] \, .$$

Find a value k such that the probability that X is greater than k is three times the probability that X is less than k.

3.16. Show that the following function can represent a probability density function only if $a = b$:

$$f(X) = \frac{2(b - X)}{ab} \qquad 0 \leqslant X \leqslant b.$$

Hence find $P[0 \leqslant X \leqslant b/2]$.

Calibrate the circumference of a 'pointer' experiment that would give rise to such a distribution.

3.17. Find the value for k in the 'rectangular' distribution:

$$f(X) = k \qquad 0 \leqslant X \leqslant 4.$$

What is the expected value of X?

3.18. Show that, for any random variable X, the expression $E(X - k)^2$ is minimised for $k = EX$.

3.19. Under what circumstances does a random variable X satisfy $E(X^2) = (EX)^2$? Illustrate your results by means of a particular example. Use this example to investigate whether $E(1/X)$ equals $1/EX$.

3.20. Prove results (3.33) and (3.34).

3.21. Suppose two random variables X and Y are related by $X = a - Y$. Show that var $X =$ var Y and that cov $(X, Y) = -$var Y. Substituting these results in (3.45) show that var $(X + Y) = 0$. Interpret this result.

3.22. Using the joint probability distribution as given in table 3.6, verify that (3.46) holds for independent random variables.

3.23. Suppose the experiment of section 3.4 is modified so that the pack of cards has the king of hearts missing but is the same in all other respects. Show that X and Y are no longer independent. Calculate their covariance and their correlation. Verify that results (3.44), (3.45) and (3.50) hold for this example.

3.24. Let X and Y be jointly distributed random variables and define Z in terms of X and Y by

$$Z = aX + bY$$

where a and b are constants. Show that

$$EZ = aEX + bEY$$

and

$$\text{var } Z = a^2 \text{ var } X + 2ab \text{ cov } (X, Y) + b^2 \text{ var } Y.$$

Hence show that, for independent random variables,

$$E(X - Y) = EX - EY$$

and

$$\text{var } (X - Y) = \text{var } X + \text{var } Y.$$

4

Proportions

4.1 Introduction

This chapter begins the application of the ideas and concepts developed in the previous two chapters to problems of applied economics. In the introduction to chapter 3 we discussed the general problem of making inferences about the characteristics of a population from the characteristics of a sample of the population. We also discussed the necessity of knowing the accuracy of such inferences. Apart from in one illustrative example, the terms 'population', 'sample', 'characteristic' and 'accuracy' were left undefined. (In the one example, the 'population' was all the households in a particular area, and the 'characteristic' was household income.) We continue to use the term 'population' in a general sense; in particular applications the population may be the set of all households in a particular area, the set of all Chinese people in the world, the set of all companies specialising in making furniture, the set of all black dogs with pink eyes, or the set of all colour television sets. The analysis of this and subsequent chapters is equally applicable to any of these populations, though of course we will be more concerned with populations of relevance to economics. The term 'characteristic' can also continue to be used in a general sense; in particular applications the characteristic may be income, eye colour, profit, number of bones or the frequency of breakdown, though not all characteristics may be applicable to all populations!

In this chapter our analysis will be concerned with a particular type of characteristic — those that are 'all-or-nothing' characteristics. By this we mean that the characteristic of interest is such that each member of the population either has the characteristic or does not have the characteristic; that is, the characteristic is such that it is impossible to have varying amounts of it. Consider, for example, the population of all households in a particular area; the characteristic of 'owner-occupier' is an 'all-or-nothing' characteristic — either the occupant owns the accommodation he lives in, or he does not (if it is mortgaged to a savings and loan association he is still the legal owner; he just has not paid for it). For the population of all Chinese people in the world, the characteristic 'has been out of China' is an 'all-or-nothing' characteristic. For the furniture companies, 'a positive pre-tax profit in the financial year 1972/73' is such a characteristic (whether they have fiddled the books or not!). Obviously there is a wide variety of characteristics of this particular type. In economics we are often interested in the possession or otherwise, by house-holds or individuals, of certain amenities or durable goods: hot-and-cold running water, washing machines, cars and the like. We are interested in the proportion of the population with such characteristics, and in compar-ing the proportions in different areas of the country and in different social and educational classes. This chapter is concerned with estimating such proportions and making comparisons between proportions on the basis of sample evidence. In addition, it lays the statistical foundation for the study of more general characteristics.

Before we consider these problems in a general framework we consider one simple example of the results to be derived in section 4.3. Suppose that in a particular area 20 per cent of the households have central heating, while 80 per cent do not. The households are each given a number, and into a box we put numbered slips of paper, one for each household. Consider the experiment of mixing the slips of paper well and drawing out one slip at random. The outcomes of this experiment are either drawing a household with central heating or drawing one without central heating. The respective probabilities on any one repetition of the experiment are 1/5 and 4/5. Suppose we repeat the experiment four times, making sure that, after every repetition, we replace the slip we have just drawn. Define the random variable X as the number of households with central heating drawn out of the box in these four drawings. Obviously X can take the values 0, 1, 2, 3, and 4; we now calculate the probabilities of X taking these five values. Anticipating future terminology we will call 'drawing a household with central heating' a 'success' and denote it by 'S'; and 'drawing a household without central heating' we will call a 'failure' and denote by 'F'. In this terminology the random variable X is the number of successes in the four repetitions of the experiment. By a sequence such as 'SFFS' we mean that the first draw was a 'success', the second and third 'failures', and the final draw a 'success'.

Note very carefully that the way we have designed the experiment and its four repetitions is such that the outcome on any one repetition does not depend on the outcome of any other repetition. (This would not be so if we did not replace the slips already drawn before the next draw took place; we return to this point later.) Thus, outcomes on successive repetitions are independent. We can thus use the multiplication rule for independent events to evaluate the probability distribution of X. Consider first the probability that X takes the value zero, that is that we get no 'successes' in the four repetitions.

$$P[X = 0] \equiv P[0] = P[FFFF]$$

$$= P[F]P[F]P[F]P[F] \quad \text{(independence)}$$

$$= \frac{4}{5} \cdot \frac{4}{5} \cdot \frac{4}{5} \cdot \frac{4}{5} = \frac{256}{625}.$$

Similarly $P[1] = P[\text{SFFF or FSFF or FFSF or FFFS}]$

$$= P[\text{SFFF}] + P[\text{FSFF}] + P[\text{FFSF}] + P[\text{FFFS}]$$

(mutually exclusive events)

$$= 4P[\text{S}]P[\text{F}]P[\text{F}]P[\text{F}] \quad \text{(independence)}$$

$$= 4 \cdot \frac{1}{5} \cdot \frac{4}{5} \cdot \frac{4}{5} \cdot \frac{4}{5} = \frac{256}{625}.$$

Similarly $P[2] = P[\text{SSFF or SFSF or SFFS or FSSF or FSFS or FFSS}]$

$$= 6P[\text{S}]P[\text{S}]P[\text{F}]P[\text{F}]$$

$$= 6 \cdot \frac{1}{5} \cdot \frac{1}{5} \cdot \frac{4}{5} \cdot \frac{4}{5} = \frac{96}{625}.$$

Continuing in this way, we get the probability distribution of X (the number of households with central heating in a sample of size 4) as given in table 4.1. This shows that $EX = 4/5$; that is, *on average*, out of the four households in the sample we get 0.8 of a household with central heating. Now 0.8 of a household represents one-fifth of the sample; and so, *on average*, one-fifth of the households in the sample have central heating. The proportion of households with central heating in the whole population is also one-fifth; we see therefore that, *on average*, the sample proportion equals the population proportion. As we will show in the next section, this is a general result, and accords well with intuition.

Figure 4.1 shows the probability density histogram and probability density polygon of X. It can easily be verified that the area under each

Table 4.1   Distribution of number of 'successes' in four repetitions

X	$P[X]$ $(\times 625)$	$XP[X]$ $(\times 625)$	$X - EX$ $(\times 5)$	$(X - EX)^2$ $(\times 25)$	$(X - EX)^2 P[X]$ $(\times 15,625)$
0	256	0	−4	16	4,096
1	256	256	1	1	256
2	96	192	6	36	3,456
3	16	48	11	121	1,936
4	1	4	16	256	256
Σ	625	500			10,000

$$EX = \frac{500}{625} = \frac{4}{5} \qquad\qquad \text{var } X = \frac{10,000}{15,625} = 0.64$$

$$\text{s.d. } X = 0.8$$

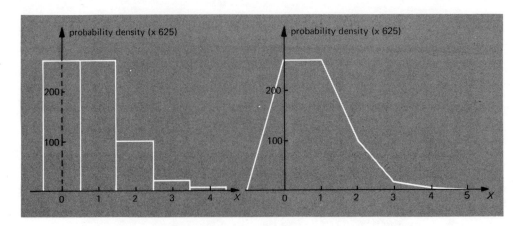

Fig. 4.1   Distribution of number of 'successes' in four repetitions

'curve' is unity. Anticipating again future developments, we will stand-ardise X. Table 4.1 shows that $EX = 4/5$ and var $X = 0.64$. Define, therefore, Z in terms of X, using (3.56), by

$$Z = \frac{X - EX}{\sqrt{\text{var } X}} = \frac{X - 4/5}{\sqrt{0.64}} = \frac{X - 4/5}{4/5}.$$

Table 4.2 shows the values of Z corresponding to each value of X. It also verifies that $EZ = 0$ and var $Z = 1$. The Z values are all $5/4$ apart, so the probability densities (the last column of table 4.2) are given by the probabilities divided by $5/4$. The probability density polygon of Z is drawn in figure 4.2. Note that the area under it is, of course, unity, and it has the same shape as the density polygon of X, although its mean and variance are different.

The probability distribution of X (given in table 4.1) not only tells us that, on average, the proportion of the households in the sample who have central heating is the same as the population proportion, but also that in almost 82 per cent (that is, 512 out of 625) of all samples of size four drawn from this population the sample will contain one or no households with central heating. Further, in only 2.78 per cent (that is, 17 out of 625) of all samples will we find three or more households with central heating, and a minute 0.16 per cent (that is, 1 out of 625) of all samples will consist entirely of centrally heated households. This last result implies that if we randomly select four households from some population (whose incidence of central heating ownership is unknown), and if we find that all four households in the sample have central heating, then it is extremely unlikely that in the population from which we drew the sample only one-fifth of the households have central heating. It is *possible* (as table 4.1

Table 4.2   Standardisation of number of 'successes' in four repetitions

X	$P[X]$ (×625)	$Z = (X - EX)/\sqrt{\mathrm{var}\,X}$ (×4)	$P[Z]$ (×625)	$ZP[Z]$ (×2,500)	$Z^2P[Z]$ (×10,000)	$P[Z]/w$ (×500)
0	256	−4	256	−1024	4096	256
1	256	1	256	256	256	256
2	96	6	96	576	3456	96
3	16	11	16	176	1936	16
4	1	16	1	16	256	1
Σ	625		625	0	10,000	
				$EZ = 0$	$\mathrm{var}\,Z = 1$	

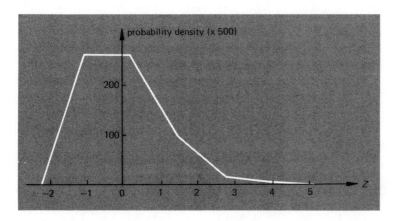

Fig. 4.2   Standardised distribution of number of 'successes' in four repetitions

shows), but it is extremely unlikely. This argument gives some insight into the methods we will develop for inference of population characteristics from sample characteristics. In section 4.5 we begin the development of these methods.

In conclusion, we note that the probability distribution given in table 4.1 is not peculiar to this particular example. If we have any experiment for which we divide the outcomes into two categories ('success' or 'failure'), and for which the probability of 'success' is 1/5, then the random variable defined as the number of 'successes' in four independent repetitions of the experiment has the probability distribution of table 4.1. Obviously we can generalise the above example, by considering different numbers of repetitions, and different probabilities of 'success'. Before we consider the most general case (in section 4.3), a few preliminaries are necessary to avoid duplication of tedious algebraic derivations. These preliminaries are considered in the next section.

4.2   Permutations and Combinations

Suppose we have n objects of some kind, and wish to select r of the n objects ($r \leqslant n$, of course) where the order of selection of the objects is important. We calculate the number of different ways the selection can be made. The first selection can be any one of the n objects. Having made this first selection there remain $(n-1)$ objects from which to choose for the second selection, and to *each* of the n possibilities on the first choice there are $(n-1)$ possibilities for the second. Thus, the first two objects can be chosen in $n(n-1)$ ways. For the third selection there remain $(n-2)$ objects, for the fourth $(n-3)$ remain. When we come to make the rth selection there remain $(n-r+1)$ objects from which to choose. Thus the total number of ways we can select r objects from n, where the order of selection is important, is

$$n(n-1)(n-2) \ldots (n-r+1).$$

We term this the number of *permutations* of n objects taken r at a time, and denote it by $_nP_r$. Thus

$$_nP_r = n(n-1)(n-2) \ldots (n-r+1). \tag{4.1}$$

To illustrate, suppose we have four oarsmen (call them A, B, C and D) and suppose we wish to select two to row in a pairs event. The order of the crew is important (that is, who is to be stroke and who is to be bow). The stroke seat can be filled in four ways; having done this, the bow seat can be filled in three ways. We thus get $12 (= 4 \times 3)$ ways of choosing the crew: AB, AC, AD, BA, BC, BD, CA, CB, CD, DA, DB, DC. Using (4.1) we have $n = 4$ and $r = 2$; thus $_4P_2 = 4 \times 3 = 12$.

To save writing, we use the symbol ! (meaning factorial) defined as follows:

$$n! = n(n-1)(n-2) \ldots 3 \times 2 \times 1.$$

That is, $n!$ (n factorial) is the product of all positive integers equal to, or less than n. For example,

$$3! = 3 \times 2 \times 1 = 6$$
$$6! = 6 \times 5 \times 4 \times 3 \times 2 \times 1 = 720$$
$$10! = 10 \times 9 \times 8 \times 7 \times 6 \times 5 \times 4 \times 3 \times 2 \times 1 = 3,628,800, \text{ etc.}$$

By convention we take $0! = 1$.

Using this notation (4.1) can be written

$$_nP_r = \frac{n(n-1)(n-2) \ldots 2 \times 1}{(n-r)(n-r-1) \ldots 2 \times 1} = \frac{n!}{(n-r)!} \tag{4.2}$$

and so, for example

$$_4P_2 = \frac{4!}{(4-2)!} = \frac{4!}{2!} = \frac{24}{2} = 12.$$

Generalising the above illustration, suppose we have six oarsmen and wish to select a crew to row in the fours event. Again the order of the crew is important. (4.2) gives the number of ways of selecting the crew as:

$$_6P_4 = \frac{6!}{(6-4)!} = \frac{6!}{2!} = \frac{720}{2} = 360.$$

Putting $r = n$ we see that $_nP_n$ counts the number of different ways that n objects can be ordered. Using (4.2) $_nP_n = n!$ (since $0! = 1$)

Suppose now that the *order* of selection is unimportant, that is that we are interested only in which objects are selected, not in their order of selection. The permutations formula (4.2) counts as different all different orderings of the same r objects. As we have seen above, having selected r objects there are $_rP_r = r!$ ways of arranging them in order. Thus the number of ways of selecting r objects from n, where the order of selection is not important, known as the number of *combinations* of n objects taken r at a time is:

$$_nC_r = \frac{_nP_r}{r!} = \frac{n!}{r!(n-r)!}. \tag{4.3}$$

For illustration, consider the number of different brag hands (three cards) that can be dealt from any ordinary pack of 52 cards. Obviously the order in which the cards are dealt is unimportant — all that matters is which three cards are in the hand. Thus, formula (4.3) is appropriate, which gives the total number of different hands as

$$_{52}C_3 = \frac{52!}{3!(52-3)!} = \frac{52!}{3!49!} = 22{,}100$$

Formulae (4.2) and (4.3) are useful in calculating various probabilities, such as the probability of getting a brag hand with all aces, the probability of a singleton in bridge, and so on. We are however mainly interested in their application to the generalisation of the example discussed in the introduction to this chapter. Consider therefore the number of ways of arranging n objects in order. If the objects are all distinguishable (as has been tacitly assumed above) the number of arrangements is $_nP_n = n!$ Suppose, however, that the objects are not distinguishable, and in fact there are just two different types of object, there being r of one type and $(n - r)$ of the other. (For example, $n = 5$ balls, $r = 2$ of which are red, $n - r = 3$ are blue). Thus, for any ordering of the n objects, we can rearrange the r of one type among themselves, leaving the overall ordering unchanged, and we can rearrange the $(n-r)$ of the other type among themselves, again leaving the overall ordering unchanged. The number of ways we can do these internal reorderings are $_rP_r = r!$ and $_{n-r}P_{n-r} = (n-r)!$ respectively. Thus the total number of ways n objects, of which r are of one type and $(n-r)$ are of another, can be ordered is

$$\frac{n!}{r!(n-r)!}. \tag{4.4}$$

You should work out why this is just the same as the number of combinations of n objects taken r at a time.

To illustrate (4.4), consider the example noted above; finding the number of ways two red and three blue balls can be arranged in order; from (4.4) it is

$$\frac{5!}{2!3!} = 10$$

(namely: RRBBB, RBRBB, RBBRB, RBBBR, BRRBB, BRBRB, BRBBR, BBRRB, BBRBR and BBBRR).

Thus, if we are repeating an experiment n times, for which there are just two outcomes, S ('success') and F ('failure'), the number of different ways we can get exactly r successes out of the n repetitions is

$$\frac{n!}{r!(n-r)!} = {_nC_r}.$$

We now use this result to generalise the example in the introduction.

4.3 The Binomial Distribution

Consider any experiment for which there are two outcomes of interest, termed 'success' and 'failure' respectively, and denoted by S and F. Suppose the probability of success on a single repetition is

$$P[S] = p$$

then

$$P[F] = 1 - p \tag{4.5}$$

Denote

$$q \equiv 1 - p.$$

Suppose we now carry out n *independent* repetitions of the experiment, and define the random variable X as the number of 'successes' obtained in the n repetitions. X takes the values 0, 1, 2, . . . , n. Consider the probability that $X = r$. Now, $X = r$ if we get r 'successes' and $(n-r)$

'failures'. As we have seen in section 4.2, the number of ways of getting exactly r 'successes' is $_nC_r$. Consider just one way:

$$\underbrace{SS \ldots S}_{r} \ \underbrace{FF \ldots F}_{(n-r)}$$

As successive repetitions are independent, it follows that

$$P[\underbrace{SS \ldots S}_{r} \ \underbrace{FF \ldots F}_{n-r}] = \underbrace{P[S]P[S] \ldots P[S]}_{r} \ \underbrace{P[F]P[F] \ldots P[F]}_{n-r}$$

$$= \underbrace{pp \ldots p}_{r} \ \underbrace{qq \ldots q}_{n-r}.$$

Thus

$$P[\underbrace{SS \ldots S}_{r} \ \underbrace{FF \ldots F}_{n-r}] = p^r q^{n-r}.$$

Now, each of the $_nC_r$ ways of getting r 'successes' has the same probability, and as all $_nC_r$ ways are mutually exclusive, it follows that

$$P[X = r] = {}_nC_r \, p^r q^{n-r}. \tag{4.6}$$

Thus the probability distribution of X, the number of 'successes' in n independent repetitions of an experiment for which the probability of a 'success' in a single repetition is p, is:

$$P[X] = \frac{n!}{X!(n-X)!} \, p^X q^{n-X} \quad (X = 0, 1, \ldots, n). \tag{4.7}$$

To satisfy the requirements of a probability distribution, the sum of the probabilities over all values of X should be one. We can see this as follows. Consider

$$(p + q)^m = \underbrace{(p + q)(p + q) \ldots (p + q)}_{m}.$$

There are $(m + 1)$ types of term in the expansion of $(p + q)^m$, namely

$$q^m, pq^{m-1}, p^2 q^{m-2}, \ldots, p^{m-2} q^2, p^{m-1} q, p^m$$

or, more succinctly,

$$p^r q^{m-r} \text{ for } r = 0,1,2, \ldots, (m-2), (m-1), m.$$

Now, the term in $p^r q^{m-r}$ in the expansion of $(p + q)^m$ consists of all products of the p in r of the m brackets with the q in the remaining $(m - r)$ brackets. Obviously there are $_mC_r$ ways of getting such products. Thus the coefficient of $p^r q^{m-r}$ in the expansion of $(p + q)^m$ is $_mC_r = m!/r!(m - r)!$

Thus

$$(p + q)^m = \sum_{r=0}^{m} \frac{m!}{r!(m-r)!} p^r q^{m-r} \qquad (4.8)$$

This is known as the *binomial theorem*.
Now, from (4.7), we have

$$\sum_{0}^{n} P[X] = \sum_{X=0}^{n} \frac{n!}{X!(n-X)!} p^X q^{n-X}$$

but from (4.8), putting $m = n$ and $r = X$ we have

$$\sum_{X=0}^{n} \frac{n!}{X!(n-X)!} = (p + q)^n$$

Thus

$$\sum_{0}^{n} P[X] = (p + q)^n = 1 \quad \text{since } q = 1 - p, \text{ from (4.5)}$$

Thus, the distribution given by (4.7) satisfies the requirement of a probability distribution in that the sum of the probabilities over all values of X is one.

The distribution defined by (4.7) is known as the *binomial distribution*. We now find its mean and variance. First, its mean:

$$EX = \sum_{0}^{n} XP[X] = \sum_{1}^{n} XP[X] \quad \text{(since for } X = 0, XP[X] = 0)$$

$$= \sum_{1}^{n} X \frac{n!}{X!(n-X)!} p^X q^{n-X} \quad \text{from (4.7)}$$

$$= np \sum_{1}^{n} \frac{(n-1)!}{(X-1)!(n-X)!} p^{X-1} q^{n-X}.$$

Put $r \equiv X - 1$, $m \equiv n - 1$; then

$$EX = np \sum_{r=0}^{m} \frac{m!}{r!(m-r)!} p^r q^{m-r}$$

$$= np(p + q)^m \quad \text{from (4.8)}$$

Thus

$$EX = np. \qquad (4.9)$$

Similarly, it can be shown (see Exercise 4.1) that

$$\text{var } X = npq. \qquad (4.10)$$

An alternative way of deriving these results is as follows. Define the n random variables $Y_1, Y_2, \ldots, Y_n$ by

$$Y_i = \begin{cases} 1 & \text{if } i\text{th repetition is a 'success'} \\ 0 & \text{if } i\text{th repetition is a 'failure'} \end{cases} \quad \text{(all } i\text{)}.$$

Then X (the number of 'successes' in the n repetitions) is simply

$$X = Y_1 + Y_2 + \ldots + Y_n. \tag{4.11}$$

Now, the random variables $Y_1, Y_2, \ldots Y_n$ are independent (since the repetitions are independent), and for each i, the probability distribution of Y_i is:

$$P[1] = p \qquad P[0] = q$$

Thus, for each i,

$$EY_i = 1 \times p + 0 \times q = p$$

and

$$\text{var } Y_i = (1-p)^2\, p + (0-p)^2\, q = pq \quad \text{(since } p + q = 1\text{)}.$$

Thus, using results (3.44) and (3.46) for independent random variables,

$$EX = E(Y_1 + Y_2 + \ldots + Y_n) = EY_1 + EY_2 + \ldots + EY_n = np$$

$$\text{var } X = \text{var } (Y_1 + Y_2 + \ldots + Y_n) = \text{var } Y_1 + \text{var } Y_2 + \ldots + \text{var } Y_n = npq.$$

Suppose we define a new random variable p' by

$$p' = \frac{X}{n}. \tag{4.12}$$

Thus, p' is the *proportion* of 'successes' in the n repetitions. Since p' is just a (constant) fraction of X, its probability distribution is given by (4.7), with X replaced by p'. Using our results on transformations ((3.54) and (3.55) with $a = 0$, $b = 1/n$), it follows from (4.9) and (4.10) that

$$\left. \begin{array}{l} Ep' = p \\ \\ \text{var } p' = \dfrac{pq}{n}. \end{array} \right\} \tag{4.13}$$

The first of these accords with intuition: the average proportion of 'successes' in the n repetitions equals p, the probability of success on a single repetition. The second is also intuitively reasonable as it shows that the dispersion of p', around p gets smaller as the number of repetitions increases.

The results derived so far in this section are completely general, and can be applied to any kind of experiment for which there are two outcomes of interest, and which we repeat n times independently. Consider, for illustration, picking five households at random from an area in which 40 per cent of all households own cars. If we sample with replacement, so that successive repetitions are independent, then X, the number of car-owning households in the sample of five, has a binomial distribution as given by (4.7) with $n = 5$ and $p = 0.4$. Thus, for example, the probability that three of the households in the sample own cars is

$$P[3] = \frac{5!}{3!(5-3)!} \left(\frac{4}{10}\right)^3 \left(\frac{6}{10}\right)^{5-3} = 10 \times \frac{4^3 \times 6^2}{10^5} = 0.23040.$$

The probability distribution of X is given in table 4.3. This also verifies (4.9) and (4.10); that is

$$EX = np = 5(0.4) = 2$$

and

$$\text{var } X = npq = 5(0.4)(0.6) = 1.2.$$

Table 4.3   Distribution of car-owning households in sample of size 5

X	$P[X]$	$XP[X]$	$(X - EX)^2$	$(X - EX)^2 P[X]$
0	0.07776	0.0	4	0.31104
1	0.25920	0.2592	1	0.25920
2	0.34560	0.6912	0	0.0
3	0.23040	0.6912	1	0.23040
4	0.07680	0.3072	4	0.30720
5	0.01024	0.0512	9	0.09216
Σ	1.0	$EX = 2.0$		var $X = 1.2$

Table 4.4 gives the distribution of p' ($=X/n$), the proportion of car-owning households in the sample, and verifies (4.13); that is

$$Ep' = p = 0.4$$

$$\text{var } p' = \frac{pq}{n} = 0.048.$$

Thus, if 40 per cent of households in the population own cars, then, *on average*, 40 per cent of the households in the sample own cars. (4.13) states this generally; that is, *on average*, the sample proportion equals the population proportion.

We are now in a position to consider any binomially distributed random variable. By inspection of (4.7) we see that the values of n and p completely determine the distribution. If we know that a random variable is binomially distributed, and if we know the values of n and p, we can completely describe its distribution. Alternatively, as inspection of (4.9)

Table 4.4   Distribution of proportion of car-owning
households in sample of size 5

p'	$P[p']$	$p'P[p']$	$(p' - Ep')^2$	$(p' - Ep')^2 P[p']$
0	0.07776	0.0	0.16	0.0124416
0.2	0.25920	0.05184	0.04	0.0103680
0.4	0.34560	0.13824	0.0	0.0
0.6	0.23040	0.13824	0.04	0.0092160
0.8	0.07680	0.06144	0.16	0.0122880
1.0	0.01024	0.01024	0.36	0.0036864
Σ	1.0	$Ep' = 0.4$		var $p' = 0.048$

and (4.10) shows, knowledge of its mean and of its variance (or standard deviation) determines the values of n and p and thus enables us to describe completely its distribution. For example, suppose we are told that a variable has a binomial distribution with mean 6 and variance 2, then (using (4.9) and (4.10)) $np = 6$ and $npq = 2$; thus $q = npq/np = 1/3$ and so $p = 1 - q = 2/3$ and $n = 9$.

We use the notation 'X is $B(np,npq)$' to mean that X is binomially distributed with mean np and variance npq.

However, although (4.7) allows us to calculate the probability distribution of a binomially distributed variable for any n and p, if n is at all large, the actual calculations needed to derive the probability distribution tend to become rather tedious. Fortunately, for large n we can approximate the binomial distribution by another distribution for which tables are available giving areas under its density curve.

Suppose the random variable X is binomially distributed with parameters n and p, that is, X is $B(np,npq)$. Consider the standardisation of X; that is

$$Z = \frac{X - EX}{\sqrt{\text{var } X}} = \frac{X - np}{\sqrt{npq}}.$$

Z, of course, has zero mean and unit variance. We have already seen, in figure 4.2, how we may plot the standardised distribution of X. In figure 4.2 we had $n = 4$ and $p = 0.2$ (see section 4.1). We can obviously plot the standardised distribution of X for any values of n and p. In figure 4.3 we draw the standardised distribution of X, where X is $B(np,npq)$, for the three cases:

(a) $n = 2$ $p = 0.5$

(b) $n = 4$ $p = 0.5$

(c) $n = 6$ $p = 0.5$.

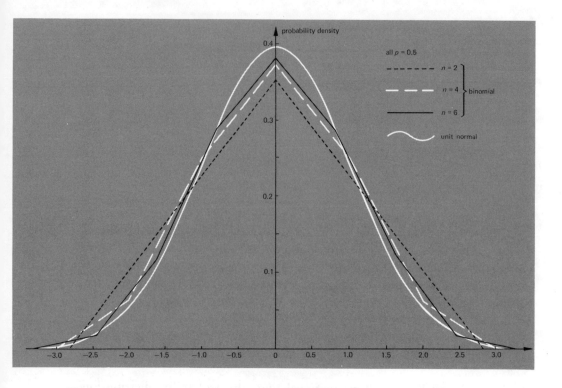

Fig. 4.3   Standardised binomial distributions for different n ($p = 0.5$)

In figure 4.4 we draw the standardised distribution of X for the three cases:

(d) $n = 2$ $p = 0.4$

(e) $n = 4$ $p = 0.4$

(f) $n = 6$ $p = 0.4.$

We notice, in both figures, that as n gets bigger the standardised binomial distribution gets closer to the smooth curve labelled 'unit normal'. The approximation of the standardised binomial distribution by the 'unit normal' improves as n increases, though for any n, the approximation is better for $p = 0.5$ than for $p = 0.4$. It can be shown that for n and p such that $np > 5$ and $nq > 5$ the approximation is good enough for areas under the standardised binomial distribution to be approximated by the corresponding area under the 'unit normal' curve, though of course the approximation improves as n gets bigger still. In the limit, as $n \to \infty$, the standardised binomial coincides with the 'unit normal'. Tables giving areas under the 'unit normal' are presented in the appendix. In the next section

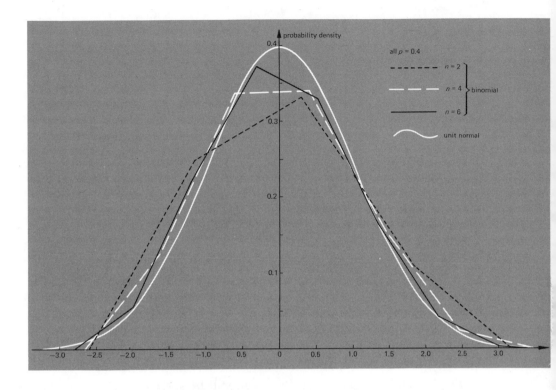

Fig. 4.4   Standardised binomial distributions for different n ($p = 0.4$)

we discuss the use of this approximation, as well as describing the characteristics of the 'normal' distribution itself.

First, however, a qualification must be made concerning the results derived above. The results were derived under the assumption that successive repetitions of the experiment were independent. We stressed that, in order for this condition to be satisfied when we took a sample of some population, we needed to sample 'with replacement'. Thus, if we were sampling by drawing numbered slips out of a box, we stressed that, after each drawing, the slip drawn must be replaced before the next is drawn. Suppose we did not replace it. Then, supposing the population size is N, and in the population the proportion of 'successes' is p, then on the first draw the probability of a 'success' is p. If we did not replace the first slip drawn before the second drawing, the probability of a 'success' on the second draw would be either $(pN - 1)/(N - 1)$ (if the first draw was a 'success') or $(pN)/(N - 1)$ (if the first draw was a 'failure'). Since $(pN - 1)/(N - 1) < p < (pN)/(N - 1)$, the probability of a 'success' on the second draw would be different from the probability of a 'success' on the first draw. Thus, if we sample without replacement, the results derived above are not applicable. Fortunately, we can still derive some useful results: we can still show that

$$EX = np \quad \text{and} \quad Ep' = p \tag{4.14}$$

but now

$$\text{var } X = npq\left(\frac{N-n}{N-1}\right) \quad \text{and} \quad \text{var } p' = \frac{pq}{n}\left(\frac{N-n}{N-1}\right) \tag{4.15}$$

that is the variance of X (and p') is smaller when we sample without replacement than when we sample with replacement. This should be intuitively reasonable. You may like to verify (4.14) and (4.15) yourself.

The factor $(N-n)/(N-1)$, multiplying the variances in the 'without replacement' case, is called the *finite population correction*. Note the relevance of the word 'finite' as

$$\frac{N-n}{N-1} \to 1 \quad \text{as} \quad N \to \infty.$$

In other words, if the population is infinite, then the same results hold for both the 'with' and the 'without replacement' cases. You should consider the reason for this. If N is large relative to n, then $(N-n)/(N-1)$ is very close to unity, and may be neglected.

Finally, and very importantly, it can still be shown, even if we sample without replacement, that the standardised variable $(X - EX)/\sqrt{\text{var } X}$ approaches the 'unit normal' as n gets bigger.

4.4 The Normal Distribution

The 'unit normal' distribution above is the standardised form of the *normal distribution*. Unlike the binomial distribution, which applies to discrete random variables, the normal distribution applies to continuous random variables. Besides providing a useful approximation to the binomial distribution (as well as to many other distributions), the normal distribution describes, in its own right, many distributions commonly found in economic variables. Furthermore, while many economic 'quantities' may not themselves be normally distributed, it is usually easy to obtain a normal distribution by means of a simple transformation (for example the distribution of the logarithm of household incomes in any particular week conforms very closely to the normal distribution).

The probability density function of a normally distributed random variable X is given by*

---

*exp$[Y]$ denotes e raised to the power Y (i.e. e^Y) where e is the base of the natural logarithm. The term $1/\sqrt{(2\pi\sigma^2)}$ in $f(X)$ is a scaling factor which ensures that the condition that the area under the probability density function equals one is satisfied.

$$f(X) = \frac{1}{\sqrt{(2\pi\sigma^2)}} \exp\left[-\frac{1}{2}\left(\frac{X-\mu}{\sigma}\right)^2\right] \quad -\infty \leqslant X \leqslant \infty. \tag{4.16}$$

Figure 4.5 illustrates the normal distribution.

If you are familiar with the calculus, you may like to verify the following results. If you are not, inspection of figure 4.5 will aid intuitive verification and understanding.

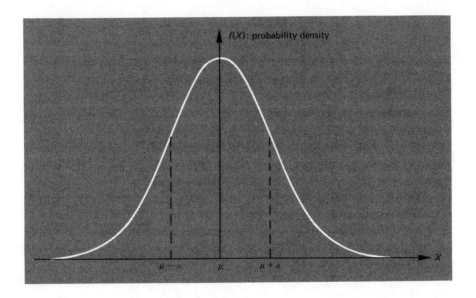

Fig. 4.5 The normal distribution

For a normally distributed random variable, with density function as given by (4.16), the following results hold:

- (a) $EX = \mu$;
- (b) var $X = \sigma^2$;
- (c) the distribution is symmetrical about the mean;
- (d) X takes values between $\pm \infty$;
- (e) $f(X)$ has points of inflexion* at $X = \mu \pm \sigma$;
- (f) the graph of the density function is bell-shaped.

As we have already noted, the standardised form of a normally distributed random variable is called the *unit normal*. Thus, if X is normally distributed with mean μ and variance σ^2 (that is, 'X is $N(\mu,\sigma^2)$'),

*A point of inflexion of a function $Y = f(X)$ is defined by $d^2Y/dX^2 = 0$. Intuitively, a point of inflexion occurs when the slope stops increasing and starts decreasing (or *vice versa*).

and we standardise as usual by

$$Z = \frac{X - EX}{\sqrt{\text{var } X}} = \frac{X - \mu}{\sigma}$$

then Z has a unit normal distribution ('Z is $N(0,1)$'). The density function of a unit normal, Z, is

$$f(Z) = \frac{1}{\sqrt{2\pi}} \exp\left(\frac{-Z^2}{2}\right) \quad -\infty \leqslant Z \leqslant \infty.$$

This was the curve labelled 'unit normal' in figures 4.3 and 4.4.

By now you should have realised that linear transformations of variables do not alter the 'shape' of the distribution. The transformation changes only the mean and the variance. (Comparison of figures 4.1 and 4.2 should convince you of this.) Thus any linear function of a normally distributed random variable is itself normally distributed. Thus, if Z is a unit normal, and X is defined in terms of Z by

$$X = \mu + \sigma Z$$

then X is also normal, with mean

$$EX = \mu + \sigma EZ = \mu \quad (\text{since } EZ = 0)$$

and variance

$$\text{var } X = \sigma^2 \text{ var } Z = \sigma^2 \ (\text{since var } Z = 1).$$

Thus, if Z is $N(0,1)$ and $X = \mu + \sigma Z$, then X is $N(\mu, \sigma^2)$. \hfill (4.17)

This, of course, is just the 'reverse' of standardisation.

As with any probability distribution, the probability of a unit normal variable lying between any two values is given by the area under the probability density curve between the two values. To facilitate calculation of these probabilities, tables of areas under a unit normal have been prepared. The table given in the appendix shows the area under a unit normal between 0 (the mean) and other values. Thus, for example, we see from the table, that if Z is a unit normal, then $P[0 \leqslant Z \leqslant 1] = 0.3413$. Other entries in the table (and the symmetry) give us other results; e.g.

$$P[-1 \leqslant Z \leqslant 1] = 0.6826$$
$$P[-2 \leqslant Z \leqslant 2] = 0.9544$$
$$P[-3 \leqslant Z \leqslant 3] = 0.9994$$
$$P[1 \leqslant Z \leqslant 2] = 0.1359 \text{ etc.}$$

We can use these tables to make probability statements about any normally distributed random variable, not just unit normals. Thus, for example, suppose X is $N(2,9)$. Then Z is $N(0,1)$ where $Z = (X - 2)/3$.

Then

$$P[2 \leqslant X \leqslant 5] = P[2 - 2 \leqslant X - 2 \leqslant 5 - 2]$$
$$= P[0 \leqslant X - 2 \leqslant 3]$$
$$= P\left[\frac{0}{3} \leqslant \frac{X-2}{3} \leqslant \frac{3}{3}\right]$$
$$= P[0 \leqslant Z \leqslant 1]$$
$$= 0.3413 \text{ from the unit normal tables.}$$

Similarly

$$P[-4 \leqslant X \leqslant 8] = P[-2 \leqslant Z \leqslant 2] = 0.9544, \text{ etc.}$$

We note that, just as with the binomial distribution, knowledge of the mean and variance of a normally distributed variable are all that are needed to determine completely its distribution.

We return now to the approximation of a standardised binomial variable by a unit normal variable. Suppose X is binomially distributed with mean np and variance npq, then $(X - np)/(\sqrt{npq})$ is approximately unit normal (for $np > 5$, $nq > 5$) and thus, using (4.17), X is approximately $N(np, npq)$. Thus,

$$\text{if } X \text{ is } B(np, npq) \quad \text{then} \quad X \text{ is approximately } N(np, npq) \qquad (4.18)$$

the approximation improving as n increases.

Note very carefully that the approximation refers to replacing 'B' by 'N', and does not refer to the mean and variance. $EX = np$ and var $X = npq$, are *exact* results. The approximation refers to the 'shape' of the distribution.

It follows from (4.18), and the definition of p', that

$$p' \text{ is approximately } N\left(p, \frac{pq}{n}\right) \qquad (4.19)$$

where, again, the approximation refers to the 'shape', not the mean and variance, which are exact results (see (4.13)).

To illustrate the use of the result (4.18) consider the probability that, in a sample of size 10 chosen with replacement from a population in which 50 per cent of all households own cars, 4 out of the 10 households in the sample own cars. Using the exact formula (4.7) with $n = 10$, $p = 0.5$, we get

$$P[4] = \frac{10!}{4!6!} (\tfrac{1}{2})^4 (\tfrac{1}{2})^6 = 210 (\tfrac{1}{2})^{10} = 0.205078125.$$

Consider now figure 4.6. The probability that X equals 4 is given by the

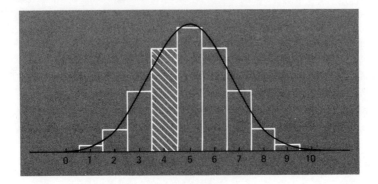

Fig. 4.6   Approximation of areas

shaded area. The approximation referred to in (4.18) means that the area under the normal curve between any two values approximates the area under the binomial curve *between the same two* values. Thus, the appropriate area under the normal curve is that lying between 3.5 and 4.5. Consider therefore $P[3.5 \leqslant X \leqslant 4.5]$ where X is $N(np, npq)$, that is where X is $N(5, 2.5)$, since $n = 10, p = 0.5$.

Now, if X is $N(5, 2.5)$

$$P[3.5 \leqslant X \leqslant 4.5] = P\left[\frac{3.5 - 5}{\sqrt{2.5}} \leqslant \frac{X - 5}{\sqrt{2.5}} \leqslant \frac{4.5 - 5}{\sqrt{2.5}}\right]$$

$$= P[-0.95 \leqslant Z \leqslant -0.32] \quad \text{where } Z \text{ is } N(0, 1)$$

$$= 0.2034 \text{ from unit normal tables.}$$

Comparing this with the exact probability (0.205078125) shows that the error of approximation is less than 1 per cent. Consider also the probability that in the sample of 10 we get 4 or less car-owning households. Using the exact formula (4.7), we have

$$P[X \leqslant 4] = P[0] + P[1] + P[2] + P[3] + P[4]$$

$$= \frac{10!}{0!10!}(\tfrac{1}{2})^0(\tfrac{1}{2})^{10} + \frac{10!}{1!9!}(\tfrac{1}{2})^1(\tfrac{1}{2})^9 + \ldots + \frac{10!}{4!6!}(\tfrac{1}{2})^4(\tfrac{1}{2})^6$$

$$= 0.0009765625 + 0.009765625 + \ldots + 0.205078125$$

$$= 0.376953125.$$

Using the approximation (4.18), we have, where X is $N(5, 2.5)$,

$$P[-0.5 \leqslant X \leqslant 4.5] = P\left[\frac{-0.5 - 5}{\sqrt{2.5}} \leqslant \frac{X - 5}{\sqrt{2.5}} \leqslant \frac{4.5 - 5}{\sqrt{2.5}}\right]$$

$$= P[-3.48 \leqslant Z \leqslant 0.32] \quad \text{where } Z \text{ is } N(0, 1)$$

$$= 0.3745 \quad \text{from unit normal tables.}$$

Again, an error of approximation of less than 1 per cent.

Obviously the use of the approximation saves a great deal of tedious computation. As a final illustration, and one for which calculation by means of the exact formula would be excessively tedious, consider the above example, but where samples of size 100 are taken instead of samples of size 10. If X is the number of car-owning households in the sample of size 100, then X is $B(np,npq)$; that is, X is $B(50,25)$, and so, using (4.18), X is approximately $N(50,25)$. For X exactly $N(50,25)$ we would have, for example

$$P[39.5 \leqslant X \leqslant 60.5] = P\left[\frac{39.5 - 50}{\sqrt{25}} \leqslant \frac{X - 50}{\sqrt{25}} \leqslant \frac{60.5 - 50}{\sqrt{25}}\right]$$
$$= P[-2.1 \leqslant Z \leqslant 2.1] \quad \text{where } Z \text{ is } N(0, 1)$$
$$= 0.9642.$$

Thus, in approximately 96 per cent of all samples of size 100 drawn with replacement from a population in which 50 per cent of households own cars, we would find between 40 and 60 car-owning households in the sample. Conversely, in less than 4 per cent of such samples would we find less than 40, or more than 60, car-owning households.

The results of the last two sections enable us to make probability statements about the number, or proportion of 'successes', in a sample given information about the proportion of 'successes' in the population as a whole. Thus, for any sample size we can calculate the probability distribution of the sample proportion given the population proportion. However, we are usually more interested in the opposite procedure: namely, inferring something about the population proportion given information on a sample proportion. This is the subject matter of the next two sections.

4.5 Estimation of the Population Proportion
So far we have been considering the problem of making probability statements about the number (or proportion) of members of a randomly chosen sample having a certain characteristic, given that we know the proportion of members of the total population with the characteristic. That is, we have used our knowledge of the population proportion, p, to *deduce* probability statements about the sample proportion p'.

However, this exercise is not of much use in itself. What is more likely to be of use in practice is the opposite procedure; that is, using our knowledge of some (randomly chosen) sample proportion, p', to *infer* something about the population proportion p. In other words, we may wish to find out what p is, but are prohibited by cost constraints or time factors from checking each individual member of the population to see whether it has the characteristic or not; we may thus be obliged to use information gained from a sample of the whole population to shed light on the value of p.

Suppose then, we have picked a random sample of n members of the population, and have found the sample proportion (with the characteristic) to be p'. How can we use this information to provide information about the value of p? The most obvious thing to do is to use p' as a *point estimate* of p. Even though p' is unlikely to be exactly equal to p in any particular sample, we do know that, *on average*, over many samples, p' will equal p. This is since $Ep' = p$ (from (4.13)). However, in any particular sample, the difference between p' and p may be considerable, and it is thus important to give some idea of the accuracy of using p' as an estimate of p. Intuitively, we can see that the larger the sample size (n) the more accurate p' will be as an estimate of p. This intuition is reinforced by our previous results that $Ep' = p$ and var $p' = pq/n$; that is, the 'spread' of p' around p is reduced as we increase n. We can incorporate these ideas by providing an *interval estimate* of p, as follows.

From (4.19) we know that

$$p' \text{ is approximately } N\left(p, \frac{pq}{n}\right)$$

and so

$$\frac{p' - p}{\sqrt{(pq/n)}} \text{ is approximately } N(0, 1).$$

This enables us to make probability statements about p'. For example, we have (approximately)

$$P\left[-1.96 \leqslant \frac{p' - p}{\sqrt{(pq/n)}} \leqslant 1.96\right] = 0.95, \text{ from the unit normal tables.}$$

that is

$$P\left[-1.96 \leqslant \frac{p' - p}{\sigma} \leqslant 1.96\right] = 0.95 \tag{4.20}$$

where $\sigma^2 = pq/n$ (if we sampled without replacement from a population of size N, σ^2 would be given by $(pq/n)[(N - n)/(N - 1)]$ (from (4.15)). The probability statement (4.20) can be expressed in two other ways:

$$P[p - 1.96\sigma \leqslant p' \leqslant p + 1.96\sigma] = 0.95 \tag{4.21}$$

or

$$P[p' - 1.96\sigma \leqslant p \leqslant p' + 1.96\sigma] = 0.95. \tag{4.22}$$

The first of these two statements (4.21) says that in 95 per cent of all samples the sample proportion lies in the interval $(p - 1.96\sigma, p + 1.96\sigma)$. The second (4.22) should be interpreted as saying that, in 95 per cent of all samples, the interval $(p' - 1.96\sigma, p' + 1.96\sigma)$ contains (or covers) p. Obviously, these are two different ways of saying the same thing, as figure 4.7 illustrates. Between $(p - 1.96\sigma)$ and $(p + 1.96\sigma)$, the area under the density function is 0.95 (the area in each of the shaded tails is 0.025). Thus, the probability that p' lies between $(p - 1.96\sigma)$ and $(p + 1.96\sigma)$ is 0.95. Now, consider the interval

$$(p' - 1.96\sigma, p' + 1.96\sigma).$$

The width of this interval is 3.92σ and is constant; however its position depends on p'. The interval is indicated in figure 4.7 for several different values of p'. In case (A) p' is contained in $(p - 1.96\sigma, p + 1.96\sigma)$ and the

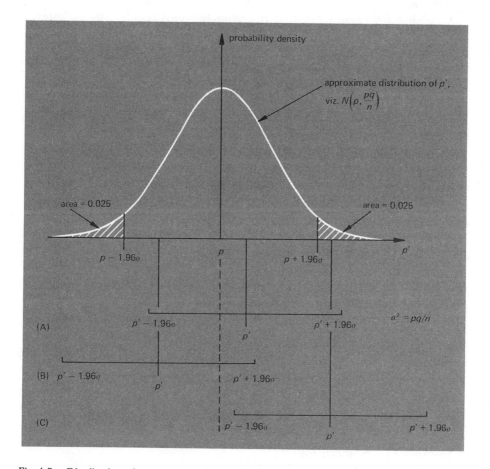

Fig. 4.7 Distribution of sample proportion

interval $(p' - 1.96\sigma, p' + 1.96\sigma)$ contains p. In case (B) p' is contained in $(p - 1.96\sigma, p + 1.96\sigma)$ and the interval $(p' - 1.96\sigma, p' + 1.96\sigma)$ contains p. In case (C) p' is *not* contained in $(p - 1.96\sigma, p + 1.96\sigma)$, and the interval $(p' - 1.96\sigma, p' + 1.96\sigma)$ does *not* contain p. These results obviously generalise: whenever p' is contained in $(p - 1.96\sigma, p + 1.96\sigma)$, and the probability of its doing so is 0.95, then the interval $(p' - 1.96\sigma, p' + 1.96\sigma)$ will contain p. Conversely, the probability that p' is not contained in $(p - 1.96\sigma, p + 1.96\sigma)$ and so $(p' - 1.96\sigma, p' + 1.96\sigma)$ does not contain p, is 0.05.

Thus the interval $(p' - 1.96\sigma, p' + 1.96\sigma)$ is such that 95 per cent of the time it will cover p. Thus, if we have observed a particular sample value of p', and if we form the interval $(p' - 1.96\sigma, p' + 1.96\sigma)$ based on this observed sample value, we can be 95 per cent confident that the interval so formed contains p. The interval $(p' - 1.96\sigma, p' + 1.96\sigma)$ is therefore naturally called a 95 per cent *confidence interval* for p; it provides an *interval estimate* for the population proportion.

Note that we cannot say that the probability of p falling in the interval is 0.95; p is a fixed number — not a random variable; it is meaningless to talk as if it had a probability. We can say either of two things:

(1) If we always construct 95 per cent confidence intervals in this way, the statement that p lies within the confidence interval will be correct 95 per cent of the time.

(2) The probability that the interval covers p is 0.95.

Even this last statement has to be interpreted carefully. What it means is that the probability of the interval $(p' - 1.96\sigma, p' + 1.96\sigma)$ covering p is 0.95. However, when we substitute an actual sample value for p', the interval so formed is then fixed (no longer random) and it either covers p or does not — but of course we do not know which.

We can of course construct other intervals with different degrees of confidence. For example, a 99 per cent interval for p is $(p' - 2.58\sigma, p' + 2.58\sigma)$.

One problem that may have occurred to you is that unless we know p we cannot find σ, and thus cannot form the interval. However, we can estimate $\sigma^2 = [p(1-p)]/n$ by substituting p' for p, and using $[p'(1-p')]/n$ as an estimate of σ^2. This naturally introduces an approximation; however, as exercise 4.21 shows, the loss of accuracy incurred is relatively small.

To illustrate, suppose in a sample of 100 households drawn with replacement, we found that 50 owned cars. Then $p' = 0.5$, and σ^2 can be estimated as

$$\sigma^2 \doteqdot \frac{p'(1-p')}{n} = \frac{(0.5)(0.5)}{100} = 0.0025$$

Thus

$$\sigma \doteqdot 0.05$$

and so, a 95 per cent confidence for p is

$$(p' - 1.96\sigma, p' + 1.96\sigma)$$

that is,

$$(0.5 - (1.96)(0.05), 0.5 + (1.96)(0.05)$$

that is,

$$(0.402, 0.598).$$

Hence we can be 95 per cent confident that between 40.2 and 59.8 per cent of all households in the population own cars. The 99 per cent confidence interval is given by

$$(0.5 - (2.58)(0.05), 0.5 + (2.58)(0.05))$$

that is $(0.371, 0.629)$. Naturally, increasing the level of confidence inevitably means a wider interval.

However, increasing the sample size increases the accuracy. Thus, if 50 per cent of households in a sample of size 400 owned cars, then σ^2 is estimated as

$$\sigma^2 \doteqdot \frac{p'(1-p')}{n} = \frac{(0.5)(0.5)}{400} = 0.000625$$

and so

$$\sigma \doteqdot 0.025.$$

In this case, the 95 per cent confidence interval is given by

$$(p' - 1.96\sigma, p' + 1.96\sigma)$$

that is

$$(0.5 - (1.96)(0.025), 0.5 + (1.96)(0.025))$$

that is, $(0.451, 0.549)$.

Thus, quadrupling the sample size halves the width of the confidence interval (for any given level of confidence), and thus doubles the accuracy.

This can easily be seen in general: the width of the 95 per cent confidence interval (for example) is

$$\text{width} = 3.92\sigma = 3.92\sqrt{(pq/n)} \tag{4.23}$$

and so the width is inversely proportional to the square root of the sample size.

Another useful result is that σ^2 (and therefore the width of the confidence interval for any given level of confidence) is greatest (for any value of n) when $p = 0.5$. This can be shown as follows:

$$\sigma^2 = \frac{pq}{n} = \frac{p(1-p)}{n}$$

Thus

$$\frac{d(\sigma^2)}{dp} = \frac{1-2p}{n} \quad \text{and} \quad \frac{d(\sigma^2)}{dp} = 0 \text{ when } p = 0.5.$$

Further,

$$\frac{d^2(\sigma^2)}{dp^2} = \frac{-2}{n}$$

which is negative; thus σ^2 is indeed maximised for $p = 0.5$.

This result is useful in two ways. First, if we do not wish to use the approximation

$$\sigma^2 \doteq \frac{p'(1-p')}{n}$$

we can give p the value 0.5 in $\sigma^2 = pq/n$, and we are then certain that the correct confidence interval is no wider than the one we obtain. Secondly, if we wish to determine the size of sample needed to achieve a certain degree of accuracy, and we have no knowledge at all of the value of p, we can estimate σ^2 using $p = 0.5$, and then we can be certain that we will achieve our desired level of accuracy. Suppose, for example, that we want our 95 per cent confidence interval to have width 0.2, then we choose n such that

$$0.2 = 3.92\sqrt{(pq/n)} \quad \text{(from (4.23))}$$

that is

$$n = pq\left(\frac{3.92}{0.2}\right)^2 = 384.16\,pq$$

so the largest sample size we would need to achieve this level of accuracy is

$$n = 384.16\,(0.5)(0.5) = 96.04.$$

We now generalise the discussion, and show the derivation of a general confidence interval. We have, from (4.19), that

p' is approximately $N(p,\sigma^2)$

where

$$\sigma^2 = \begin{cases} \dfrac{pq}{n} & \text{if we sample with replacement} \\[2em] \dfrac{pq}{n}\left(\dfrac{N-n}{N-1}\right) & \text{if we sample without replacement (from (4.14))} \end{cases}$$

and so

$\dfrac{p'-p}{\sigma}$ is approximately $N(0, 1)$.

Thus

$$P\left[-Z_{\alpha/2} \leqslant \frac{p'-p}{\sigma} \leqslant Z_{\alpha/2}\right] = 1 - \alpha \tag{4.24}$$

where Z_β is defined as that value for which $P[Z \geqslant Z_\beta] = \beta$ where Z is a unit normal.

Now (4.24) can be expressed as:

$$P[p' - Z_{\alpha/2}\sigma \leqslant p \leqslant p' + Z_{\alpha/2}\sigma] = 1 - \alpha$$

and thus a $100(1-\alpha)$ per cent confidence interval for p is

$$(p' - Z_{\alpha/2}\sigma, p' + Z_{\alpha/2}\sigma). \tag{4.25}$$

We restate the interpretation of the phrase '$100(1-\alpha)$ per cent confidence interval for p'. It means that for $100(1-\alpha)$ per cent of all samples, the interval as given by (4.25) will cover p; for any *particular* sample whether the interval covers p or not we do not know. This interpretation is crucially important, and it is vital that you fully understand it before proceeding.

Note that (4.25) is just one of many possible $100(1-\alpha)$ per cent confidence intervals, since there are many possible values, Z_1 and Z_2, that satisfy

$$P\left[Z_1 \leqslant \frac{p'-p}{\sigma} \leqslant Z_2\right] = 1 - \alpha.$$

The particular values chosen above are such that $Z_1 = -Z_2$, and give rise to a confidence interval *centred* on p'. You should verify that the following intervals are all 95 per cent confidence intervals for p; you should also provide some more examples yourself.

(a) $(p' - 2.33\sigma, p' + 1.75\sigma)$

(b) $(p' - 2.05\sigma, p' + 1.88\sigma)$

(c) $(p' - 1.96\sigma, p' + 1.96\sigma)$

(d) $(p' - 1.88\sigma, p' + 2.05\sigma)$

(e) $(p' - 1.75\sigma, p' + 2.33\sigma)$

(f) $(0, p' + 1.64\sigma)$

(g) $(p' - 1.64\sigma, 1)$

(The last two are somewhat tricky and require a little thought.) In general, unless we have some prior information, we will use centred confidence intervals (as in (c) above), as they convey the fact that we have no reason, one way or the other, to bias our interval estimates.

One fact that is probably fairly apparent, although we have only tacitly mentioned it, is that the results derived in this chapter are all conditional on the assumption that the samples are chosen *randomly*. A random sample is one for which every member of the population has an equal chance of being selected. If there is bias in the selection of the sample, there will, of course, be bias in the inferences based on the sample evidence. Thus, when we use the word 'sample' we always mean a 'random sample'.

We close this section with an illustration based on the 1970 *Survey of Consumer Finances*. In Table 4–10 of the *Survey*, the distribution by income of family purchases of new cars is given. For example, out of 472 families sampled with incomes above $15,000, 118 purchased a new car in 1969. For all families, 335 out of the total sample of 2,567 families purchased a new car in 1969. We use these figures to derive confidence intervals for the proportion of families purchasing a new car in all $15,000 plus income families, and in all families. First however we must note that the *Survey* chose their sample without replacement, and so, strictly, we ought to apply the finite population correction. In practice, as the sample size was small relative to the population size, we can ignore this correction. We consider first families with incomes greater than $15,000. Here we have $n = 472$ and $p' = 0.25$. Thus the 95 per cent confidence interval for the proportion of all families (with incomes greater than $15,000) who purchased new cars is, from (424),

$$(p' - 1.96\sigma, p' + 1.96\sigma)$$

We estimate $\sigma^2 = [p(1-p)]/n$ in two ways; first by approximating p by p' and then, to allow absolute safety in the calculation of the confidence interval, by putting $p = 0.5$. Replacing p by p',

$$\sigma^2 \doteqdot \frac{(0.25)\,(0.75)}{472}$$

and so, $\sigma \doteqdot 0.0199$. Our 95 per cent confidence interval is thus

$$(0.25 - (1.96)(0.0199), 0.25 + (1.96)(0.0199))$$

i.e.

$$(0.211, 0.289).$$

We can thus be 95 per cent confident that between 21.1 and 28.9 per cent of all families with incomes above $15,000 purchased a new car in 1969. Now, putting $p = 0.5$,

$$\sigma^2 \doteq \frac{(0.5)(0.5)}{472}$$

and so $\sigma \doteq 0.0230$, and so our 95 per cent confidence interval is (0.205, 0.295). Note the very small difference using the two approximations.

For the U.S. as a whole we calculate the 99 per cent confidence interval

$$(p' - 2.58\sigma, p' + 2.58\sigma)$$

where $n = 2,576$ $p' = 335/2,576 = 0.13$, and we estimate σ^2 by $\sigma^2 \doteq [p'(1 - p')]/n = [(0.13)(0.87)] / 2,576$, and so $\sigma \doteq 0.0066$. Thus, the 99 per cent confidence interval for the proportion of all U.S. families purchasing new cars in 1969 is (0.113, 0.147).

4.6 Testing Hypotheses about the Population Proportion

Up to now we have been concerned solely with the problem of estimating the population proportion using the information contained in a sample. We may however approach this problem from a slightly different point of view. In constructing economic models the 'specification' that we arrive at from economic theory may lead us to expect certain parameters to have particular values (or to lie within a certain range of values). The economist will, therefore, be largely concerned with testing the 'specification' of his model.

In general the initial specification of the model is called the *null hypothesis*, the object being to test whether the evidence from a sample is consistent with this null hypothesis. On the basis of the test we must decide whether to accept or reject the null hypothesis. If only a sample from the population is available, there is always a possibility that any decision made on the basis of our tests will be wrong. Furthermore, there will always be costs associated with making the decision – so that the most natural approach to 'decision-making' in this context would be to attempt to reduce the 'risk' (in terms of the expected loss) to a minimum.

Since we either reject or accept the null hypothesis there are obviously two kinds of mistakes that can be made, and, in general, the costs involved for each kind will not be the same; we can reject the null hypothesis when, in fact, it is correct (we call this a *Type I error*), or we can accept the null hypothesis when in fact it is incorrect (a *Type II error*). It is important to note that, if we are making our decision on the evidence of some sample, we can never be absolutely sure, except in certain trivial cases, that we have made the correct decision (in other words, the probability of a Type I error, or the probability of a Type II error, is never zero, except in certain trivial cases).

Suppose our *a priori* hypothesis about the value of p, the population proportion, states that it is p_0. This is our null hypothesis. Suppose we wish to test this null hypothesis against the alternative hypothesis that p is p_1. As will become apparent, we cannot prove that the null hypothesis holds in *absolute* terms, but can only show that it is *relatively* better than some alternative. Formally, we are testing H_0 against H_1 where:

$$H_0 : p = p_0 \quad (p_0 < p_1, \text{ say})$$
$$H_1 : p = p_1$$

Now suppose we draw a sample of size n (with replacement) from the population, and, as before, call p' the sample proportion. We wish to use this sample information to test whether it supports H_0 or H_1.

We already know (from (4.19)) that if H_0 is correct, then p' will be approximately normally distributed with mean p_0 and variance $[p_0(1 - p_0)]/n$. Similarly, we know (from (4.19)) that if H_1 is correct then p' will be approximately normal with mean p_1 and variance $[p_1(1 - p_1)]/n$. Formally,

If H_0 is correct then

$$p' \text{ is (approximately) } N(p_0, \sigma_0{}^2) \text{ where } \sigma_0{}^2 = \frac{p_0(1 - p_0)}{n}$$

If H_1 is correct then

$$p' \text{ is (approximately) } N(p_1, \sigma_1{}^2) \text{ where } \sigma_1{}^2 = \frac{p_1(1 - p_1)}{n}$$

(4.26)

Figure 4.8 illustrates these two statements.

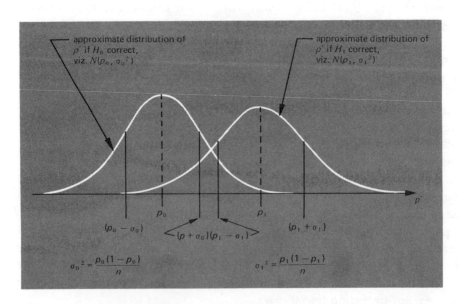

Fig. 4.8   Distribution of p' under two hypotheses

If our actual sample value p' is less than p_0, it seems reasonable to decide that H_0 is more likely to be correct. Similarly, if our actual sample value p' is greater than p_1, it seems reasonable to decide that H_1 is more likely to be correct. However, we have a problem in deciding which is more likely to be correct if our sample value p' lies between p_0 and p_1. Let us investigate the consequences if we choose some *critical value* p^* and decide to accept H_0 (reject H_1) if the observed p' is less than or equal to p^*, and to accept H_1 (reject H_0) if the observed p' is greater than p^*. Thus consider the decision rule:

$$\left.\begin{array}{l} \text{If } p' \leqslant p^* \text{ then accept } H_0 \quad (\text{reject } H_1) \\[2mm] \text{If } p' > p^* \text{ then reject } H_0 \quad (\text{accept } H_1) \end{array}\right\} \quad (4.27)$$

The consequences of this decision rule are given in table 4.5. Suppose we choose the critical value p^* as illustrated in figure 4.9. Now, if H_0 is actually correct, so that p' has the left-hand distribution in figure 4.9, then p' can be greater than p^*, and the probability that p' is greater than

Table 4.5   Consequences of decision rule (4.27)

	If H_0 correct	If H_1 correct
If $p' \leqslant p^*$ then accept H_0	Correct decision	Type II error
If $p' > p^*$ then accept H_1	Type I error	Correct decision

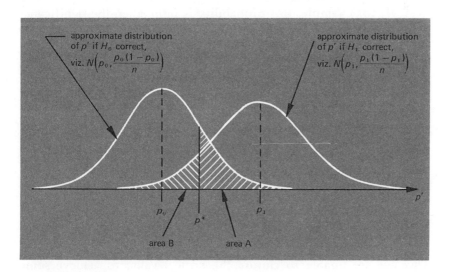

Fig. 4.9   Probabilities of errors for given critical value

$p*$ is given by area A. Thus:

$$P[p' > p* \mid H_0 \text{ correct}] = \text{area A}.$$

Hence, if our decision rule is given by (4.27), we see immediately that the probability of making a Type I error is given by area A. Similarly

$$P[p' \leqslant p* \mid H_1 \text{ correct}] = \text{area B}$$

and so, the probability of making a Type II error is given by area B. We see that moving $p*$ to the right increases area B, and thus increases the probability of a Type II error, and at the same time decreases area A and thus decreases the probability of a Type I error. The opposite happens if we move $p*$ to the left. Thus, we have (for a given sample size) a trade-off between the probability of a Type I error and the probability of a Type II error. Increasing the sample size (for a given critical value) reduces the variance of p' under both hypotheses, and thus reduces both the probability of a Type I error and the probability of a Type II error.

Where we actually fix $p*$ in any particular test will depend on the relative consequences of making the two kinds of error. Thus, for example, if making a Type I error is twice as costly as making a Type II error, we should (if we wish to equate the expected cost of a Type I error with that of a Type II error) choose the critical value so that the probability of making a Type I error is half the probability of making a Type II error, (that is, choose $p*$ to make area B twice as large as area A). The size of the sample will depend on the cost of sampling relative to the cost of making errors. In the illustrative example at the end of this section we consider the problem of choosing the optimal sample size.

The probability of making a Type I error is called the *significance level* of the test. If, for example, this probability is 0.05, then the significance level of the test is 5 per cent. We call the test *significant* (at the α per cent level) if the null hypothesis is rejected (at the α per cent significance level). The *power* of the test is defined as the probability of rejecting H_0 when it is wrong, that is

$$\text{power} = 1 - P[\text{Type II error}].$$

In many cases, no *specific* alternative hypothesis is proposed, and the test is between H_0 and H_1 where

either (1) $H_0 : p = p_0$ $H_1 : p > p_0$

or   (2) $H_0 : p = p_0$ $H_1 : p < p_0$

or   (3) $H_0 : p = p_0$ $H_1 : p \neq p_0$.

Obviously, in such cases we cannot calculate the probability of a Type II error. Thus we are unable to choose the critical value to achieve some desired trade-off between the two kinds of error (unless we are prepared to state a specific $p \neq p_0$). In such cases, it is usual to choose the critical value to achieve some desired significance level.

Thus, in case (1) we choose the critical value $p^*(>p_0)$ such that $P[p' > p^* \mid H_0$ correct] equals the desired significance level, rejecting H_0 if $p' > p^*$, and accepting H_0 if $p' \leqslant p^*$.

In case (2) we choose the critical value p^* ($<p_0$) such that $P[p' < p^* \mid H_0$ correct] equals the desired significance level, rejecting H_0 if $p' < p^*$, and accepting H_0 if $p' \geqslant p^*$.

In case (3) we choose *two* critical values p_1^* and p_2^* (where $p_1^* < p_0 < p_2^*$) such that $P[p' < p_1^*$ or $p' > p_2^* \mid H_0$ correct] equals the desired significance level, rejecting H_0 if $p' < p_1^*$ or $p' > p_2^*$, and accepting H_0 if $p_1^* \leqslant p' \leqslant p_2^*$.

For obvious reasons, cases (1) and (2) are called *one-tailed tests*, while case (3) is a *two-tailed test*.

Consider the following extended illustration. Suppose an electrician, at present earning \$10,000 net a year, is considering whether to set up a colour television repair service in a particular area. He calculates that if 30 per cent of the households in the area own colour television sets he will earn a net income of \$7,500 a year, but if 50 per cent of the households own colour sets he will earn a net income of \$11,250 a year. He thus concludes that, if the 30 per cent figure is correct, he is better off staying in his present job, whereas if the 50 per cent figure is correct, he will do better to give up his present job and to set up the repair service (we suppose that he knows that one of these two proportions is correct, but not which one). His choice commits him for one year. He can gain information by sampling households; each household sampled incurs a cost of \$2.50. We are asked to determine the optimal sample size, and the optimal decision to take in the light of the sample evidence.

We have

$$H_0 : p = 0.3$$
$$H_1 : p = 0.5.$$

Denote, as usual, by p' the proportion of households in a sample of size n with colour sets. Then, from (4.19),

$$p' \text{ is (approximately) } N\left(p, \frac{p(1-p)}{n}\right).$$

Suppose, first, we consider a sample of size 50. Then

if H_0 correct p' is (approx) $N(0.3, 0.0042)$

if H_1 correct p' is (approx) $N(0.5, 0.0050)$.

These alternative distributions are pictured in figure 4.10. Now, consider the costs of making errors. If he decides that H_1 is correct when H_0 is in fact true he will earn \$2,500 less than he would have done if he had stayed in his present job. If he decides H_0 is correct when H_1 is in fact true he will earn \$1,250 less than he would have done if he had set up the repair service. Thus, the cost of a Type I error (\$2,500) is twice as large as the cost of a Type II error (\$1,250). Therefore, we want to choose the critical value p^* so that the probability of a Type I error is half the probability of a Type II error*. Suppose we set $p^* = 0.4$, then

$$
\begin{aligned}
P[\text{Type I error}] &= P[p' > 0.4 \mid H_0 \text{ correct}] \\
&= P[p' > 0.4 \mid p' \text{ is } N(0.3, 0.0042)] \\
&= P\left[Z > \frac{0.4 - 0.3}{\sqrt{0.0042}} \mid Z \text{ is } N(0, 1) \right] \\
&= 0.0618
\end{aligned}
$$

and

$$
\begin{aligned}
P[\text{Type II error}] &= P[p' < 0.4 \mid H_1 \text{ correct}] \\
&= P[p' < 0.4 \mid p' \text{ is } N(0.5, 0.0050)] \\
&= P\left[Z < \frac{0.4 - 0.5}{\sqrt{0.0050}} \mid Z \text{ is } N(0, 1) \right] \\
&= 0.0793.
\end{aligned}
$$

Manifestly $p^* = 0.4$ does not achieve the desired trade-off, and the optimal p^* will be greater than 0.4. Actual determination of the optimal p^* is a question of a trial and error. Such a process yields the value $p^* = 0.408$ which gives

$$P[\text{Type I error}] \doteq 0.048 \quad P[\text{Type II error}] = 0.096.$$

Now, this critical value was derived on the assumption of a sample of size 50. Is this sample size optimal?

$$
\begin{aligned}
\text{The expected loss} &= (\text{cost of Type I error})P[\text{Type I error}] \\
&= 2,500 \times 0.048 = 120 \\
&= (\text{cost of Type II error})P[\text{Type II error}] \\
&= 1,250 \times 0.096 = 120.
\end{aligned}
$$

The expected loss is \$120. The cost of taking 50 observations is \$125. If we take one more observation, the cost of taking observations will rise by

*This assumes that we wish to equate the expected cost of a Type I error and the expected cost of a Type II error; this assumes in turn constant marginal utility of income.

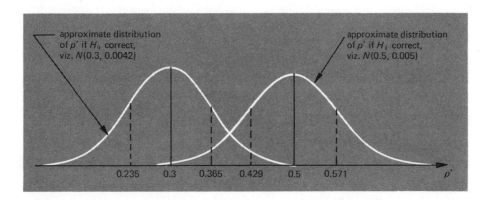

Fig. 4.10   The electrician's problem

$2.50; at the same time the expected loss will decrease. To find by how much the expected loss will decrease we would have to repeat the above analysis for a sample of size 51 (this would necessitate determining a new value for p^*). If the expected loss decreased by more than $2.50 it would be worth taking the extra observation. Conversely, it would be worth taking one less observation (that is only 49) if the saving of observation cost ($2.50) was more than the increase in expected loss. Thus the optimal * sample size is that size at which the marginal cost of taking observations equals the (negative of the) marginal expected loss. You may like to find what the optimal size is in this problem. Suppose though that a sample of size 50 was optimal; then if the proportion in a sample of size 50 was greater than 0.408 we would advise the electrician to set up the repair service, but if the sample proportion was less than .408 we would advise him to stay in his present job.

Consider now the rather simpler example of testing whether the incidence of two-car-ownership among families with heads in the 55–64 age group is different from among families in the U.S. as a whole. From the 1970 *Survey of Consumer Finances*, the proportion of two-car-owning families in the sample with heads in the 55–64 age group and in the whole of the sample were 0.27 and 0.29 respectively. Assume for the moment that the latter figure was not a sample proportion, but that 0.29 was the population proportion in the whole of the U.S.

We wish to test $H_0: p = 0.29$ against $H_1: p \neq 0.29$, where p is the population proportion in families with heads. Now, if H_0 is correct, the sample proportion p' (based on a sample size $n = 426$) is (approximately)

---

*This assumes we wish to minimize cost plus expected loss; which again assumes constant marginal utility of income.

N (0.29) (0.29) (0.71)]/426); that is, if H_0 is correct p' is (approx) N (0.29, 0.0004833). Suppose we chose a 5 per cent significance level. The test above is two-tailed, so we need two critical values p_1* and p_2* which satisfy $P[p' < p_1$* or $p' > p_2$* | H_0 correct] = 0.05. Thus

$$p_1* = 0.29 - 1.96\sqrt{0.0004833} = 0.247$$

$$p_2* = 0.29 + 1.96\sqrt{0.0004833} = 0.333.$$

As the observed value of p' (=0.27) does lie between p_1* and p_2*, the test is not significant at the 5 per cent level, and so we accept the hypothesis that the proportion of two-car-owning families with heads aged 55–64 is the same as in the U.S. as a whole.

If we carried out the one-tailed test, H_0 : p = 0.29, against H_1 : $p < 0.29$ and used a 1 per cent significance level, then our critical value would be $p*$ = 0.29 − 2.33$\sqrt{0.0004833}$ = 0.239 and our test would again be insignificant, rejecting H_1 in favour of H_0.

If we wish to compare two proportions, both of which are based on sample evidence, we need some new theory. This is considered in the next section.

4.7 The Difference Between Two Proportions

From chapter 3 we know that if X and Y are independent random variables, then

$$E(X - Y) = EX - EY$$

and

$$\text{var } (X - Y) = \text{var } X + \text{var } Y.$$

Further, it can be shown (by methods outside the scope of this book) that if X and Y are independently and *normally* distributed, then the difference $(X - Y)$ is also *normally* distributed. Combining these results we get:

if X is $N(\mu_X, \sigma_X^2)$ and Y is $N(\mu_Y, \sigma_Y^2)$, and X and Y are independent, then $(X - Y)$ is $N(\mu_X - \mu_Y, \sigma_X^2 + \sigma_Y^2)$. $\qquad$ (4.28)

Consider, then, two populations where the proportion in the populations having a certain characteristic are p_1 and p_2 respectively. Suppose we take a sample of size n_1 from the first population, and a sample of size n_2 from the second population. Denote the respective sample proportions by p_1' and p_2'. We have, from (4.19) that

$$p_1' \text{ is (approximately) } N(p_1, \sigma_1^2) \quad \text{where} \quad \sigma_1^2 = \frac{p_1(1-p_1)}{n_1}$$

and

$$p_2' \text{ is (approximately) } N(p_2, \sigma_2{}^2) \quad \text{where} \quad \sigma_2{}^2 = \frac{p_2(1-p_2)}{n_2}.$$

(We are assuming that sampling is done with replacement; if it is done without replacement, both variances need to be multiplied by the appropriate finite population correction.)

If the two samples are collected independently, then p_1' and p_2' are independent, and we can use (4.28) to show

$$(p_1' - p_2') \text{ is (approximately) } N(p_1 - p_2, \sigma_1{}^2 + \sigma_2{}^2) \qquad (4.29)$$

Denoting

$$\left. \begin{array}{l} p' = p_1' - p_2' \\ p = p_1 - p_2 \\ \text{and} \\ \sigma^2 = \sigma_1{}^2 + \sigma_2{}^2 \end{array} \right\} \qquad (4.30)$$

(4.29) can be written:

$$p' \text{ is (approximately) } N(p, \sigma^2). \qquad (4.31)$$

This result enables us to form confidence intervals for, and test hypotheses about, the difference between two population proportions.

To illustrate, we find a 95 per cent confidence interval for the difference between the proportion of two-car-owning families with heads under 25 and the proportion of two-car-owning families with heads aged 25–34. The relevant sample observations (once again obtained from the *Survey* for 1970) are:

$$p_1' = 0.15 \quad n_1 = 237 \text{ (heads aged under 25)}$$
$$p_2' = 0.26 \quad n_2 = 471 \text{ (heads aged 25–34)}$$

Estimating the variances as usual, we get

$$\sigma_1{}^2 \doteqdot 0.0005379 \quad \sigma_2{}^2 \doteqdot 0.0004084$$

and so σ^2 (the variance of $p_1' - p_2'$) is estimated by

$$\sigma^2 \doteqdot 0.0009463.$$

Also, $p' = p_1' - p_2' = -0.11$. Hence, a 95 per cent confidence interval for $p = p_1 - p_2$ is $(p' - 1.96\sigma, p' + 1.96\sigma)$ that is, $(-0.170, -0.050)$, and so we can be 95 per cent confident that in families with heads under 25 between 5.0 and 17.0 per cent less of the families own two cars than in families with heads aged 25–34.

Finally, suppose we wished to test $H_0 : p_1 = p_2$ against $H_1 : p_1 < p_2$. We could rephrase this test as being $H_0 : p = 0$ against $H_1 : p < 0$ where

$p = p_1 - p_2$. Now, if H_0 is correct, we see from (4.30) and (4.31) that

p' is (approximately) $N(0, \sigma^2)$ \hfill (4.32)

where

$$p' = p_1' - p_2' \quad \text{and} \quad \sigma^2 = \sigma_1^2 + \sigma_2^2 = \frac{p_1(1 - p_1)}{n_1} + \frac{p_2(1 - p_2)}{n_2}.$$

If we choose a 5 per cent significance level, then, as the test is one-tailed, we need a critical value p^* such that

$P[p' < p^* \mid H_0 \text{ correct}] = 0.05.$

Inspection of (4.32) and the normal area tables show that

$p^* = 0 - 1.64\sigma = -1.64\sigma.$ \hfill (4.33)

Now, we do not know σ, and therefore must estimate it. To be consistent, however, we must estimate it on the assumption that H_0 is correct, that is $p_1 = p_2$. Now, for families with heads aged under 25, 39 families out of the 257 in the sample owned two cars, while for families with heads aged 25–34, 122 out of the 471 owned two cars; thus in the two sets combined, 161 families owned two cars out of a total sample of 728—a proportion of 0.221. We estimate σ^2 using $p_1 = p_2 \doteq 0.221$, giving

$$\sigma^2 \doteq \frac{(0.221)(0.779)}{257} + \frac{(0.221)(0.779)}{471} = 0.0010353$$

Thus the critical value p^* is, from (4.33),

$p^* = -0.053.$

The observed value of $p' = p_1' - p_2' = -0.11$, which is less than p^* and so we reject the null hypothesis at the 5 per cent level: the incidence of two-car ownership in families with heads aged under 25 *is* less than in families with heads aged 25–34.

4.8 Summary

In this chapter we began by considering the distribution of the number (and proportion) of members of a sample with a certain characteristic, given a known proportion in the population with the characteristic. This distribution is called the binomial distribution. We showed that, *on average*, the sample proportion equals the population proportion, and that the variance of the sample proportion is inversely proportional to the sample size. We also showed that the binomial distribution can be approximated by the normal distribution; this approximation facilitates the calculation of probabilities. We then changed our angle of approach from that of making deductions about the sample proportion, given knowledge of the population proportion, to that of making inferences about the population proportion, given knowledge of the sample proportion. The fact that, *on average*, the sample proportion equals the population proportion enabled us to use the former as a point estimate of the latter. Interval estimates were obtained, in the form of confidence intervals, by considering the distribution of the sample proportion about its mean, the population proportion. The same considerations allowed us to test hypotheses about the population proportion using sample evidence. We saw that the choice of the sample size and the significance level (or confidence level for interval estimates) depended on the costs of sampling and on the use to which the results were to be put. The close links between estimation and hypothesis testing, although not explicitly emphasised, should have become apparent. The chapter concluded by considering the generalisation of the above results to the difference between two proportions.

4.9 Exercises

4.1. Verify directly (by using its probability distribution) that the variance of a binomially distributed variable is npq.

4.2. Calculate the probability distribution of the number of '6's obtained when a fair die is rolled six times. Verify that its mean is np and its variance is npq.

4.3. Suppose that 60 per cent of all households in the U.S. have washing machines. Define X as the number of households with washing machines in a random sample of size 100. What is the distribution of X? Find its mean and variance.

4.4. Use the data given in exercise 2.1 to find the mean and variance of the proportion of people in a random sample (taken from the whole U.S. population) of size 2,500 that are 10 years of age or less.

4.5. Suppose that 20 per cent of middle-class people over the age of 50 have been to university, while only 10 per cent of working-class people over 50 have been to university. Find the mean and variance of the proportion of university-educated people in a sample of 100 middle-class people over 50. Find the mean and variance of the proportion of university-educated people in an (independent) random sample of 400 working-class people over 50. Then find the mean and variance of the difference between these two proportions.

4.6. Given that a binomially distributed random variable has a mean 8 and variance 6, find n, p and q.

4.7. Given that a binomially distributed random variable has mean 20 and standard deviation 4, find n, p and q.

4.8. Determine the smallest sample size, from a population with 30 per cent coloured people, needed to ensure that the probability of having at least one coloured person in the sample is

    (a) not less than 0.50;
    (b) not less than 0.95.

4.9. A theatre has three vacant seats which may be booked by telephone. The management knows that, on average, 50 per cent of telephone bookings are not taken up, and has therefore decided to accept more than three reservations. Given that the opportunity cost of each empty seat is $2.50 and the value placed by the management on the goodwill lost by each over-sale (that is, each person who arrives to find his seat already taken) is $5, the management, being a cost minimiser, wishes to know how many bookings to take. What advice would you give? Discuss the assumptions you used to formulate your advice.

4.10. Given that the random variable X is normally distributed with mean 10 and variance 4 (that is, X is $N(10,4)$), calculate the following probabilities:

    (a) X is greater than 12;
    (b) X is greater than 11;
    (c) X is less than 9;
    (d) X lies between 9 and 12;
    (e) X equals 9;
    (f) X lies between 8 and 11;
    (g) X equals 11.

Evaluate the same probabilities for the case where X is $N(12,16)$.

4.11. Suppose your standardised mark in an examination was 0.8. If the marks were normally distributed, what percentage of the students would be expected to have done better than you?

4.12. A fair coin is tossed eight times. Find the probability, both exactly, by the binomial distribution formula, and approximately, by means of the normal curve approximation, of getting

    (a) 5 heads;
    (b) at least 6 heads.

4.13. Suppose that 10 per cent of 'everlasting' tights develop faults before their guarantee expires. By means of the normal curve approximation calculate the probability that a shop which has sold 100 such tights will have to
  (a) replace more than 16 of them;
  (b) replace between 5 and 15 of them.
Discuss, in the light of your answers, how much the firm ought to 'charge' for the guarantee.

4.14.
(a) What is meant by random sampling?
(b) Why are random samples important?
(c) How would you choose a random sample of size 100 from the students at your university?
(d) How would you choose a random sample of 100 shoppers in your home town?
(e) How would you choose a random sample of 100 motorists on your nearest trunk road?

4.15. Your local Students' Union wishes to estimate the percentage of students who favour a new union constitution:
(a) It proposes to select a random sample of 200 students. If the results of this poll yielded $p' = 0.6$, how accurate is this estimate of the true proportion likely to be?
(b) How large a sample should the Union plan to take if an estimate accurate to within 0.04 with 95 per cent confidence is desired? For 99 per cent confidence?

4.16. If a random sample of size 100 had been taken. and p' is found to be 0.5, with what confidence can one assert that p' is not more than 0.04 from the population proportion?

4.17 The following information on ownership of two or more cars was obtained from tables 1-6 and 4-17 of the 1970 *Survey of Consumer Finances*.

Education of Head of family	Proportion owning 2 or more cars in 1970	Number of families in sample
0–8 grades	0.14	596
9–11 grades	0.27	449
12 grades	0.30	483
Some college	0.35	685
College degree	0.41	347

Calculate the following confidence intervals for the proportion of families owning two or more cars:

(a) a 95 per cent confidence interval for the '0–8 grades'.
(b) a 90 per cent confidence interval for the '12 grades'.
(c) a 99 per cent confidence interval for the 'college degree'.

4.18 Using the data in exercise 4.17 test the following hypotheses:
(a) that not less than 30% of the '9–11 grades' own 2 or more cars.
(b) that 40% of the 'college degree' families own 2 or more cars.
(c) that not more than 30% of the 'some college' families own 2 or more cars.

4.19 Using the data of exercise 4.17,
(a) find a 95 per cent confidence interval for the difference between the incidence of two-car ownership in any two of the educational categories.
(b) for any two of the educational categories test the hypothesis that there is no difference in the incidence of two-car ownership.

4.20 A worker is, at present, paid a constant weekly wage of $200. His employers offer him the chance of changing to a piece-work rate, initially for one year. He performs two different tasks A and B, and at the beginning of each week is allocated his task, which lasts for the duration of that week. Job A is paid at a lower rate than B, and the worker knows that the probability, p, of getting job A in a week (independent of other weeks) is either 0.6 or 0.4; but the management (for obvious reasons) will not say which is the correct value. If $p = 0.6$ his average weekly earnings on piece-rate would be $150, and if $p = 0.4$ it would be $250.

He was allocated job A 18 times in the last 40 weeks. What would you advise him to do?

Indicate how you might decide whether it would be worth waiting a few weeks before taking the decision.

4.21. Solve each of the inequalities in the expression:

$$p' - Z\sqrt{\left(\frac{p(1-p)}{n}\right)} \leqslant p \leqslant p' + Z\sqrt{\left(\frac{p(1-p)}{n}\right)}$$

for p. This will involve the solution of a quadratic equation. Use your results to obtain a confidence interval for p that does not contain p in its limits. Use the results to find an exact confidence interval for 4.17 (a), (b) or (c). Compare and comment.

4.22. Discuss briefly the following statements (note: they are not all correct).

(a) 100 per cent confidence intervals for the population proportion are either pointless or worthless.
(b) In hypothesis testing the effect of making the Type I error zero is to render the test worthless.
(c) There is no connection between confidence intervals and hypothesis testing.
(d) If, when testing a null hypothesis, we find the test significant at the 5 per cent level, we can conclude that the probability of the null hypothesis being correct is 5 per cent or less.

4.23. If you rolled a die 240 times and obtained 50 '6's, would you decide that the die favoured '6's?

4.24. Let the null hypothesis be $p = 0.5$. How large a sample is needed for a sample value $p' = 0.45$ to be significant at the 5 per cent level?

4.25. Let the null hypothesis be $p = 0.5$. For a significance level of 5 per cent, find the probability of a Type II error (for a sample size 100) when the alternative hypothesis is
(a) $p = 0.6$;
(b) $p = 0.75$;
(c) $p \neq 0.5$.

5

Means and Variances

5.1 Introduction

In chapter 4 we introduced the concepts and methods involved in making inferences about population characteristics from sample evidence. We showed also how the accuracy of such inferences could be assessed. The analysis of chapter 4 was, however, restricted to a certain type of characteristic, namely those of an 'all-or-nothing' nature. In economics we are also interested in making inferences about more general types of characteristics (such as income, expenditure, investment and price, for example) which do not have this 'all-or-nothing' nature, but can take a whole range of values. This chapter therefore, broadens the scope of the analysis of chapter 4, to enable us to make inferences about these more general types of characteristics.

Although the scope of the analysis is broadened, however, the concepts and methodology are the same. Our inferences are again based on knowledge of the probability distributions of the sample characteristics. Interval estimates of population characteristics, in the form of confidence intervals, are derived in a similar fashion, and their interpretation is the same. The methods used to test *a priori* hypotheses about population characteristics using sample evidence are also the same as in chapter 4.

We are considering therefore characteristics that can take a whole range of values, such as income, expenditure, investment and price. We are interested in the distribution of such characteristics in some population, and in particular in the mean and variance of the population distribution. We wish to make inferences about population means and population variances from sample evidence. For example, we may wish to estimate mean household income in a particular region, or we may wish to test whether mean household income in one region is different from that in another region. Also, we may wish to compare the dispersion of incomes in two regions, or at two different points in time, to see whether there are any differences in the distribution of income. Other examples, which indicate the scope of this chapter, include the estimation of mean *per capita* expenditure on certain goods, and the inter-regional or international comparison of the mean expenditure; the estimation of the dispersion of prices of a particular good; testing whether the mean proportion of income devoted to bread is smaller for high-income groups than for low-income groups; and comparing the dispersion of the price of 'risky' shares with the dispersion of the price of 'safe' shares.

The first step in our analysis is, as before, to deduce probability statements about sample characteristics, given knowledge of population characteristics. Suppose, first, that we are interested in making inferences about the population mean; it would seem reasonable, particularly in view of the fact that we found that knowledge of the distribution of the sample proportion enabled us to make inferences about the population proportion, to investigate the distribution of the sample mean. This we do in the next section.

5.2   Distribution of the Sample Mean

Suppose we are interested in a particular characteristic, or variable, which we denote by X. The variable X takes particular values for each member of the population. We denote by μ the mean value of X over the whole population, and we denote by σ^2 the variance of X in the population.

Suppose we pick at random one member of the population. Denote by X_1 the value of X for this member. Obviously X_1 is a random variable as its value depends on which member is picked. The probability distribution of X_1 must be the same as the frequency distribution of X in the population, and so the expected value of X_1 is μ and the variance of X_1 is σ^2; that is

$$EX_1 = \mu \quad \operatorname{var} X_1 = \sigma^2.$$

Suppose now we pick a random sample of size 2 (with replacement) from the population, denoting by X_1 and X_2 the observed values of X for the first and second members of the sample respectively. The probability distribution of X_1 must be the same as the frequency distribution of X in the population. Similarly, the probability distribution of X_2 must be the same as the frequency distribution of X in the population. Hence

$$EX_1 = EX_2 = \mu \quad \operatorname{var} X_1 = \operatorname{var} X_2 = \sigma^2.$$

By repeating the above argument, we can see that, if we pick a random sample of size n (with replacement) from the population, and denote by X_i the value of X for the ith member of the sample ($i = 1, 2, \ldots, n$), then, for each i, the probability distribution of X_i is the same as the frequency distribution of X in the population, and so

$$EX_i = \mu \quad \operatorname{var} X_i = \sigma^2 \quad (i = 1, 2, \ldots, n). \tag{5.1}$$

It follows also, since we have sampled with replacement, that, for $i \neq j$, the value that X_i takes is not influenced by the value that X_j takes. Thus

$$X_i \text{ and } X_j \text{ are independent } (i \neq j; i, j = 1, 2, \ldots, n). \tag{5.2}$$

Denote (as usual) the mean of the n sample observations by $\bar{X}$; thus, the sample mean is

$$\bar{X} = \frac{1}{n}(X_1 + X_2 + \ldots + X_n) = \frac{1}{n}\sum_1^n X_i. \tag{5.3}$$

Consider the mean and variance of $\bar{X}$. First, its mean:

$$E\bar{X} = E\left[\frac{1}{n}(X_1 + X_2 + \ldots + X_n)\right]$$

$$= \frac{1}{n} E(X_1 + X_2 + \ldots + X_n) \qquad \text{from (3.54)}$$

$$= \frac{1}{n}(EX_1 + EX_2 + \ldots + EX_n) \quad \text{from (3.44)}$$

$$= \frac{1}{n}\underbrace{(\mu + \mu + \ldots + \mu)}_{n} \qquad\qquad \text{from (5.1)}$$

Thus,

$$E\bar{X} = \mu \tag{5.4}$$

and so, *on average*, the sample mean equals the population mean. This directly parallels our earlier result for proportions that, *on average*, the sample proportion equals the population proportion. Now,

$$\text{var } \bar{X} = \text{var}\left[\frac{1}{n}(X_1 + X_2 + \ldots + X_n)\right]$$

$$= \frac{1}{n^2}\text{var }(X_1 + X_2 + \ldots + X_n). \quad \text{from (3.55)}$$

Now $X_1, X_2, \ldots, X_n$ are all independent (from (5.2)), and so we can use the result (3.46) that the variance of the sum of independent random variables equals the sum of their variances. Thus

$$\text{var } \bar{X} = \frac{1}{n^2}(\text{var } X_1 + \text{var } X_2 + \ldots + \text{var } X_n)$$

$$= \frac{1}{n^2}\underbrace{(\sigma^2 + \sigma^2 + \ldots + \sigma^2)}_{n} \quad \text{from (5.1)}$$

Thus

$$\text{var } \bar{X} = \frac{\sigma^2}{n}. \tag{5.5}$$

Taken in conjunction with (5.4), this result shows that the dispersion of the sample mean around the population mean gets smaller as the sample size increases. This is intuitively reasonable. Note, once again, the parallel with our result for proportions that the dispersion of the sample proportion around the population proportion gets smaller as the sample size increases. In fact, both the variance of the sample mean and the variance of the sample proportion are inversely proportional to the sample size.

Note, also, that (5.5) shows that the larger the dispersion of X in the population (that is, the larger is σ^2), the larger is the dispersion of the sample mean around the population mean.

It is important not to confuse the two distributions: (a) the distribution of the variable (X) in the population; this has mean μ and variance σ^2; (b) the distribution of the mean $(\bar{X})$ of a sample of size n taken (with replacement) from the population; this has mean μ and variance σ^2/n.

To illustrate results (5.1), (5.2), (5.4) and (5.5) consider the following very simple example. Suppose there are just five members in the population, for which the variable X takes the values

3, 9, 15, 21 and 27.

The population mean and variance are respectively

$\mu = 15$ $\sigma^2 = 72$.

Suppose we take samples of size $n = 2$ with replacement. There are 25 different possible samples we may obtain, each with probability 1/25. Table 5.1 lists these 25 possible samples and gives also the value of $\bar{X}$ for

Table 5.1   List of samples of size 2 with replacement

Sample number	X_1	X_2	$\bar{X}$	Sample number	X_1	X_2	$\bar{X}$
1	3	3	3	14	15	21	18
2	3	9	6	15	15	27	21
3	3	15	9	16	21	3	12
4	3	21	12	17	21	9	15
5	3	27	15	18	21	15	18
6	9	3	6	19	21	21	21
7	9	9	9	20	21	27	24
8	9	15	12	21	27	3	15
9	9	21	15	22	27	9	18
10	9	27	18	23	27	15	21
11	15	3	9	24	27	21	24
12	15	9	12	25	27	27	27
13	15	15	15				

Each sample has probability 1/25.

Table 5.2   Distribution of X_2

Sample numbers	X_2	$P[X_2]$ (×5)	$X_2 P[X_2]$ (×5)	$(X_2 - EX_2)^2 P[X_2]$ (×5)
1, 6, 11, 16, 21	3	1	3	144
2, 7, 12, 17, 22	9	1	9	36
3, 8, 13, 18, 23	15	1	15	0
4, 9, 14, 19, 24	21	1	21	36
5, 10, 15, 20, 25	27	1	27	144
Σ		5	75	360

$$EX_2 = \frac{75}{5} = 15 \qquad \text{var } X_2 = \frac{360}{5} = 72$$

each sample. It should be fairly apparent from inspection of table 5.1 that X_1 takes the values, 3, 9, 15, 21 and 27 each with probability 1/5, so that $EX_1 = 15 \ (= \mu)$ and var $X_1 = 72 \ (= \sigma^2)$. Similarly, $EX_2 = 15 \ (= \mu)$ and var $X_2 = 72 \ (= \sigma^2)$. However, to be absolutely certain, we present in table 5.2 the probability distribution of X_2 and derive its mean and variance. Thus our example verifies (5.1). You should verify (5.2) yourself by deriving the joint probability distribution of X_1 and X_2, and checking the independence condition (3.37). The distribution of $\bar{X}$ is given in table 5.3, and this verifies (5.4) and (5.5) since

$$E\bar{X} = 15 = \mu$$

and

$$\text{var } \bar{X} = 36 = 72/2 = \sigma^2/n.$$

Table 5.3   Distribution of $\bar{X}$ for samples of size 2 with replacement

Sample numbers	Number of samples	$\bar{X}$	$P[\bar{X}]$ (×25)	$\bar{X}P[\bar{X}]$ (×25)	$(\bar{X} - E\bar{X})^2 P[\bar{X}]$ (×25)
1	1	3	1	3	144
2, 6	2	6	2	12	162
3, 7, 11	3	9	3	27	108
4, 8, 12, 16	4	12	4	48	36
5, 9, 13, 17, 21	5	15	5	75	0
10, 14, 18, 22	4	18	4	72	36
15, 19, 23	3	21	3	63	108
20, 24	2	24	2	48	162
25	1	27	1	27	144
Σ	25		25	375	900

$$E\bar{X} = \frac{375}{25} = 15 \qquad \text{var } \bar{X} = \frac{900}{25} = 36$$

Table 5.4   Distribution of $\bar{X}$ for
samples of size 3 with replacement

$\bar{X}$	$P[\bar{X}]$ (×125)	$\bar{X}P[\bar{X}]$ (×125)	$(\bar{X}-E\bar{X})^2\, P[\bar{X}]$ (×125)
3	1	3	144
5	3	15	300
7	6	42	384
9	10	90	360
11	15	165	240
13	18	234	72
15	19	285	0
17	18	306	72
19	15	285	240
21	10	210	360
23	6	138	384
25	3	75	300
27	1	27	144
Σ	125	1875	3000

$$E\bar{X} = \frac{1875}{125} = 15 \qquad \text{var } \bar{X} = \frac{3000}{125} = 24$$

In table 5.4 we give the probability distribution of the sample mean for samples of size $n = 3$ from the same population. You should check the entries in the table. This again verifies (5.4) and (5.5) since

$$E\bar{X} = 15 = \mu$$

and

$$\text{var } \bar{X} = 24 = 72/3 = \sigma^2/n.$$

We now consider whether the results derived above need to be modified if sampling is carried out without replacement. The first thing we notice is that (5.2) no longer holds if sampling is done without replacement; that is, X_i and X_j are no longer independent. However, we notice also that this independence condition (5.2) was not needed to prove that $E\bar{X} = \mu$. Thus this result holds whether sampling is done with or without replacement. We see, however, that to prove that var $\bar{X} = \sigma^2/n$ we did require the independence condition (5.2). If we sample without replacement, the variance of the sample mean is no longer given by (5.5).

We can show that when sampling is done without replacement from a population of size N, the variance of the sample mean is given by

$$\text{var } \bar{X} = \frac{\sigma^2}{n}\left(\frac{N-n}{N-1}\right) \tag{5.6}$$

that is, the 'without replacement' variance is the 'with replacement' variance multiplied by the factor $(N - n)/(N - 1)$, which we recognise, of course, as the finite population correction.

The proof of (5.6) can be carried out in two stages. First, it can be shown that, if we sample without replacement from a population of size N, then

$$\text{cov}\,(X_i, X_j) = \frac{-\sigma^2}{N - 1} \quad \text{for all } i, j; i \neq j. \tag{5.7}$$

This stage is left as an exercise for the reader (see exercise 5.4). We now generalise result (3.45) to obtain

$$\left. \begin{aligned} \text{var}\,(X_1 + X_2 + \ldots + X_n) &= \text{var}\, X_1 + \text{var}\, X_2 + \ldots + \text{var}\, X_n \\ &+ 2\, \text{cov}\,(X_1, X_2) + 2\, \text{cov}\,(X_1, X_3) + \ldots + 2\, \text{cov}\,(X_1, X_n) \\ &\qquad + 2\, \text{cov}\,(X_2, X_3) + \ldots + 2\, \text{cov}\,(X_2, X_n) \\ &\qquad\qquad + \ldots \\ &\qquad\qquad\qquad + \ldots \\ &\qquad\qquad\qquad\qquad + 2\, \text{cov}\,(X_{n-1}, X_n) \end{aligned} \right\} \tag{5.8}$$

Each variance term in (5.8) is σ^2, from (5.1). Each covariance term is $-\sigma^2/(N - 1)$ from (5.7). There are* $(n - 1) + (n - 2) + \ldots + 2 + 1 = \frac{1}{2}(n - 1)n$ such covariance terms in (5.8). Thus

$$\text{var}\,(X_1 + X_2 + \ldots + X_n) = n\sigma^2 + 2\{\tfrac{1}{2}(n - 1)n\}\left(\frac{-\sigma^2}{N - 1}\right)$$

$$= n\sigma^2 - \frac{n(n - 1)}{N - 1}\,\sigma^2$$

$$= n\left(\frac{N - n}{N - 1}\right)\sigma^2$$

and so the variance of the sample mean is

$$\text{var}\, \bar{X} = \text{var}\left(\frac{X_1 + X_2 + \ldots + X_n}{n}\right) = \frac{1}{n^2}\, \text{var}\,(X_1 + X_2 + \ldots + X_n)$$

that is,

$$\text{var}\, \bar{X} = \frac{\sigma^2}{n}\left(\frac{N - n}{N - 1}\right).$$

*To evaluate $S = (n - 1) + (n - 2) + \ldots + 2 + 1$ note that S can also
be written $\quad S = \quad 1 \quad + \quad 2 \quad + \ldots + (n - 2) + (n - 1)$
Thus adding: $2S = \underbrace{\quad n \quad + \quad n \quad + \ldots + \quad n \quad + \quad n \quad}_{(n - 1)} = (n - 1)n$

Table 5.5   List of samples of size 2 without replacement

Sample number	X_1	X_2	$\bar{X}$	Sample number	X_1	X_2	$\bar{X}$
1	3	9	6	11	15	21	18
2	3	15	9	12	15	27	21
3	3	21	12	13	21	3	12
4	3	27	15	14	21	9	15
5	9	3	6	15	21	15	18
6	9	15	12	16	21	27	24
7	9	21	15	17	27	3	15
8	9	27	18	18	27	9	18
9	15	3	9	19	27	15	21
10	15	9	12	20	27	21	24

Each sample has probability 1/20.

We illustrate this result, using the population introduced earlier. Table 5.5 lists all possible samples of size $n = 2$ where sampling is done without replacement. Inspection of table 5.5 verifies that (5.1) still holds, but that (5.2) (that X_1 and X_2 are independent) no longer holds. You should verify the validity of (5.7) concerning the covariance of X_1 and X_2. Table 5.6 gives the probability distribution of the sample mean, and

Table 5.6   Distribution of $\bar{X}$ for samples of size 2 without replacement

Sample numbers	Number of samples	$\bar{X}$	$P[\bar{X}]$ (× 10)	$\bar{X}P[\bar{X}]$ (× 10)	$(\bar{X} - E\bar{X})^2 P[\bar{X}]$ (× 10)
1, 5	2	6	1	6	81
2, 9	2	9	1	9	36
3, 6, 10, 13	4	12	2	24	18
4, 7, 14, 17	4	15	2	30	0
8, 11, 15, 18	4	18	2	36	18
12, 19	2	21	1	21	36
16, 20	2	24	1	24	81
Σ	20		10	150	270

$$E\bar{X} = \frac{150}{10} = 15 \qquad \text{var } \bar{X} = \frac{270}{10} = 27$$

verifies that (5.4) still holds; that is $E\bar{X} = 15 \ (= \mu)$. It verifies (5.6) also; that is,

$$\text{var } \bar{X} = 27 = \frac{72}{2}\left(\frac{5-2}{5-1}\right) = \frac{\sigma^2}{n}\left(\frac{N-n}{N-1}\right).$$

Collecting results obtained so far we have:

if sampling with replacement $E\bar{X} = \mu$ var $\bar{X} = \sigma^2/n$

i. sampling without replacement $E\bar{X} = \mu$ var $\bar{X} = \dfrac{\sigma^2}{n}\left(\dfrac{N-n}{N-1}\right).$ (5.9)

We continue the discussion assuming that sampling is done with replacement, but subsequent results are equally applicable to the 'without replacement' case, as long as we apply the modifications noted above.

We have found the mean and variance of the distribution of the sample mean; we now consider the 'shape' of the distribution of the sample mean. If the frequency distribution of X in the population is normal, then each X_i $(i = 1,2, \ldots, n)$ will also be normally distributed, since the probability distribution of each X_i must necessarily be the same as the frequency distribution of X in the population. Now, generalising the result stated in (4.28), that the sum of two independent normally distributed random variables is also normal, it follows that $X_1 + X_2 + \ldots + X_n$ is normally distributed, and hence $\bar{X} = (X_1 + X_2 + \ldots + X_n)/n$ is also normally distributed. Thus, combining this result with (5.4) and (5.5) we have

if X is $N(\mu, \sigma^2)$ then $\bar{X}$ is $N(\mu, \sigma^2/n)$ (5.10)

or, standardising,

if X is $N(\mu, \sigma^2)$ then $\dfrac{\bar{X} - \mu}{\sigma/\sqrt{n}}$ is $N(0, 1).$ (5.11)

Now, if the distribution of the population is not normal, then neither is the distribution of each X_i. However, particularly since the results derived in this section parallel so closely the results of section 4.3, you may have anticipated that the standardised distribution of $\bar{X}$ (that is, $(\bar{X} - \mu)/(\sigma/\sqrt{n})$) can be approximated by the unit normal distribution for n sufficiently large, with the approximation improving as n gets larger. Inspection of tables 5.3 and 5.4 shows that, even for very small n, there is evidence of normality in the distribution of $\bar{X}$. For example, table 5.4 shows that the probability that $\bar{X}$ lies within one standard deviation of its mean (between $15 \pm \sqrt{24}$) is 0.68; for a normally distributed random variable this probability is 0.6826; also, that the probability that $\bar{X}$ lies within two standard deviations of its mean (between $15 \pm 2\sqrt{24}$) is 0.936; the corresponding probability for a normally distributed variable is 0.9544. Thus, even for n as small as 3, the distribution of the sample mean $\bar{X}$ is fairly well approximated by the normal distribution. Note, also, that the population from which the samples were drawn in our illustrative example was rectangularly distributed.

Our illustrations thus give intuitive insight into the general result that, whatever the 'shape' of the population distribution, the distribution of the sample mean is approximately normal, with the approximation improving as n increases. This result is one of the famous *central limit theorems*, and gives further justification of the extensive use of the normal distribution in applied statistics. Combining this with our previous results (5.4) and (5.5) we have:

> for any population distribution with mean μ and variance σ^2, $\bar{X}$ is approximately $N(\mu, \sigma^2/n)$, with the approximation improving as n increases; $\qquad$ (5.12)

or, standardising,

> for any population distribution with mean μ and variance σ^2, $(\bar{X} - \mu)/(\sigma/\sqrt{n})$ is approximately $N(0,1)$, with the approximation improving as n increases. $\qquad$ (5.13)

Notice, very carefully, that the approximation referred to in (5.12) and (5.13) applies to the 'shape' of the distribution of $\bar{X}$. The approximation does *not* refer to the mean and variance of $\bar{X}$, which are *exact* results. We note, of course, that (5.12) and (5.13) are appropriate for the case when sampling is done with replacement; if sampling is done without replacement, the variance of $\bar{X}$ needs to be multiplied by the finite population correction.

We collect our results together in table 5.7. To illustrate these results we consider the population of all families in the United States. In the *Statistical Abstract of the United States 1972* the distribution of pre-tax income for this population (for 1970) is given in table 525. The size of the population is $N = 51,948,000$, and the mean income (in dollars) is approximately $\mu = 10,773$. The standard deviation of income is approximately $\sigma = 1,163$ (dollars). If the 6,868 distribution of income was normal (which inspection of table 525 in the *Statistical Abstract* shows is untrue), this information would enable us to say, for example, that the probability that a person picked at random from this population has an income within one standard deviation of the mean (that is, between $3,905 and $17,641) is 0.6826 (from the normal area tables).

However, for sufficiently large n the central limit theorem shows that $\bar{X}$ is approximately normal. Samples of size $n = 100$ should certainly be 'sufficiently large' (cf. $n = 3$ in table 5.4). For such a large population relative to this sample size, whether we sample with or without replacement is immaterial; since, for $N = 51,948,000$ and $n = 100$, the finite population correction is $(N - n)/(N - 1) = 0.999998$. Now, for

Table 5.7   Summary of results of section 5.2

If the population distribution of X has mean μ and variance σ^2, and if we take a random sample of size n from this population, then the sample mean $\bar{X}$ has the following properties:

	With replacement	Without replacement	
$E\bar{X} =$	μ	μ	
var $\bar{X} =$	$\dfrac{\sigma^2}{n}$	$\dfrac{\sigma^2}{n}\left(\dfrac{N-n}{N-1}\right)$	
If population is normal	$\dfrac{\bar{X}-\mu}{\sigma\sqrt{n}}$	$\dfrac{\bar{X}-\mu}{\dfrac{\sigma}{\sqrt{n}}\sqrt{\left(\dfrac{N-n}{N-1}\right)}}$	is $N(0,1)$
If population is non-normal	$\dfrac{\bar{X}-\mu}{\sigma\sqrt{n}}$	$\dfrac{\bar{X}-\mu}{\dfrac{\sigma}{\sqrt{n}}\sqrt{\left(\dfrac{N-n}{N-1}\right)}}$	is approximately $N(0,1)$

samples of size 100, we have, from table 5.7,

$$E\bar{X} = \mu = 10{,}773 \quad \text{var } \bar{X} = \frac{\sigma^2}{n} = \left(\frac{\sigma}{\sqrt{n}}\right)^2 = \left(\frac{6{,}868}{\sqrt{100}}\right)^2 = (686.8)^2$$

and so $(\bar{X} - 10{,}773)/686.8$ is approximately $N(0,1)$. Thus, from the unit normal area tables, we have, approximately,

$$P\left[-1.96 \leqslant \frac{\bar{X}-10{,}773}{686.8} \leqslant 1.96\right] = 0.95$$

and so

$$P[\,9{,}427 \leqslant \bar{X} \leqslant 12{,}119\,] = 0.95.$$

Thus, in 95 per cent of all samples of size 100 taken from this population, the mean income of the families in the sample will be between \$9,427 and \$12,119.

If we took samples of size 10,000 (and we can again ignore the finite population correction which is 0.9998), then,

$$E\bar{X} = \mu = 10{,}773 \quad \text{var } \bar{X} = \left(\frac{\sigma}{\sqrt{n}}\right)^2 = (68.68)^2$$

and so, $(\bar{X} - 10{,}773)/\,68.68$ is approximately $N(0,1)$, giving

$$P\left[-1.96 \leqslant \frac{\bar{X}-10{,}773}{68.68} \leqslant 1.96\right] = 0.95$$

that is,

$$P[10,638 \leqslant \bar{X} \leqslant 10,908] = 0.95$$

and so in 95 per cent of all samples of size 10,000 the sample mean income will be between $10,638 and $10,908. By similar means, we can show that in 99 per cent of such samples, the sample mean income will be between $10,596 and $10,950.

Thus the results obtained in this section, and summarised in table 5.7, enable us to deduce probability statements about the sample mean, given knowledge of the population mean (and population variance). The next section considers how we may make inferences about the population mean, given knowledge of a sample mean.

5.3   Inferences about Population Means I

We now consider how we may use the results of section 5.2 to make inferences about some unknown population mean, given information on the mean of a sample drawn from the population. Throughout this section we will be assuming that the sample is drawn with replacement; our results can, of course, be modified for sampling without replacement. Also, by 'sample' we will of course be meaning 'random sample'; by now the importance of random sampling should be so apparent that repeated qualifications of this nature are unnecessary. As a final introductory remark, we repeat our earlier statement that the concepts and method-ology of this chapter are identical to those used in chapter 4; thus, our discussion in this chapter will be relatively brief to avoid repetition of material already presented. If you find the exposition too brief, you may well find it helpful to reread the corresponding sections of the previous chapter.

From table (5.7) we see that $(\bar{X} - \mu)/(\sigma/\sqrt{n})$ has either exactly, if the population is normal, or approximately, by the central limit theorem if the population is non-normal, a unit normal distribution. To avoid repetition in the subsequent discussion we will assume a normally distributed population. The results are understood to hold approximately for non-normally distributed populations. Thus,

$$P\left[-Z_{\alpha/2} \leqslant \frac{\bar{X} - \mu}{\sigma/\sqrt{n}} \leqslant Z_{\alpha/2}\right] = 1 - \alpha \tag{5.14}$$

where Z_β is defined as that value for which $P[Z \geqslant Z_\beta] = \beta$ where Z is a unit normal (for example, $Z_{0.025} = 1.96$, $Z_{0.05} = 1.64$, $Z_{0.01} = 2.33$, $Z_{0.005} = 2.58$, etc.). The probability statement (5.14) can be expressed in two other forms:

$$P[\mu - Z_{\alpha/2}\,\sigma/\sqrt{n} \leqslant \overline{X} \leqslant \mu + Z_{\alpha/2}\,\sigma/\sqrt{n}] = 1 - \alpha \tag{5.15}$$

$$P[\overline{X} - Z_{\alpha/2}\,\sigma/\sqrt{n} \leqslant \mu \leqslant \overline{X} + Z_{\alpha/2}\,\sigma/\sqrt{n}] = 1 - \alpha. \tag{5.16}$$

Figure 5.1 illustrates these two statements. (Compare this with figure 4.7 illustrating probability statements (4.21) and (4.22).) Statement (5.15) merely says that the area under the probability density curve of $\overline{X}$ between $\mu - Z_{\alpha/2}\,\sigma/\sqrt{n}$ and $\mu + Z_{\alpha/2}\,\sigma/\sqrt{n}$ is $(1 - \alpha)$, (the area in each shaded tail is $\alpha/2$). Consider now the interval

$$(\overline{X} - Z_{\alpha/2}\,\sigma/\sqrt{n},\ \ \overline{X} + Z_{\alpha/2}\,\sigma/\sqrt{n}).$$

Its width $(2Z_{\alpha/2}\,\sigma/\sqrt{n})$ is constant, but its position is random. We see that whenever $\overline{X}$ lies in the interval $(\mu - Z_{\alpha/2}\,\sigma/\sqrt{n},\ \mu + Z_{\alpha/2}\,\sigma/\sqrt{n})$ then the interval $(\overline{X} - Z_{\alpha/2}\,\sigma/\sqrt{n},\ \overline{X} + Z_{\alpha/2}\,\sigma/\sqrt{n})$ covers (or contains) μ. (Cases A

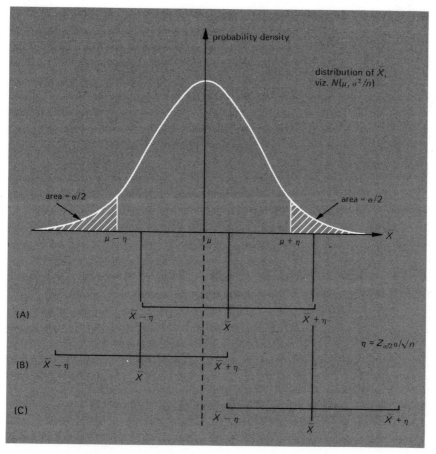

Figure 5.1   Distribution of sample mean

and B.) Whenever $\bar{X}$ lies outside the interval $(\mu - Z_{\alpha/2}\sigma/\sqrt{n},$ $\mu + Z_{\alpha/2}\sigma/\sqrt{n})$, then the interval $(\bar{X} - Z_{\alpha/2}\sigma/\sqrt{n}, \bar{X} + Z_{\alpha/2}\sigma/\sqrt{n})$ does not cover μ. (Case C.) Probability statement (5.16) states that the interval $(\bar{X} - Z_{\alpha/2}\sigma/\sqrt{n}, \bar{X} + Z_{\alpha/2}\sigma/\sqrt{n})$ covers μ with probability $(1 - \alpha)$ since $\bar{X}$ lies in $(\mu - Z_{\alpha/2}\sigma/\sqrt{n}, \mu + Z_{\alpha/2}\sigma/\sqrt{n})$ with probability $(1 - \alpha)$. Thus for $100(1 - \alpha)$ per cent of all samples, the interval $(\bar{X} - Z_{\alpha/2}\sigma/\sqrt{n},$ $\bar{X} + Z_{\alpha/2}\sigma/\sqrt{n})$ covers (or contains) μ. Thus, for any particular sample, (that is, for any observed value of $\bar{X}$) we can be $100(1 - \alpha)$ per cent confident that the interval

$$(\bar{X} - Z_{\alpha/2}\sigma/\sqrt{n}, \quad \bar{X} + Z_{\alpha/2}\sigma/\sqrt{n}) \tag{5.17}$$

covers μ. This interval is, thus, a $100(1 - \alpha)$ per cent *confidence interval* for μ, and it provides an interval estimate of μ. (Obviously a *point* estimate for μ is provided by $\bar{X}$ itself, and, *on average*, we will be correct using $\bar{X}$ as a point estimate of μ since $E\bar{X} = \mu$.)

Notice that we say that (5.17) is 'a' $100(1 - \alpha)$ per cent confidence interval for μ; there are obviously many others, the one presented in (5.17) is the one centred on $\bar{X}$, and is the most suitable if we have no reason, one way or the other, to bias the confidence interval.

Inspection of (5.17) shows that we have the same kind of problem that arose in chapter 4: if we do not know σ we cannot form the confidence interval. In chapter 4 we saw that there were ways round this problem when forming confidence intervals for the population proportion. Unfortunately, such ways round the problem cannot be used when forming confidence intervals for the population mean. For the moment, therefore, we assume that we have knowledge of the value of σ. This is obviously an extreme assumption (if we do not know μ, it is even more unlikely that we know σ); however, we will maintain it for the remainder of this section. In section 5.9 we relax this assumption; the intervening sections cover the theory necessary to enable us to do so.

Once again, it is very important that the interpretation of the phrase '$100(1 - \alpha)$ per cent confidence interval for μ' is fully understood. It means that, if we always form $100(1 - \alpha)$ per cent confidence intervals for μ in this way, then the statement that the confidence interval contains μ will be correct $100(1 - \alpha)$ per cent of the time. Whether it does for any particular sample, we do not know.

We illustrate by forming an interval estimate for the mean family income in 1969 for families with an operative as its head. We obtain the data, once again, from the 1970 *Survey of Consumer Finances*. Table 1-6 of the *Survey* shows that the mean (1969) family income of the 360 families in the sample (with operatives as heads) was $9,910. Suppose we know, from some other source, that the standard deviation of family income (for families with operative heads) was $\sigma = 4,400$.

We have:

$$X = 9,910 \quad n = 360 \quad \sigma = 4,400$$

As the distribution of income in the population is not normal, the following results are understood to hold approximately. For 95 per cent confidence, $\alpha = 0.05$, and so $Z_{\alpha/2} = Z_{0.025} = 1.96$. Substituting these values in (5.17), we get the following 95 per cent confidence interval for μ. the mean family income for the population of *all* families with operative heads, $(9,910 - 1.96(4,400/\sqrt{360}), 9,910 + 1.96(4,400/\sqrt{360}))$, that is $(9,455, 10,365)$, and so we can be 95 per cent confident that the mean family income for families with operative heads in 1969 was between \$9,455 and \$10,365.

From the same table, we have for the U.S. as a whole $\bar{X} = 10,420$, $n = 2,576$, and suppose we know $\sigma = 7,800$. For a 99 per cent confidence interval, $Z_{0.005} = 2.58$; thus a 99 per cent confidence interval for the U.S. mean family income is $(10,420 - 2.58(7,800/\sqrt{2,576}), 10,420 + 2.58(7,800/\sqrt{2,576}))$, that is $(10,024, 10,816)$.

We can also use the results contained in table 5.7 to test hypotheses about the population mean. Suppose we wish to test whether sample evidence supports the hypothesis H_0 or the hypothesis H_1 where

$$H_0 : \mu = \mu_0 \qquad (\mu_0 < \mu_1, \text{ say})$$
$$H_1 : \mu = \mu_1 .$$

Now, we know that $(\bar{X} - \mu)/(\sigma/\sqrt{n})$ is $N(0,1)$ (or approximately so if the population is not normal; this qualification is understood to apply throughout the following discussion), that is, $\bar{X}$ is $N(\mu, \sigma^2/n)$. Hence:

if H_0 is correct $\bar{X}$ is $N(\mu_0, \sigma^2/n)$

if H_1 is correct $\bar{X}$ is $N(\mu_1, \sigma^2/n)$

where we assume (in this section) that the population variance σ^2 is known and is therefore not subject to test. Figure 5.2 illustrates the distribution of $\bar{X}$ under the two hypotheses about the population mean. If the observed value of $\bar{X}$ is less than μ_0, it seems reasonable to conclude that H_0 is more likely to be correct; and if the observed value of $\bar{X}$ is greater than μ_1 it seems reasonable to conclude that H_1 is more likely to be correct. Suppose we choose some critical value X^*, and use the decision rule

if $\bar{X} < X^*$ then accept H_0 (reject H_1)

if $\bar{X} > X^*$ then accept H_1 (reject H_0)

The probability of a Type I error (rejecting H_0, the null hypothesis, when

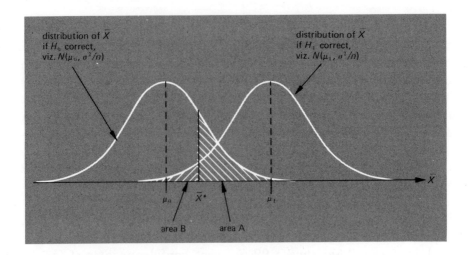

distribution of $\bar{X}$
if H_0 correct,
viz. $N(\mu_0, \sigma^2/n)$

distribution of $\bar{X}$
if H_1 correct,
viz. $N(\mu_1, \sigma^2/n)$

μ_0 $\bar{X}^*$ μ_1

area B   area A

Figure 5.2   Distribution of $\bar{X}$ under two hypotheses about the population mean

in fact it is correct) is given therefore by

$$P[\text{Type I error}] = P[\bar{X} > X^* \mid H_0 \text{ correct}]$$
$$= P[\bar{X} > X^* \mid \bar{X} \text{ is } N(\mu_0, \sigma^2/n)]$$
$$= \text{area A in figure 5.2.}$$

Similarly

$$P[\text{Type II error}] = \text{area B in figure 5.2.}$$

Naturally, again we get a 'trade-off' (for a given sample size) between the probability of a Type I error and the probability of a Type II error. Thus, once again, the choice of the optimal critical value X^* depends on the relative consequences of the two kinds of error. We can of course reduce both probabilities (for a given critical value) by increasing the sample size, as this reduces the variance of $\bar{X}$ under both hypotheses. Here again, the choice of the optimal sample size depends on the cost of obtaining observations relative to the cost of making mistakes.

If the alternative hypothesis (H_1) is of a non-specific form, we are of course unable to calculate the probability of a Type II error. In such cases the choice of critical value will be determined by the desired significance level (the probability of a Type I error). Consider the following three cases:

(1) $H_0 : \mu = \mu_0$ $H_1 : \mu > \mu_0$

(2) $H_0 : \mu = \mu_0$ $H_1 : \mu < \mu_0$

(3) $H_0 : \mu = \mu_0$ $H_1 : \mu \neq \mu_0$.

Suppose in each case an α per cent significance level is desired. Then, in case (1), we choose the critical value X^* $(>\mu_0)$ such that $P[\bar{X} > X^* \mid H_0$ correct$] = \alpha$. But, if H_0 is correct, $\bar{X}$ is $N(\mu_0, \sigma^2/n)$. Thus X^* is chosen so that $P[\bar{X} > X^* \mid \bar{X}$ is $N(\mu_0, \sigma^2/n]] = \alpha$, that is

$$X^* = \mu_0 + Z_\alpha \sigma/\sqrt{n} \qquad (5.18)$$

where Z_α is defined as before. Our decision rule is to accept H_0 if $\bar{X} \leqslant X^*$ and to reject H_0 if $\bar{X} > X^*$. In case (2) we choose the critical value X^* $(<\mu_0)$ such that $P[\bar{X} < X^* \mid H_0$ correct$] = \alpha$. Thus

$$X^* = \mu_0 - Z_\alpha \sigma/\sqrt{n} \qquad (5.19)$$

and our decision rule is to accept H_0 if $\bar{X} \geqslant X^*$ and to reject H_0 if $\bar{X} < X^*$.

In case (3) we choose two critical values X_1^* and X_2^* $(X_1^* < \mu_0 < X_2^*)$ such that $P[\bar{X} < X_1^*$ or $\bar{X} > X_2^* \mid H_0$ correct$] = \alpha$. Thus (for a symmetric test)

$$\left. \begin{array}{l} X_1^* = \mu_0 - Z_{\alpha/2}\sigma/\sqrt{n} \\ X_2^* = \mu_0 + Z_{\alpha/2}\sigma/\sqrt{n} \end{array} \right\} \qquad (5.20)$$

and our decision rule is to accept H_0 if $X_1^* \leqslant \bar{X} \leqslant X_2^*$ and reject H_0 if $\bar{X} < X_1^*$ or $\bar{X} > X_2^*$.

We illustrate by testing whether the evidence in the 1970 *Survey of Consumer Finances* supports the view that mean family income in black families is lower than in the U.S. as a whole. Suppose, for the purposes of illustration, that the U.S. mean income figure of $10,420 in 1969 is the population mean, rather than a sample mean. Denoting mean family income in black families by μ, we wish to test $H_0 : \mu = 10,420$ against $H_1 : \mu < 10,420$. If H_0 is correct, the sample mean income in a sample of size n will be (approximately) $N(10,420, \sigma^2/n)$. Suppose a 1 per cent significance level is desired, then we choose our critical value X^* by (from (5.19))

$$X^* = 10,420 - 2.33\sigma/\sqrt{n}.$$

From black families $(n=)$ 279 families were included in the sample. We suppose that the standard deviation of family income in black families is $\sigma = 4,000$. Thus $X^* = 10,420 - 2.33(4,000/\sqrt{279}) = 9,862$. Now, the observed value of $\bar{X}$ (the sample mean income in black families) was $6,870. This is less than X^*. Thus we find the test significant at the 1 per cent level, and thus conclude that mean income in black families *is* lower than in the U.S. as a whole.

Finally in this section, we consider how we may make inferences about the difference between two population means. This section closely parallels section 4.7, so we will be brief. Suppose a sample of size n_1 is taken from a population with mean μ_1, and variance $\sigma_1{}^2$. Then, denoting by $\bar{X}_1$ the sample mean, it follows (from table 5.7) that

$$\bar{X}_1 \text{ is } N(\mu_1, \sigma_1{}^2/n_1) \quad \text{(or approximately so)} \tag{5.21}$$

Similarly if $\bar{X}_2$ is the mean of a sample of size n_2 taken *independently* from another population with mean μ_2 and variance $\sigma_2{}^2$, then

$$\bar{X}_2 \text{ is } N(\mu_2, \sigma_2{}^2/n_2) \quad \text{(or approximately so)} \tag{5.22}$$

Since $\bar{X}_1$ and $\bar{X}_2$ are independent, we use the results of section 4.7 to show

$$\bar{X}_1 - \bar{X}_2 \text{ is } N\left(\mu_1 - \mu_2, \frac{\sigma_1{}^2}{n_1} + \frac{\sigma_2{}^2}{n_2}\right) \tag{5.23}$$

Result (5.23) enables us to form confidence intervals for, and test hypotheses about, $\mu_1 - \mu_2$, given sample evidence (and, of course, still assuming we know the population variances).

As a brief illustration we test whether there is any significant difference between mean family income for families with operative heads and those with clerical heads. We are testing $H_0 : \mu_1 - \mu_2 = 0$ against $H_1 : \mu_1 - \mu_2 \neq 0$ where subscripts 1 and 2 refer to the operative heads and to clerical heads respectively. If H_0 is correct $\bar{X}_1 - \bar{X}_2$ is (approximately) $N(0, \sigma_1{}^2/n_1 + \sigma_2{}^2/n_2)$. This is a two-tailed test and so, for a 5 per cent significance level, we have, from (5.20), the two critical values for $(\bar{X}_1 - \bar{X}_2)$:

$$0 - 1.96\sqrt{\left(\frac{\sigma_1{}^2}{n_1} + \frac{\sigma_2{}^2}{n_2}\right)} \quad \text{and} \quad 0 + 1.96\sqrt{\left(\frac{\sigma_1{}^2}{n_1} + \frac{\sigma_2{}^2}{n_2}\right)}$$

Now

$$\sigma_1 = 4{,}400 \qquad \sigma_2 = 5{,}000$$
$$n_1 = 360 \qquad n_2 = 287$$

Thus, the critical values for $(\bar{X}_1 - \bar{X}_2)$ are -735 and 735. The observed values of $\bar{X}_1$ and $\bar{X}_2$ are respectively 9,910 and 10,740. Thus, the observed value of $(\bar{X}_1 - \bar{X}_2)$ is -830 which lies outside the critical values. Therefore the test is significant at the 5 per cent level, and so we conclude that, on the basis of the sample evidence, there *is* a significant difference between mean family income in the two types of family.

The whole of the analysis of this section was based on the assumption that we knew the values of the relevant population variances. In the next section we consider how we may estimate the population variance when it is not known.

5.4 Estimation of the Population Variance

Suppose that the variable X has population mean μ and population variance σ^2. Suppose that we pick a random sample of size n from this population. Denote the values of X in the sample as before by $X_1, X_2, \ldots, X_n$. As will become apparent, the results in this and subsequent sections require that X_i and X_j are independent for all i, j $(i \neq j)$. We will therefore assume that sampling is carried out with replacement, or, equivalently (and you should convince yourself that it is), that sampling is carried out without replacement from an *infinite population*. Thus we have:

$$EX_i = \mu \quad \text{var } X_i = \sigma^2 \qquad (i = 1, 2, \ldots, n)$$
$$X_i \text{ and } X_j \text{ are independent} \quad (i \neq j; i, j = 1, 2, \ldots, n). \qquad \left.\right\} \quad (5.24)$$

Suppose that both μ and σ^2 are unknown. We consider in this section how we may estimate σ^2 using sample evidence.

We have already seen that the sample proportion provides a (point) estimate of the population proportion, and that the sample mean provides a (point) estimate of the population mean. We have seen also, for both the proportion and the mean, that using the sample characteristic to estimate the corresponding population characteristic provides an estimate that is correct *on average*; that is

$$Ep' = p \quad E\bar{X} = \mu.$$

It would seem reasonable, therefore, to investigate whether the sample variance can be used as a (point) estimate of the population variance, and to investigate whether such an estimate would be correct *on average*.

Denote the sample variance (that is, the variance of the n observations in the sample) by d^2. Thus,

$$d^2 = \frac{1}{n} \sum_1^n (X_i - \bar{X})^2. \qquad (5.25)$$

We now find the average value of d^2 (over all possible samples) to see whether, *on average*, d^2 equals σ^2, that is, to see whether Ed^2 equals σ^2. From (5.25) we can express d^2 as

$$d^2 = \frac{1}{n} \sum_1^n \{(X_i - \mu) - (\bar{X} - \mu)\}^2$$

$$= \frac{1}{n} \sum_1^n \{(X_i - \mu)^2 - 2(\bar{X} - \mu)(X_i - \mu) + (\bar{X} - \mu)^2\}$$

Thus

$$d^2 = \frac{1}{n} \sum_1^n (X_i - \mu)^2 - (\bar{X} - \mu)^2. \qquad (5.26)$$

Consider the expected value of the ith term in the sum in (5.26):

$$E(X_i - \mu)^2.$$

Since $EX_i = \mu$ from (5.24), $E(X_i - \mu)^2 = E(X_i - EX_i)^2 = \text{var } X_i$, from (3.30). Thus

$$E(X_i - \mu)^2 = \sigma^2 \quad \text{from (5.24)} \tag{5.27}$$

Also, since $E\bar{X} = \mu$ and $\text{var } \bar{X} = \sigma^2/n$ (from table 5.7), then

$$E(\bar{X} - \mu)^2 = E(\bar{X} - E\bar{X})^2 = \text{var } \bar{X} = \sigma^2/n. \tag{5.28}$$

Thus, substituting (5.27) and (5.28) into (5.26), we have

$$Ed^2 = \frac{1}{n} \underbrace{(\sigma^2 + \sigma^2 + \ldots + \sigma^2)}_{n} - \sigma^2/n.$$

That is,

$$Ed^2 = \left(\frac{n-1}{n}\right)\sigma^2 \tag{5.29}$$

and so, *on average*, d^2 is *not* equal to σ^2, and if we use d^2 to estimate σ^2 we will, *on average*, be underestimating σ^2. However (5.29) enables us to find a way of estimating σ^2 that *will* be correct on average. Define s^2 by

$$s^2 = \left(\frac{n}{n-1}\right)d^2 = \frac{1}{n-1}\sum_1^n (X_i - \bar{X})^2 \quad \text{(from (5.25))}. \tag{5.30}$$

Then, combining (5.30) and (5.29), we have

$$Es^2 = \sigma^2 \tag{5.31}$$

and so, if we use s^2 to estimate σ^2, we will, *on average*, be correct. s^2 differs from d^2 (the sample variance) only in that the sum of squares of the deviations of the X_i from their mean is divided by $(n-1)$ instead of by n.

We illustrate results (5.29) and (5.31) using the population defined in section 4.2 and for which $\sigma^2 = 72$. The list of all possible samples of size $n = 2$ (with replacement) was given in table 5.1. For sample number 5, for example, $X_1 = 3, X_2 = 27$ and so $\bar{X} = 15$. Thus

$$d^2 = \frac{1}{n}\sum_1^n (X_i - \bar{X})^2 = \frac{1}{2}\{(3 - 15)^2 + (27 - 15)^2\} = 144$$

and

$$s^2 = \frac{1}{n-1}\sum_1^n (X_i - \bar{X})^2 = 288.$$

Table 5.8   Distributions of d^2 and s^2 $(n = 2)$

Sample numbers	d^2	$P[d^2]$ (×25)	$d^2P[d^2]$ (×25)	s^2	$P[s^2]$ (×25)	$s^2P[s^2]$ (×25)
1, 7, 13, 19, 25	0	5	0	0	5	0
2, 6, 8, 12, 14, 18, 20, 24	9	8	72	18	8	144
3, 9, 11, 15, 17, 23	36	6	216	72	6	432
4, 10, 16, 22	81	4	324	162	4	648
5, 21	144	2	288	288	2	576
	Σ	25	900	Σ	25	1,800

$$Ed^2 = \frac{900}{25} = 36 \qquad Es^2 = \frac{1,800}{25} = 72$$

Table 5.9   Distributions of d^2 and s^2 $(n = 3)$

d^2	$P[d^2]$ (×125)	$d^2P[d^2]$ (×125)	s^2	$P[s^2]$ (×125)	$s^2P[s^2]$ (×125)
0	5	0	0	5	0
8	24	192	12	24	288
24	18	432	36	18	648
32	18	576	48	18	864
56	24	1,344	84	24	2,016
72	12	864	108	12	1,296
96	6	576	144	6	864
104	12	1,248	156	12	1,872
128	6	768	192	6	1,152
Σ	125	6,000	Σ	125	9,000

$$Ed^2 = \frac{6,000}{125} = 48 \qquad Es^2 = \frac{9,000}{125} = 72$$

Repeating this for all samples we derive the probability distributions of d^2 and s^2, which are given in table 5.8. This table verifies (5.29) and (5.31) since

$$Ed^2 = 36 = \left(\frac{2-1}{2}\right)72 = \left(\frac{n-1}{n}\right)\sigma^2 \quad \text{and} \quad Es^2 = 72 = \sigma^2 .$$

Table 5.9 gives the probability distributions of d^2 and s^2 when we take samples of size $n = 3$ (with replacement) from the same population. Once again (5.29) and (5.31) are verified, since as table 5.9 shows

$$Ed^2 = 48 = \left(\frac{3-1}{3}\right)72 = \left(\frac{n-1}{n}\right)\sigma^2 \quad \text{and} \quad Es^2 = 72 = \sigma^2 .$$

You should check the entries in this table.

By inspection of (5.29) (and the illustrative examples) it can be seen that, although d^2 underestimates σ^2 on average, the amount of underestimation decreases as n increases. However for small n the amount of underestimation can be quite large.

Our results show that if we estimate the population variance σ^2 by s^2 (a modification of the sample variance, as given by (5.30)) we will, *on average*, be correct. The value of s^2 in a *particular* sample may, however, be a long way from the value of σ^2, especially for small n, as tables 5.8 and 5.9 show very clearly. For large n, as we will see, the dispersion of s^2 around σ^2 will be sufficiently small for us to approximate σ^2 by s^2 without introducing too large an error of approximation. This is, in fact, what we did in section 4.3 to arrive at an approximate value for σ^2. Thus, the values we used for σ were obtained from table 1–6 of the 1970 *Survey,* and were the appropriate value of s (rounded to the nearest dollar).

Thus, for sufficiently large n (it will become apparent in section 5.9 how large is 'sufficiently large'), we may approximate σ by s when making inferences about the population mean. For small n we need to find the distribution of s^2 around σ^2 to enable us to avoid large errors of approximation when making inferences about the population mean. In section 5.6 we derive the distribution of s^2. First we introduce a new theoretical distribution.

5.5 The chi-square distribution

We have so far introduced two theoretical distributions, the binomial and the normal. We now introduce a third, the *chi-square distribution*, which, as we will see, is based on the normal distribution. Before formally defining the chi-square distribution, we go on a small digression which, hopefully, will help intuitive understanding of the chi-square distribution.

Suppose we have k random variables which we denote by Z_i $(i = 1, 2, \ldots, k)$. Suppose further, though it is not strictly necessary for the assertion we are about to make, that the k variables are independent. Define a new random variable Y in terms of the Z_i by

$$Y = g(Z_1, Z_2, \ldots, Z_k) \qquad\qquad (5.32)$$

where g is any function. Then, as we know the probability distributions of the Z_i $(i = 1, 2, \ldots, k)$, we can derive the probability distribution of Y.

To illustrate this general result, suppose $k = 2$, and that the probability distributions of Z_1 and Z_2 are the same, and are as given in table 5.10. You should check that

$$EZ_1 = EZ_2 = 0 \quad \text{var } Z_1 = \text{var } Z_2 = 1.$$

Table 5.10   Distribution of Z_i $(i = 1, 2, \ldots, k)$

Z_i	$P[Z_i]$
-2	1/8
0	3/4
2	1/8
Σ	1

Suppose we define (for example)

$$Y = g(Z_1, Z_2) = \frac{Z_1^{\ 3}}{(Z_2 + 3)^2}$$

Thus if $Z_1 = -2$ and $Z_2 = 0$ then

$$Y = \frac{(-2)^3}{(0 + 3)^2} = \frac{-8}{9}$$

and since Z_1 and Z_2 are independent

$$P\left[Y = \frac{-8}{9}\right] = P[Z_1 = -2, Z_2 = 0] = \left(\frac{1}{8}\right)\left(\frac{3}{4}\right) = \frac{3}{32}$$

Further,

$$P[Y = 0] = P[Z_1 = 0, Z_2 = -2] + P[Z_1 = 0, Z_2 = 0] + P[Z_1 = 0, Z_2 = 2]$$

$$= \left(\frac{3}{4}\right)\left(\frac{1}{8}\right) + \left(\frac{3}{4}\right)\left(\frac{3}{4}\right) + \left(\frac{3}{4}\right)\left(\frac{1}{8}\right) = \frac{3}{4}.$$

Continuing in this way we derive the probability distribution of Y as given in the last two columns of table 5.11. Alternatively, we could have defined Y by

$$Y = g(Z_1, Z_2) = Z_1^{\ 2} + Z_2^{\ 2}.$$

In this case, the probability distribution of Y is given in table 5.12; as usual, you will find it helpful to check the entries. Note that $EY = 2$, a result you should have been able to anticipate given the definition of Y, and the distributions of Z_1 and Z_2.

Suppose now we introduce a third random variable Z_3, with the same distribution as Z_1 and Z_2, as given in table 5.10. Suppose we define Y in terms of these three random variables by

$$Y = Z_1^{\ 2} + Z_2^{\ 2} + Z_3^{\ 2}$$

(you should be able to show that $EY = 3$). As before, we can derive the distribution of Y; it is given in table 5.13.

Table 5.11   Distribution of $Y = Z_1^3/(Z_2 + 3)^2$

Z_1	Z_2	$P[Z_1, Z_2]$ (×64)	Y (×225)	Y (×225)	$P[Y]$ (×64)
-2	-2	1	-1,800	-1,800	1
-2	0	6	-200	-200	6
-2	2	1	-72	-72	1
0	-2	6	0	0	48
0	0	36	0	72	1
0	2	6	0	200	6
2	-2	1	1,800	1,800	1
2	0	6	200		
2	2	1	72		
Σ		64		Σ	64

Table 5.12   Distribution of $Y = Z_1^2 + Z_2^2$

Z_1	Z_2	$P[Z_1, Z_2]$ (×64)	Y	Y	$P[Y]$ (×64)
-2	-2	1	8	0	36
-2	0	6	4	4	24
-2	2	1	8	8	4
0	-2	6	4		
0	0	36	0		
0	2	6	4		
2	-2	1	8		
2	0	6	4		
2	2	1	8		
Σ		64		Σ	64

Table 5.13   Distribution of $Y = Z_1^2 + Z_2^2 + Z_3^2$

Y	$P[Y]$ (×64)
0	27
4	27
8	9
12	1
Σ	64

These examples illustrate that, if we know the probability distributions of k (independent) random variables Z_1, Z_2, ..., Z_k, then it is a straightforward matter to derive the probability distribution of any function of the Z_i ($i = 1, 2, \ldots, k$).

Consider then the following distribution. Suppose Z_1, Z_2, ..., Z_k are k *independent unit normal* random variables. Define the random variable Y_k by

$$Y_k = Z_1^2 + Z_2^2 + \ldots + Z_k^2. \tag{5.33}$$

As we know the probability distribution of each Z_i, we can derive the probability distribution of Y_k, for any k. Obviously, for different values of k, Y_k has a different distribution. Although the derivation of the probability distribution of Y_k is conceptually straightforward, the actual calculations necessary are obviously somewhat tedious. Fortunately, since variables of the form of Y_k occur frequently in statistics, the distribution of Y_k, for any k, has already been calculated and tabulated. The distribution of the variable Y_k, defined by (5.33), is known as the *chi-square distribution with k degrees of freedom*. Tables of the chi-square distribution for different degrees of freedom (that is, different values of k) are given in the appendix. The term 'degrees of freedom' probably seems somewhat obscure at this stage; however its meaning should become apparent later in the book. As far as we are concerned at the moment, the number of degrees of freedom of a chi-square distribution is just the number of independent unit normals that have been squared and added together to obtain the variable.

From the tables of the chi-square distribution we can see, for example, that if Y_3 has a chi-square distribution with 3 degrees of freedom, then $P[Y_3 \geqslant 4.11] = 0.25$, $P[0.352 \leqslant Y_3 \leqslant 7.81] = 0.90$, etc. For Y_{20}, a chi-square with 20 degrees of freedom, we have $P[Y_{20} \geqslant 37.6] = 0.01$, $P[12.44 \leqslant Y_{20} \leqslant 28.4] = 0.80$, etc.

We note that a variable with a chi-square distribution is necessarily positive, and that

$$E(Y_k) = k \tag{5.34}$$

(This follows since $EZ_i^2 = E(Z_i - EZ_i)^2 = \text{var } Z_i = 1$, since Z_i is $N(0,1)$, all i.)

Since a chi-square variable with 1 degree of freedom is simply one unit normal squared (put $k = 1$ in (5.33)), we can 'check' the chi-square table against the unit normal table. We have $Y_1 = Z_1^2$ where Y_1 has a chi-squared distribution with 1 degree of freedom, and Z_1 is a unit normal. Thus, for example,

$$P[Y_1 \leqslant 3.84] = P[Z_1^2 \leqslant 3.84] = P[-\sqrt{3.84} \leqslant Z_1 \leqslant \sqrt{3.84}]$$
$$= P[-1.96 \leqslant Z_1 \leqslant 1.96] = 0.95$$

from the unit normal tables.

Consulting the chi-square table, we see that, for a chi-square variable with 1 degree of freedom, the probability of it being less than or equal to 3.84 is indeed 0.95. You may find it helpful to 'check' other entries yourself.

By inspection of the definition (5.33), it follows immediately that, if Y_k has a chi-square distribution with k degrees of freedom, and if Y_m has an independent chi-square distribution with m degrees of freedom, then the variable $(Y_m + Y_k)$ has a chi-square distribution with $(m + k)$ degrees of freedom. We use the terminology 'Y_k is χ_k^2' to mean that Y_k has a chi-square distribution with k degrees of freedom. Therefore we can write the above result as:

$$\left. \begin{array}{l} \text{if } Y_k \text{ is } X_k{}^2, \text{ and } Y_m \text{ is } X_m{}^2, \text{ and } Y_k, Y_m \text{ are independent,} \\ \text{then } (Y_k + Y_m) \text{ is } X_{k+m}{}^2. \end{array} \right\} \quad (5.35)$$

The import of all the above is that, if we wish to make probability statements about some random variable, and if we can express it in the form (5.33), then the variable has a chi-square distribution, and thus we can use the chi-square tables to make probability statements about it.

To illustrate the use of this distribution, consider the following (rather surrealistic!) example. Suppose two men are fighting a rather unusual duel. They are both standing at point A. The first duellist walks a distance Z_1 yards north (if $Z_1 > 0$) or south (if $Z_1 < 0$) where Z_1 is a unit normal random variable. The second walks a distance Z_2 yards east (if $Z_2 > 0$) or west (if $Z_2 < 0$) where Z_2 is a unit normal random variable (Z_1 and Z_2 are independent). When they have both finished walking they fire at each other with pistols which have a range of $\sqrt{6}$ yards. What is the probability that they are out of range of each other?

The distance between them when they fire is given by Y where $Y^2 = Z_1{}^2 + Z_2{}^2$ (by Pythagoras' theorem). Define $Y_2 = Y^2$. Then, since Z_1, Z_2 are independent unit normals, $Y_2 = Z_1{}^2 + Z_2{}^2$ has a chi-square distribution with 2 degrees of freedom (from (5.33)). Now, the question asks what is $P[Y \geqslant \sqrt{6}]$

$$\begin{aligned} \text{but} \quad P[Y \geqslant \sqrt{6}] &= P[Y^2 \geqslant 6] \\ &= P[Y_2 \geqslant 6] \\ &\doteqdot 0.05 \end{aligned}$$

from the tables of the chi-square distribution. Hence the probability is 0.05 that they are out of range.

We return to our main theme in the next section, and consider how we may use the material of this section to determine the distribution of s^2 (our estimator of σ^2).

5.6 Distribution of s^2

We investigate the distribution of s^2. From (5.30) s^2 is defined by

$$s^2 = \frac{1}{n-1} \sum_{1}^{n} (X_i - \bar{X})^2.$$

Consider

$$(n-1)\frac{s^2}{\sigma^2} = \frac{1}{\sigma^2} \sum_{1}^{n} (X_i - \bar{X})^2$$

$$= \frac{1}{\sigma^2}\left\{\sum_1^n (X_i - \mu)^2 - n(\bar{X} - \mu)^2\right\} \text{ as shown earlier (cf. (5.26))}.$$

$$= \sum_1^n \left(\frac{X_i - \mu}{\sigma}\right)^2 - \left(\frac{\bar{X} - \mu}{\sigma/\sqrt{n}}\right)^2$$

Thus,

$$(n-1)\frac{s^2}{\sigma^2} + \left(\frac{\bar{X} - \mu}{\sigma/\sqrt{n}}\right)^2 = \left(\frac{X_1 - \mu}{\sigma}\right)^2 + \left(\frac{X_2 - \mu}{\sigma}\right)^2 + \ldots + \left(\frac{X_n - \mu}{\sigma}\right)^2.$$

$$(5.36)$$

Now, *if the population is normally distributed*, with mean μ and variance σ^2, then

$$X_i \text{ is } N(\mu, \sigma^2) \qquad (i = 1, 2, \ldots, n)$$

and so

$$\left(\frac{X_i - \mu}{\sigma}\right) \text{ is } N(0, 1) \qquad (i = 1, 2, \ldots, n)$$

and, of course, as we are sampling with replacement, X_i and X_j are independent, all $i \neq j$. Thus, the expression on the right-hand side of (5.36) is the sum of n independent unit normals squared, and thus has (from (5.33)) a chi-square distribution with n degrees of freedom. We know also, from table 5.7, that $(\bar{X} - \mu)/(\sigma/\sqrt{n})$ is $N(0,1)$, and thus $\{(\bar{X} - \mu)/(\sigma/\sqrt{n})\}^2$ is X_1^2. Now it can be shown (though by methods outside the scope of this book) that s^2 and $\bar{X}$ are independent. (Intuitively it can be seen that s^2 measures dispersion *around* $\bar{X}$; the dispersion of a set of data is not determined by where the set of data is centred). Thus $(n-1)s^2/\sigma^2$ and $\{(\bar{X} - \mu)/(\sigma/\sqrt{n})\}^2$ are independent. Therefore, we know that the right-hand side of (5.36) is X_n^2, the second term on the left-hand side of (5.36) is X_1^2, and the two terms on the left-hand side of (5.36) are independent. It follows, using the converse of (5.35), that the first term on the left-hand side of (5.36) is X_{n-1}^2. Thus,

$$\text{if } X \text{ is } N(\mu, \sigma^2) \quad \text{then} \quad (n-1)\frac{s^2}{\sigma^2} \text{ is } X_{n-1}^2. \tag{5.37}$$

Since

$$(n-1)\frac{s^2}{\sigma^2} \equiv \frac{1}{\sigma^2}\sum_1^n (X_i - \bar{X})^2 \equiv \frac{nd^2}{\sigma^2}$$

it follows immediately from (5.37) that

$$\text{if } X \text{ is } N(\mu, \sigma^2) \quad \text{then} \quad \frac{nd^2}{\sigma^2} \text{ is } X_{n-1}^2. \tag{5.38}$$

We note that (5.37) taken in conjunction with (5.34) proves (5.31): from (5.34), if Y_k is $X_k{}^2$ then $E Y_k = k$; but from (5.37) $(n-1)s^2/\sigma^2$ is $X_{n-1}{}^2$; hence $E(n-1)s^2/\sigma^2 = (n-1)$; that is $E s^2 = \sigma^2$.

Results (5.37) and (5.38) enable us to make probability statements about the values of s^2 and d^2 in samples drawn from a normally distributed population with known variance. For example, suppose the population is normally distributed with variance $\sigma^2 = 4$. Then, if we take samples of size $n = 17$ from this population,

$$(n-1)\frac{s^2}{\sigma^2} = 16\frac{s^2}{4} = 4s^2 \text{ is } X_{16}{}^2 \quad \text{(from (5.37))}.$$

Thus, from chi-square tables, we have, for example,

$$P[4s^2 \geqslant 26.3] = 0.05 \qquad \text{i.e. } P[s^2 \geqslant 6.575] = 0.05$$

$$P[6.91 \leqslant 4s^2 \leqslant 28.8] = 0.95 \quad \text{i.e. } P[1.7275 \leqslant s^2 \leqslant 7.2] = 0.95, \text{ etc.}$$

Thus, in 95 per cent of all samples of size $n = 17$ drawn from a normally distributed population with variance $\sigma^2 = 4$, the value of s^2 will be between 1.7275 and 7.2.

If we took samples of size $n = 41$ from the same population, then, from (5.37),

$$(n-1)\frac{s^2}{\sigma^2} = 40\frac{s^2}{4} = 10s^2 \text{ is } X_{40}{}^2$$

and so, for example,

$$P[10s^2 \geqslant 55.8] = 0.05 \qquad \text{i.e. } P[s^2 \geqslant 5.58] = 0.05$$

$$P[24.4 \leqslant 10s^2 \leqslant 59.3] = 0.95 \quad \text{i.e. } P[2.44 \leqslant s^2 \leqslant 5.93] = 0.95, \text{ etc.}$$

Comparing this with the corresponding result for $n = 17$ illustrates our previous assertion that the dispersion of s^2 around σ^2 decreases as n increases. (For $n = 101$, we would have $P[2.968 \leqslant s^2 \leqslant 4.724] = 0.95$.)

5.7 Inferences about Population Variances

Result (5.37) also enables us to make inferences about some unknown variance (of a normally distributed population), given sample evidence. From (5.37) we have

$$P\left\{Y_{n-1,1-\alpha/2} \leqslant (n-1)\frac{s^2}{\sigma^2} \leqslant Y_{n-1,\alpha/2}\right\} = 1 - \alpha \tag{5.39}$$

where $Y_{k,\beta}$ is defined as that value for which $P[Y_k \geqslant Y_{k,\beta}] = \beta$ where Y_k is $X_k{}^2$. We can express the probability statement (5.39) as

$$P\left[\frac{(n-1)s^2}{Y_{n-1,\alpha/2}} \leqslant \sigma^2 \leqslant \frac{(n-1)s^2}{Y_{n-1,1-\alpha/2}}\right] = 1 - \alpha$$

and so a $100(1-\alpha)$ per cent confidence interval for σ^2 is

$$\left(\frac{(n-1)s^2}{Y_{n-1,\alpha/2}}, \frac{(n-1)s^2}{Y_{n-1,1-\alpha/2}}\right). \tag{5.40}$$

For example, suppose we collected a sample of size $n = 12$ from a normally distributed population with unknown variance σ^2. Suppose our 12 observations were as given in table 5.14. Then $\bar{X} = 6.1$ and $11s^2 = 74.18$. For a 90 per cent confidence interval the appropriate entries in the chi-square table are

$$Y_{n-1,\,1-\alpha/2} = Y_{11,\,0.95} = 4.57$$

$$Y_{n-1,\,\alpha/2} = Y_{11,\,0.05} = 19.68.$$

Thus, from (5.40), a 90 per cent confidence interval for σ^2 is

$$\left(\frac{74.18}{19.68}, \frac{74.18}{4.57}\right)$$

that is, we can be 90 per cent confident that the interval $(3.77, 16.23)$ covers σ^2.

Table 5.14   Hypothetical sample

i	X_i	$X_i - \bar{X}$	$(X_i - \bar{X})^2$
1	5.4	-0.7	0.49
2	4.9	-1.2	1.44
3	6.4	0.3	0.09
4	1.8	-4.3	18.49
5	4.8	-1.3	1.69
6	3.6	-2.5	6.25
7	5.7	-0.4	0.16
8	11.2	5.1	26.01
9	10.0	3.9	15.21
10	6.6	0.5	0.25
11	5.0	-1.1	1.21
12	7.8	1.7	2.89
Σ	73.2	0.0	74.18

$$\bar{X} = 6.1 \qquad s^2 = \frac{74.18}{11} \doteq 6.744$$

Suppose we wish to use the sample evidence of table 5.14 to test the hypothesis $H_0 : \sigma^2 = 4$ against $H_1 : \sigma^2 > 4$. Now, from (5.37), $(n-1)s^2/\sigma^2$ is χ^2_{n-1}, and so with $n = 12$ observations $11s^2/\sigma^2$ is $\chi_{11}{}^2$.

Thus, if H_0 is correct then $11s^2/4$ is X_{11}^2. We choose a critical value C^* for the variable $11s^2/4$, to achieve the desired significance level (say 1 per cent); thus C^* must satisfy

$$P\left[\frac{11s^2}{4} > C^* \mid H_0 \text{ correct}\right] = 0.01$$

that is

$$P\left[\frac{11s^2}{4} > C^* \mid \frac{11s^2}{4} \text{ is } X_{11}^2\right] = 0.01.$$

From the chi-square tables, $C^* = 24.7$. Our observed value of $11s^2/4$ is 18.545 which is less than C^*, and so the test is not significant at the 1 per cent level, and the sample evidence is consistent with the hypothesis that the (normally distributed) population from which it was drawn has variance $\sigma^2 = 4$. Figure 5.3 illustrates this procedure.

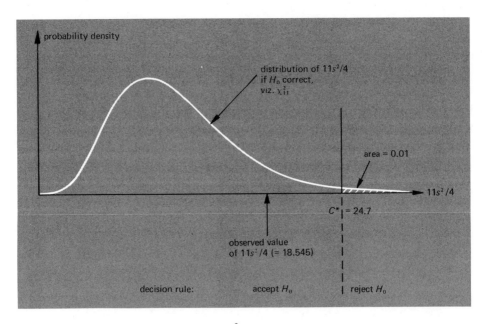

Figure 5.3   Testing a hypothesis about σ^2

In this section we have determined the distribution of s^2. In section 5.9 we consider how this knowledge improves our inferences about the population mean when the population variance is unknown. First we introduce another theoretical distribution.

5.8 The *t*-distribution

In section 5.6 we discussed at length how knowledge of the probability distribution of a set of random variables enabled us to derive the probability distribution of any function of the random variables. You may find it helpful to reread this before proceeding.

Suppose that Y_k has a chi-square distribution with k degrees of freedom, and that Z has a unit normal distribution. Suppose further that Y_k and Z are independent. Define the random variable t_k by

$$t_k = \frac{Z}{\sqrt{(Y_k/k)}}. \tag{5.41}$$

Then, as we know the probability distributions of Z and Y_k for any k, it follows that we can derive the probability distribution of t_k for any k. Obviously, for different k the distribution of t_k will be different. Once again, although the derivation of the distribution of t_k is conceptually straightforward, in practice the computations necessary are somewhat tedious. However, as this distribution occurs frequently in statistics, its probability distribution has already been calculated and tabulated. The variable as defined by (5.41) has what is called a *t-distribution* with k *degrees of freedom*. Tables are provided in the appendix. Inspection of (5.41) shows that (for any k) the *t*-distribution is symmetrical about zero. Inspection of the tables of the *t*-distribution also shows that its shape is similar to the unit normal, but is more spread out. As k increases the *t*-distribution approaches the unit normal.

To illustrate the use of the tables, suppose we know that a variable (call it t_5 for consistency) has a *t*-distribution with 5 degrees of freedom. From the table we get

$$P[t_5 \geqslant 2.015] = 0.05$$
$$P[-2.015 \leqslant t_5 \leqslant 2.015] = 0.90$$
$$P[0.92 \leqslant t_5] = 0.20 \quad \text{etc.}$$

Once again, the import of the above is that, if we wish to make probability statements about some random variable, and if we can express the variable in the form of (5.41), then it has a *t*-distribution, and we can thus use the *t*-tables to make probability statements about it.

To illustrate, we consider further the unusual duel described in section 5.5. Suppose that one of the two men is walking along a horizontal beam which can be raised or lowered relative to the other man, who walks along a fixed horizontal beam. Suppose that initially the two beams are level, but during the period before they fire the moving beam is raised by Z yards (if $Z > 0$) or lowered by Z yards (if $Z < 0$) where Z is a unit normal random variable, independent of Z_1 and Z_2. Suppose further that their 'pistols' are mounted to their respective beams and have a limited range of elevation. Suppose that the maximum elevation (or declination) of each 'pistol' is 1.334. What is the probability that the elevation of their 'pistols' is sufficient to be on target?

The actual elevation when firing is about to commence is Z/Y (Z being the vertical distance between them and Y being the horizontal distance between them). Call this $W \equiv Z/Y$. We want to find $P[-1.334 \leqslant W \leqslant 1.334]$. Now what is the distribution of W? We know that Y^2 is a chi-square with 2 degrees of freedom and Z is an independent unit normal. Hence $Z/(\sqrt{(Y^2/2)})$ has a t-distribution with 2 degrees of freedom (from (5.41)). But,

$$\frac{Z}{\sqrt{(Y^2/2)}} = \left(\frac{Z}{Y}\right)\sqrt{2} = W\sqrt{2}.$$

Therefore $W\sqrt{2}$ has a t-distribution with 2 degrees of freedom. Hence call $W\sqrt{2} \equiv t_2$;

$$\therefore \quad P[-1.334 \leqslant W \leqslant 1.334] = P[-1.334\sqrt{2} \leqslant W\sqrt{2} \leqslant 1.334\sqrt{2}]$$
$$= P[-1.886 \leqslant t_2 \leqslant 1.886]$$
$$= 0.80 \text{ from the tables of the}$$
$$t\text{-distribution}$$

Hence the probability is 0.8 that the elevation of their 'pistols' is sufficient.

5.9   Inferences about Population Means II

We are now in a position to form confidence intervals for, and to test hypotheses about, the population mean, even when we do not know the population variance. Previously, we had to rely on an approximation, namely replacing σ^2, whenever it appeared, by s^2. This approximation was reasonable if n, the sample size, was large, but for small n, the approximation could introduce a large error.

Consider the variable $(\bar{X} - \mu)/(s/\sqrt{n})$ where, of course, $\bar{X} = \Sigma X_i/n$ and $s^2 = \Sigma_1^n (X_i - \bar{X})^2/(n-1)$. We will show that, if the population is normally distributed, then $(\bar{X} - \mu)/(s/\sqrt{n})$ has a t-distribution with $(n-1)$ degrees of freedom. We do this by showing it can be expressed in the form of (5.41):

$$\frac{\bar{X} - \mu}{s\sqrt{n}} \equiv \left(\frac{\bar{X} - \mu}{\sigma/\sqrt{n}}\right) \bigg/ s/\sigma \equiv \left(\frac{\bar{X} - \mu}{\sigma/\sqrt{n}}\right) \bigg/ \sqrt{(s^2/\sigma^2)}$$

$$\equiv \left(\frac{\bar{X}-\mu}{\sigma/\sqrt{n}}\right) \bigg/ \sqrt{\left(\frac{(n-1)s^2/\sigma^2}{n-1}\right)}. \tag{5.42}$$

If the population is normally distributed, then the numerator $(\bar{X}-\mu)/(\sigma/\sqrt{n})$ is a unit normal, from table 5.7. The denominator $\sqrt{[\{(n-1)s^2/\sigma^2\}/(n-1)]}$ is the square root of a variable with a chi-square distribution $((n-1)s^2/\sigma^2)$ divided by its number of degrees of freedom $(n-1)$, from (5.37). Further, as we have already noted, $\bar{X}$ and s^2 are independent. Thus, $(\bar{X}-\mu)/(s/\sqrt{n})$ can indeed be expressed in the form (5.41). Thus,

if X is $N(\mu,\sigma^2)$,then $(\bar{X}-\mu)/(s/\sqrt{n})$ has a t-distribution with $(n-1)$ degrees of freedom (5.43)

Taken in conjunction with the result that $(\bar{X}-\mu)/(\sigma/\sqrt{n})$ is $N(0,1)$, and the fact that the t-distribution is more spread out than the unit normal, result (5.43) makes good intuitive sense.

Now that we know the exact distribution of $(\bar{X}-\mu)/(s/\sqrt{n})$, we can make inferences about the unknown mean of a normally distributed population from sample evidence, even when we do not know the population variance. From (5.43) we have

$$P\left[-t_{n-1,\,\alpha/2} \leqslant \frac{\bar{X}-\mu}{s/\sqrt{n}} \leqslant t_{n-1,\,\alpha/2}\right] = 1-\alpha \tag{5.44}$$

where $t_{k,\beta}$ is that value for which $P[t_k \geqslant t_{k,\beta}] = \beta$ where t_k has a t-distribution with k degrees of freedom. We can express (5.44) as:

$$P[\bar{X}-t_{n-1,\,\alpha/2}\,s/\sqrt{n} \leqslant \mu \leqslant \bar{X}+t_{n-1,\,\alpha/2}\,s/\sqrt{n}] = 1-\alpha$$

and so, in $100(1-\alpha)$ per cent of all samples, the interval

$$(\bar{X}-t_{n-1,\,\alpha/2}\,s/\sqrt{n},\,\bar{X}+t_{n-1,\,\alpha/2}s/\sqrt{n}) \tag{5.45}$$

covers μ. Hence, the interval (5.45) is a $100(1-\alpha)$ per cent confidence interval for μ. Notice the interval (5.45) can be formed using only sample evidence (and the appropriate entry from the t-tables).

As an example, consider the sample given in table 5.14, which we assume was obtained from a normally distributed population with unknown mean and unknown variance. We have

$$n = 12 \quad \bar{X} = 6.1 \quad s = \sqrt{6.744} = 2.597.$$

For a 95 per cent confidence interval for μ, the appropriate entry in the t-tables is

$$t_{n-1,\alpha/2} = t_{11,0.025} = 2.201.$$

Thus, from (5.45) a 95 per cent confidence interval for μ, is

$$\{6.1 - 2.201 \ (2.597/\sqrt{12}), 6.1 + 2.201 \ (2.597/\sqrt{12})\}$$

that is, (4.45, 7.75); and so, we can be 95 per cent confident that the mean of the (normally distributed) population from which the sample data of table 5.14 was drawn is between 4.45 and 7.75.

We can also use the data of table 5.14 to test whether the sample evidence supports $H_0 : \mu = 8$ or $H_1 : \mu < 8$.

We know that $(\bar{X} - \mu)/(s/\sqrt{n})$ has a t-distribution with $(n - 1)$ degrees of freedom (from (5.43)). Hence, if H_0 is correct, for samples of size $n = 12$, then $(\bar{X} - 8)/(s/\sqrt{12})$ has a t-distribution with 11 degrees of freedom. Figure 5.4 illustrates this. Suppose we require a 5 per cent significance level. Then we choose a critical value C^* for $(\bar{X} - 8)/(s/\sqrt{12})$ so that

$$P\left[\frac{\bar{X} - 8}{s/\sqrt{12}} < C^* \mid H_0 \ \text{correct}\right] = 0.05$$

and reject H_0 if $(\bar{X} - 8)/(s/\sqrt{12})$ is less than C^* (that is, if $\bar{X}$ is significantly below 8). C^* is defined by

$$P\left[\frac{\bar{X} - 8}{s/\sqrt{12}} < C^* \ \middle| \ \begin{array}{l} \frac{\bar{X} - 8}{s/\sqrt{12}} \ \text{has } t\text{-distribution with} \\ 11 \text{ degrees of freedom} \end{array}\right] = 0.05$$

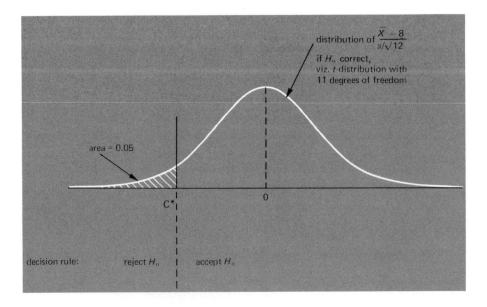

Figure 5.4   Testing a hypothesis about μ

From the tables this gives $C^* = -1.796$. Now, our observed value of $(\bar{X} - 8)/(s/\sqrt{12})$ is $(6.1 - 8)/(2.597/\sqrt{12}) = -1.424$ which is greater than C^*. So our test is not significant (at the 5 per cent level). Thus we conclude that the evidence of the sample data of table 5.14 is not inconsistent with the hypothesis that the mean of the population from which the sample was drawn is equal to 8.

Generalising, we consider the cases

$\quad$ (1) $H_0 : \mu = \mu_0 \quad H_1 : \mu > \mu_0$

$\quad$ (2) $H_0 : \mu = \mu_0 \quad H_1 : \mu < \mu_0$

$\quad$ (3) $H_0 : \mu = \mu_0 \quad H_1 : \mu \neq \mu_0$.

For an α per cent significance level in each case we get the following rules. In case (1), accept H_0 if $(\bar{X} - \mu_0)/(s/\sqrt{n}) \leqslant C^*$ and reject H_0 if $(\bar{X} - \mu_0)/(s/\sqrt{n}) > C^*$ where C^* is defined by

$$P\left[\frac{\bar{X} - \mu_0}{s/\sqrt{n}} > C^* \mid H_0 \text{ correct}\right] = \alpha.$$

That is,

$$P\left[\frac{\bar{X} - \mu_0}{s/\sqrt{n}} > C^* \,\middle|\, \begin{array}{l} \frac{\bar{X} - \mu_0}{s/\sqrt{n}} \text{ has } t\text{-distribution with } (n-1) \\ \text{degrees of freedom} \end{array}\right] = \alpha$$

that is,

$$C^* = t_{n-1,\,\alpha}.$$

Similarly, in case (2), we should accept H_0 if $(\bar{X} - \mu_0)/(s/\sqrt{n}) \geqslant C^*$ and reject H_0 if

$$\frac{\bar{X} - \mu_0}{s/\sqrt{n}} < C^* \quad \text{where} \quad C^* = t_{n-1,\,\alpha}.$$

Finally, in case (3), we accept H_0 if $C_1^* \leqslant (\bar{X} - \mu_0)/(s/\sqrt{n}) \leqslant C_2^*$ and reject H_0 if

$$\frac{\bar{X} - \mu_0}{s/\sqrt{n}} < C_1^* \quad \text{or} \quad \frac{\bar{X} - \mu_0}{s/\sqrt{n}} > C_2^*$$

where $\quad C_1^* = -t_{n-1,\,\alpha/2} ; \quad C_2^* = t_{n-1,\,\alpha/2} .$

Finally, we consider how we may make inferences about the difference between two unknown population means. Suppose a sample of size n_1 is taken from a normally distributed population which has mean μ_1 and variance $\sigma_1{}^2$. Denote, as usual, by $\bar{X}_1$ and $s_1{}^2$ the mean and modified variance of the sample observations. Suppose a sample of size n_2 is taken independently from another normally distributed population, which has a mean μ_2 and variance $\sigma_2{}^2$. Let $\bar{X}_2$ and $s_2{}^2$ be the sample mean and modified sample variance. We then have, from table 5.7,

$$\bar{X}_1 \text{ is } N(\mu_1, \sigma_1{}^2/n_1) \text{ and } \bar{X}_2 \text{ is } N(\mu_2, \sigma_2{}^2/n_2).$$

Thus, since $\bar{X}_1$ and $\bar{X}_2$ are independent

$$\bar{X}_1 - \bar{X}_2 \text{ is } N\left(\mu_1 - \mu_2, \frac{\sigma_1{}^2}{n_1} + \frac{\sigma_2{}^2}{n_2}\right)$$

and so

$$\frac{(\bar{X}_1 - \bar{X}_2) - (\mu_1 - \mu_2)}{\sqrt{\left(\dfrac{\sigma_1{}^2}{n_1} + \dfrac{\sigma_2{}^2}{n_2}\right)}} \quad \text{is } N(0, 1). \tag{5.46}$$

Also, from (5.37),

$$(n_1 - 1)\frac{s_1{}^2}{\sigma_1{}^2} \quad \text{is} \quad \chi^2_{n_1-1} \quad \text{and} \quad (n_2 - 1)\frac{s_2{}^2}{\sigma_2{}^2} \quad \text{is} \quad \chi^2_{n_2-1}.$$

Now, suppose

$$\sigma_1{}^2 = \sigma_2{}^2 = \sigma^2 \tag{5.47}$$

then

$$\frac{(n_1 - 1)s_1{}^2 + (n_2 - 1)s_2{}^2}{\sigma^2} \quad \text{is} \quad \chi^2_{n_1+n_2-2} \quad \text{(from (5.35)).} \tag{5.48}$$

Also, $\bar{X}_1$, $\bar{X}_2$, $s_1{}^2$ and $s_2{}^2$ are all independent. Thus, from (5.41)

$$\frac{(\bar{X}_1 - \bar{X}_2) - (\mu_1 - \mu_2)}{\sigma\sqrt{\left(\dfrac{1}{n_1} + \dfrac{1}{n_2}\right)}} \Bigg/ \sqrt{\left(\dfrac{\dfrac{(n_1 - 1)s_1{}^2 + (n_2 - 1)s_2{}^2}{\sigma^2}}{(n_1 + n_2 - 2)}\right)} \tag{5.49}$$

has a t-distribution with $(n_1 + n_2 - 2)$ degrees of freedom.
We can simplify (5.49) to obtain the result,

$$\left.\begin{aligned} &\frac{(\bar{X}_1 - \bar{X}_2) - (\mu_1 - \mu_2)}{s} \Bigg/ \sqrt{\left(\dfrac{n_1 n_2}{n_1 + n_2}\right)} \\[4pt] &\text{has a } t\text{-distribution with } (n_1 + n_2 - 2) \text{ degrees of freedom where} \\[4pt] &s^2 = \frac{(n_1 - 1)s_1{}^2 + (n_2 - 1)s_2{}^2}{n_1 + n_2 - 2} \end{aligned}\right\} \tag{5.50}$$

Result (5.50) enables us to make inferences about the difference between two population means using only sample evidence. Notice, however, that (5.50) was derived under the assumption (5.47) that the two population variances are equal. For unequal variances the derivation of a comparable result to (5.50) is rather complex. As we will not be using such a result, we will not investigate it further. However, it should be remembered that if n_1 and n_2 are large, then σ_1 and σ_2 can be approximated by s_1 and s_2 respectively, and the methods of section 5.3 used to make inferences about the difference between population means.

Table 5.15   Expenditure of households by household composition

Weekly household expenditure	Two adults five children	Two adults six or more children
$ 75 and under $100	1	0
$100 and under $125	7	3
$125 and under $150	11	6
$150 and under $175	5	6
$175 and under $200	10	6
$200 and under $225	3	1
$225 and under $250	5	3
$250 and under $300	3	2
$300 and under $400	4	2
$400 or more	1	1
Totals	50	30

To conclude this section we give a further example of the main results presented in it. Consider the population of all households consisting of two adults and five children. Table 5.15 presents information on weekly household expenditure of a sample of 50 households chosen from this population. (Table 5.15 is adapted from U.K. data). For this sample of 50, mean weekly household expenditure $\bar{X} = 191.50$ and the modified variance $s^2 = 5,940.3$. Assuming that the weekly expenditure of all households in this population is normally distributed with unknown mean μ, it follows from (5.43), that $(\bar{X} - \mu)/(s/\sqrt{50})$ has a t-distribution with 49 degrees of freedom. Thus, from (5.45) and the tables of the t-distribution, a 90 per cent confidence interval for μ is given by

$$(\bar{X} - 1.678 \, s/\sqrt{50}, \; \bar{X} + 1.678 \, s/\sqrt{50})$$

that is, $(173.21, 209.79)$.

Similarly, using the second column of table 5.15, we obtain the following 90 per cent confidence interval for the mean weekly expenditure of all households consisting of two adults and six or more children (here $\bar{X} = 196.25, s^2 = 6,154.625$, and $n = 30$).

$$(\bar{X} - 1.699 \, s/\sqrt{30}, \ \bar{X} + 1.699 \, s/\sqrt{30})$$

that is, $(171.91, 220.59)$.

Using subscripts 1 and 2 to refer to households consisting of two adults and five children and to households consisting of two adults and six or more children respectively, we now derive a 90 per cent confidence interval for the difference $(\mu_1 - \mu_2)$ between mean household expenditure in these two populations. From (5.50) we have (assuming that the population variances are equal, and, of course, still assuming normally distributed populations) that

$$\frac{(\bar{X}_1 - \bar{X}_2) - (\mu_1 - \mu_2)}{s} \bigg/ \sqrt{\left(\frac{n_1 n_2}{n_1 + n_2}\right)}$$

has a t-distribution with $(n_1 + n_2 - 2)$ degrees of freedom, where

$$s^2 = \frac{(n_1 - 1)s_1^2 + (n_2 - 1)s_2^2}{n_1 + n_2 - 2}.$$

Here we have $n_1 = 50, n_2 = 30$, and so the quantity

$$\frac{(\bar{X}_1 - \bar{X}_2) - (\mu_1 - \mu_2)}{s/\sqrt{18.75}}$$

has a t-distribution with 78 degrees of freedom.

Thus, (interpolating the t-tables)

$$P\left[-1.667 \leqslant \frac{(\bar{X}_1 - \bar{X}_2) - (\mu_1 - \mu_2)}{s/\sqrt{18.75}} \leqslant 1.667\right] = 0.90.$$

Thus, a 90 per cent confidence interval for $(\mu_1 - \mu_2)$ is

$$\{(\bar{X}_1 - \bar{X}_2) - 1.667 \, s/\sqrt{18.75}, \ (\bar{X}_1 - \bar{X}_2) + 1.667 \, s/\sqrt{18.75}\}.$$

Our assumption that $\sigma_1^2 = \sigma_2^2 = \sigma^2$ (that the population variances are equal) appears plausible (since $s_1^2 = 5,940.3$ and $s_2^2 = 6,154.625$). Our estimate s^2 of σ^2 is

$$s^2 = \frac{(n_1 - 1)s_1^2 + (n_2 - 1)s_2^2}{n_1 + n_2 - 2} = 6,019.975.$$

Thus, a 90 per cent confidence interval for $(\mu_1 - \mu_2)$ is $(-34.62, 25.12)$. The width of this interval clearly reflects the small sample sizes. However, on the basis of this evidence, we have to conclude that households consisting of two adults and six or more children do not appear to spend significantly more on average than households consisting of two adults and only five children.

As a final comment, it should be noted that inspection of table 5.15 does not indicate support for the assumption of a normally distributed population. We must therefore be cautious when interpeting the results derived above, and be prepared to admit that they may hold only approximately.

5.10 Summary

This chapter extended the scope of the concepts and methodology introduced in chapter 4. The chapter as a whole was concerned with making inferences about population means and population variances using sample evidence. We began by showing that the sample mean equalled, *on average*, the population mean, and that the dispersion of the sample mean around the population mean was inversely proportional to the sample size. We showed also that the distribution of the sample mean was either exactly or approximately normal.

These results enabled us to form confidence intervals for, and test hypotheses about, an unknown population mean using sample evidence. However, at this stage we had to assume knowledge of the population variance. Thus we went on to consider how we might estimate the population variance, when it was unknown, using sample evidence. We found that the sample variance underestimated, *on average*, the population variance; this led us to introduce the modified sample variance (s^2), which did, *on average*, equal the population variance. Consideration of the distribution of s^2 necessitated the introduction of a new theoretical distribution — the chi-square distribution. Then, by showing that $(n-1)$ s^2/σ^2 had a chi-square distribution (when the population had a normal distribution), we were in a position to form confidence intervals for, and test hypotheses about, an unknown population variance using sample evidence.

Returning to the problem of making inferences about an unknown population mean when the population variance was unknown led to the introduction of another new theoretical distribution — the t-distribution. We went on to show that, if the population was normally distributed, then the sample statistic $(\bar{X} - \mu)/(s/\sqrt{n})$ had a t-distribution (with $(n-1)$ degrees of freedom). This information then enabled us to form confidence intervals for, and test hypotheses about, an unknown population mean when the population variance was also unknown.

5.11 Exercises

5.1. A class contains 13 students who have sat an examination; 3 of them gained A grades, 7 B grades and the remaining 3 C grades. An A grade is awarded when the candidate has scored 3 marks, a B grade for 2 marks and a C grade for 1 mark. Let X be the variable: number of marks.

(a) Find the mean mark of the class, and the standard deviation of the marks in the class. Denote these by μ and σ respectively.

(b) Suppose we now select 3 students at random from the 13 in the class, where sampling is done with replacement. Find the probability that the 3 students selected will all have A grades. Find the probability that of the 3 students selected, 2 will have A grades, the other a B grade. Continue this process of finding probabilities for all possible combinations of A, B and C grades in the sample of 3 students. (You should find 10 different combinations, and the attached probabilities.)

(c) For each of these 10 different combinations, find the average mark $\bar{X}$ in the sample of 3.

(d) Hence find the probability distribution of $\bar{X}$, and find its mean and variance, denoted by $E\bar{X}$ and var $\bar{X}$ respectively.

(e) Is there any connection between $E\bar{X}$ and μ? Between var $\bar{X}$ and σ^2? Did you anticipate these results?

(f) How closely does the distribution of $\bar{X}$ approximate the normal distribution?

5.2. Repeat exercise 5.1 entirely for the case when sampling is done without replacement. Compare the results of the two exercises and discuss.

5.3. For the population used in section 5.2 as an illustrative example find the probability distribution of $\bar{X}$ for the samples of size 3 drawn without replacement. Verify (5.4) and (5.6). Verify also that X_1, X_2 and X_3 are not independent, and that (5.7) holds.

5.4. By explicitly finding the joint probability distribution of X_i and X_j ($i \neq j$), where X_i and X_j are the ith and jth observations in a sample of size of n drawn without replacement from a population of size N, prove that the covariance between X_i and X_j is as given in (5.7). (Hint: if $X_{(k)}$ and $X_{(m)}$ are respectively the kth and mth values of X in the population, show that, for $i \neq j$,

$$P[X_i = X_{(k)}, X_j = X_{(m)}] = \begin{cases} \dfrac{1}{N(N-1)} & k \neq m \\ 0 & k = m. \end{cases}$$

5.5. Given that the (continuous) variable X has population mean 30 and standard deviation 8, calculate the probability that the sample mean $\bar{X}$, based on a sample of size 256 drawn with replacement, will:

(a) be less than 29;
(b) be less than 30.5;
(c) exceed 29.5;
(d) lie between 29.5 and 30.5;
(e) equal 30;
(f) lie between 30.5 and 31.5.

5.6. Repeat exercise 5.5, substituting a sample size of 64 instead of one of 256.

5.7. Competitors in an athletic competition have to complete six laps of a course in 18 minutes. Experience has shown that the lengths of time (in minutes) taken by the competitors to complete a lap are statistically independent and normally distributed with mean 3 minutes and standard deviation 1 minute.

(a) What is the mean length of time left by competitors to complete the last lap?

(b) What is the probability that a competitor completes five or less laps in the time available?

(c) What additional proportion of competitors would complete the six laps if they were given an extra 3 minutes?

(d) Criticise the independence assertion made in the first paragraph.

5.8. You are told that heights of students are normally distributed with an unknown mean μ and a standard deviation 4 in. You take a sample of size 64 and find that the average height in the sample is 69 in.

(a) Find a 95 per cent and a 99 per cent confidence interval for μ.

(b) With what confidence can we say that the error of estimate is less than 1 in?

(c) How large a sample would one need to obtain an estimate of μ which is in error by less than ½ in with probability 0.9544?

5.9 By consulting the appropriate tables in a *Survey of Consumer Finances*, find an interval estimate for mean family income with a head of an educational background of your choice. (Choose any confidence level, and suppose that the population variance can be reasonably well approximated by s^2, the modified sample variance.)

5.10 Test whether the evidence of the *Survey* suggests that mean family income in the family type of your choice is significantly different from the U.S. mean family income.

5.11. Suppose you are testing the hypothesis $\mu = 70$, where you are given that $\sigma = 10$. How large a sample would you need for a value of $\bar{X} = 68$ to be significant at the 1 per cent level on a two-tailed test? On a one-tailed test? At what significance level would a value of $\bar{X} = 68$ on a sample size of 100 be significant on a two-tailed test? On a one-tailed test?

5.12. Show that the result of chapter 4, that p' is approximately normal, is a particular case of the result in this chapter that $\bar{X}$ is approximately normal.

5.13. Why can we not use the test described in section 5.3, in conjunction with the data of the *Survey* to test whether there is a significant difference between mean family income in black families and in the U.S. as a whole?

5.14. A factory specialises in making solid cubes measuring $|X|$ feet along all sides. The production process is designed so that the machines produce cubes in such a way as to make X a unit normal random variable. What proportion of the cubes have a volume less than 8 cubic feet?

    After having been made, the cubes are painted at a cost of 1 cent per square foot. What proportion of the cubes will cost more than 30 cents to paint? If the cubes are packed in (random) bundles of four, determine the median cost per bundle of the paint on the cubes.

5.15. A housing estate is being built with square gardens at the front and at the back, the front garden being $|a|$ yards square, the back garden $|b|$ yards square. a and b are independent unit normal variables. Find the probability of a particular house having a total garden area of at least 6 square yards. Determine the size under which 95 per cent of the front gardens will lie on average.

5.16. Suppose X is normally distributed with mean 30 and variance 16. Find the probability that s^2, the modified sample variance in a sample of size 17, will:

(a) be greater than 19.37;
(b) be less than 9.31;
(c) lie between 5.81 and 32.0.

5.17. Suppose the weekly consumption expenditure of households in the U.S. is normally distributed with standard deviation $35. Suppose we take a random sample of 15 old-age pensioner households, and find that the standard deviation of weekly consumption expenditure of these fifteen households is $25. Is the variation of old-age pensioner's expenditure less than that for the U.S. as a whole?

5.18. If X and Y are independent unit normal random variables, demonstrate that $Z = X/Y$ has a t-distribution with one degree of freedom. Hence find the probability that Z is positive.

Suppose we wish to test whether a variable X (which we know is normally distributed with unit variance) has a zero mean or a positive mean. Suppose, however, that we cannot observe X directly, but can only observe the variable $Z = X/Y$, where we know Y is a unit normal and is independent of X. Would an observation $Z = 10$ lead us to reject the hypothesis that X has a zero mean at the 5 per cent level? At the 1 per cent level?

5.19. A random sample of 10 students was selected, and it was found that their expenditure per year on books was as follows (in $):

75, 50, 60, 35, 35, 50, 35, 75, 50, 35.

(a) Find a 95 per cent confidence interval for the variance of the expenditure on books of all students

(b) Test the hypothesis that, on average, students spend $62.50 per year on books.

(c) Discuss how you might find the probability of a Type II error of your test, where the alternative hypothesis is that the average expenditure is $37.50 per year.

5.20. A random sample of 10 doctors was selected and it was found that their weekly incomes (in $) were as follows:

650, 950, 1450, 650, 950, 650, 650, 1150, 950, 1450.

(a) Find a 90 per cent confidence interval for the average weekly income of doctors.

(b) Test the hypothesis that the standard deviation of doctors' incomes is $200.

(c) How *might* you find the probability of a Type II error of your test where the alternative hypothesis is that the standard deviation of doctors' incomes is $400?

5.21. Suppose a consumer council wants to investigate whether the value of second-hand cars has changed. A sample of 36 cars is taken, with a mean price of $1,500 and a standard deviation of $75. Last year the mean value of second-hand cars was $1,450.

(a) At the 5 per cent level of significance test the hypothesis that, on average, car prices have not risen.

(b) What is the largest sample size for which a mean of $1,450 would *not* be significant at the 5 per cent level? (You may find it helpful to find an approximate result first.)

5.22. One play of a game is as follows: a coin is tossed; if it lands tails then two objects are selected randomly (without replacement) from a box containing three dollar pieces and two worthless metal discs (of the same shape and weight). If it lands heads, then three objects are selected from the same box. The probability of the coin landing tails is p.

(a) Derive the probability distribution of the amount paid per play.

(b) If $p = 2/3$, what is the maximum amount you would pay per play of the game?

(c) If out of n plays of the game you were paid a total of Z, how would you estimate the value of p? Explain how you might obtain confidence intervals for this estimate.

5.23. For any two occupations, or for any two different household compositions, use the data of the *Survey* to test for any significant difference in mean household income (assume the appropriate population variances are equal). Discuss how you might investigate whether this assumption is likely to be correct.

5.24. Discuss briefly the following statements (note they are not necessarily correct):

(a) Knowledge of the population distribution is unnecessary for forming estimates of, and confidence intervals for, the population mean.

(b) The variance of the mean of a random sample drawn with replacement is only approximately equal to σ^2/n, where σ^2 is the population variance and n the sample size, the approximation becoming increasingly smaller as n gets larger.

(c) When estimating a population mean, knowledge of the population variance enables a more precise estimate to be made.

(d) The *principle* of hypothesis testing relies critically on the normal distribution.

(e) The popular practice of providing point estimates of population parameters is not only misleading but also dangerous.

(f) The variance of the sample mean in a sample drawn without replacement is always smaller than in a sample drawn with replacement.

6

Properties of Estimators

6.1 Introduction

The preceding two chapters were concerned with the problems of estimating, and testing hypotheses about certain population parameters; in particular p, the population proportion of those having a certain characteristic; μ, the population mean of a variable; and σ^2, the population variance of a variable. In each case we used sample information to provide an estimator of the relevant population parameter. We used p', the sample proportion, as our estimator of p, the population proportion. We used $\bar{X}$, the sample mean, as our estimator of μ, the population mean. We used s^2, a modification of the sample variance, as our estimator of σ^2, the population variance. We chose these particular pieces of sample information as estimators since, *on average*, they equal the population parameter they are estimating; that is, since

$$Ep' = p$$
$$E\bar{X} = \mu$$

and

$$Es^2 = \sigma^2.$$

The sample variance, d^2, was rejected as an estimator of the population variance, σ^2, since, *on average*, d^2 underestimates σ^2; that is, since

$$Ed^2 = \left(\frac{n-1}{n}\right)\sigma^2 < \sigma^2.$$

This chapter investigates whether this procedure for choosing estimators is optimal, and, in particular, whether p', $\bar{X}$ and s^2 are indeed the 'best' ways of using the sample information to estimate p, μ and σ^2 respectively. We also investigate whether we can unambiguously say that s^2 is a 'better' estimator of σ^2 than is d^2.

First, we introduce some terminology. Let θ denote some population parameter we wish to estimate. Let $\hat{\theta}$ denote an estimator of θ, based on sample information. Since $\hat{\theta}$ is a sample statistic, it is necessarily a random variable. We call $\hat{\theta}$ an *unbiased estimator* of θ if, *on average*, $\hat{\theta}$ equals θ; that is, if $E\hat{\theta} = \theta$. Conversely, if $E\hat{\theta} \neq \theta$ then $\hat{\theta}$ is a *biased estimator* of θ. The *bias* of $\hat{\theta}$ as an estimator of θ is $(E\hat{\theta} - \theta)$, which, of course, is zero if $\hat{\theta}$ is unbiased.

Using this terminology we see that p', $\bar{X}$ and s^2 are unbiased estimators of p, μ and σ^2 respectively, while d^2 is a (negatively) biased estimator of σ^2. This chapter investigates whether an unbiased estimator is always to be preferred to a biased estimator. Obviously, we cannot decide in general which estimator is best, as the choice of estimator will depend on the use to which we put our estimator, and on the consequences of mis-estimation of the population parameter. The next section considers a particular estimation problem, and investigates the consequences of mis-estimation in a particular setting.

6.2   Illustration of Consequences of Mis-estimation

Let us consider a simple Keynesian multiplier model:

$$C = \alpha + \beta Y \tag{6.1}$$

$$Y = C + G \tag{6.2}$$

where C is aggregate consumption, Y aggregate income and G government expenditure (the economy is assumed closed and all investment is undertaken by the government; all variables are in real terms). The first equation is, of course, the aggregate consumption function, and the second the income identity. Substituting (6.1) in (6.2) and solving for Y in terms of G we get:

$$Y = \frac{\alpha}{1-\beta} + \frac{1}{1-\beta} G.$$

Thus a change in government expenditure of ΔG will lead to an increase in income of ΔY given by:

$$\Delta Y = \frac{1}{1-\beta} \Delta G \quad \text{or} \quad \Delta Y = k \Delta G.$$

The multiplier $k = 1/(1-\beta)$ depends on β, the marginal propensity to consume (M.P.C.).

Suppose that the government knows that next year the economy is going to be underemployed, and that the government calculates that an increase in income of \$1,000m is needed to remove the anticipated unemployment. In order to avoid this unemployment the government needs to increase its expenditure by $\Delta G = \Delta Y/k = \Delta Y (1-\beta) = 1,000 (1-\beta)$ in \$m. If the government knows exactly the value of β, it can easily evaluate the required increase in its expenditure. However, in practice, the government does not know the exact value of β. It has estimates of it based on sample information (ways of finding estimates are discussed in the next chapter). What are the consequences for the economy if the government's estimate of β is not correct? If the government under-estimates β (that is, if its estimate is lower than the true value of β), it will overestimate $(1 - \beta)$ and thus will increase government expenditure by more than that required to remove the unemployment. In terms of our simplified economic analysis the consequence is clear — inflation will result. If the government overestimates β, it will underestimate $(1 - \beta)$ and thus it will increase government expenditure by less than that required to remove the unemployment. The consequence of the mis-estimation is again clear — the unemployment, although lessened, will persist. Further, the greater the underestimation of β the worse the inflation, and the greater the overestimation the worse the unemployment. We see, therefore, that there are costs associated with mis-estimation; costs, which, in this example, increase with the magnitude of the mis-estimation. Notice also that in our example the costs of over- and underestimation are not the same. In table 6.1 we give hypothetical values for the consequences of mis-estimation.

Table 6.1   Consequences of mis-estimation of the M.P.C.

Magnitude of the mis-estimation of β	Consequences of the mis-estimation if it is an:	
	under-estimate (inflation rate, %)	over-estimate (unemployment rate, %)
0	0	0
0.05	1	0.25
0.10	2	0.50
0.15	4	1.00
0.20	7	1.75
0.25	11	2.75

The comparison of these costs will depend upon the government's social welfare function. If it considers that a 1 per cent inflation rate is worse than a ¼ per cent unemployment rate (and that a 2 per cent inflation rate is worse than a ½ per cent unemployment rate, etc.), it will prefer (if it is unable to determine β exactly) to overestimate rather than underestimate the value of β. Thus it may prefer a method of estimating β that overestimates it on average.

Consider the following hypothetical loss-of-welfare function:

If there is x per cent inflation, welfare loss = $8x$

If there is y per cent unemployment, welfare loss = $4y$.

Using this function table 6.1 can be translated into a welfare loss table, as given in table 6.2.

Suppose now that there are two different ways of using sample information to provide estimators of β. Denote these two estimators by $\hat{\beta}_1$

Table 6.2   Welfare loss table

Magnitude of the mis-estimation of β	Welfare loss of the mis-estimation if it is an:	
	underestimate	overestimate
0	0	0
0.05	8	1
0.10	16	2
0.15	32	4
0.20	56	7
0.25	88	11

Table 6.3   Distributions of two estimators of β

$\hat{\beta}_1 - \beta$	$P[\hat{\beta}_1 - \beta]$	$(\hat{\beta}_1 - \beta)P[\hat{\beta}_1 - \beta]$	$\hat{\beta}_2 - \beta$	$P[\hat{\beta}_2 - \beta]$	$(\hat{\beta}_2 - \beta)P[\hat{\beta}_2 - \beta]$
−0.10	0.1	−0.01	−0.05	0.1	−0.005
−0.05	0.2	−0.01	0	0.2	0
0	0.4	0	0.05	0.4	0.02
0.05	0.2	0.01	0.10	0.2	0.02
0.10	0.1	0.01	0.15	0.1	0.015
Σ	1	$E(\hat{\beta}_1 - \beta) = 0$	Σ	1	$E(\hat{\beta}_2 - \beta) = 0.05$

and $\hat{\beta}_2$. As we saw in chapters 4 and 5, it is usually possible to find the distributions of estimators. Suppose the distributions of $\hat{\beta}_1$ and $\hat{\beta}_2$ are as given in table 6.3. (Chapter 7 shows how we may derive the distributions of estimators of β.)

Table 6.3 shows that:

$$E(\hat{\beta}_1 - \beta) = 0$$

that is, $E\hat{\beta}_1 = \beta$, and so $\hat{\beta}_1$ is an unbiased estimator of β.

$$E(\hat{\beta}_2 - \beta) = 0.05$$

that is $E\hat{\beta}_2 = \beta + 0.05$, and so $\hat{\beta}_2$ is a (positively) biased estimator of β.

Combining the probability distributions of $\hat{\beta}_1$ and $\hat{\beta}_2$ with the welfare loss table (table 6.2), we find the probability distributions of the loss (L) using the two estimators. These are given in table 6.4.

Thus, if the government uses estimator β_1 its expected loss $EL = 3.6$. If the government uses estimator $\hat{\beta}_2$ its expected loss $EL = 2.0$. Thus, *on average*, the government will suffer a smaller welfare loss using estimator $\hat{\beta}_2$, even though, as we have already shown, $\hat{\beta}_2$, is a *biased* estimator of β.

Table 6.4   Distributions of welfare loss using
$\hat{\beta}_1$ and $\hat{\beta}_2$

using $\hat{\beta}_1$			using $\hat{\beta}_2$		
L	P[L]	LP[L]	L	P[L]	LP[L]
16	0.1	1.6	8	0.1	0.8
8	0.2	1.6	0	0.2	0
0	0.4	0	1	0.4	0.4
1	0.2	0.2	2	0.2	0.4
2	0.1	0.2	4	0.1	0.4
Σ	1	EL = 3.6	Σ	1	EL = 2.0

This result follows since we have constructed our welfare loss function in such a way that a 1 per cent inflation rate is considered to be worse than a ¼ per cent unemployment rate. Let us consider what would happen if the government considered these equally bad; that is, suppose the loss-of-welfare function is:

If there is x per cent inflation, welfare loss = x

If there is y per cent unemployment, welfare loss = $4y$

Table 6.5   Alternative welfare loss table

Magnitude of the mis-estimation of β	Welfare loss of the mis-estimation if it is an:	
	underestimate	overestimate
0	0	0
0.05	1	1
0.10	2	2
0.15	4	4
0.20	7	7
0.25	11	11

Our welfare loss table is now as given in table 6.5, and, as can be seen by inspection of this table, the losses are symmetric. Combining this new welfare loss function with the distribution of our two estimators as given in table 6.3 gives the distributions of the loss under the two estimators. These are shown in table 6.6.

Now, the expected loss using $\hat{\beta}_1$ is $EL = 0.8$, and using $\hat{\beta}_2$ the expected loss is $EL = 1.3$. In this situation the government would now prefer $\hat{\beta}_1$ to $\hat{\beta}_2$, as an estimator of β. They would prefer the unbiased estimator to the biased estimator.

Table 6.6   Distribution of welfare loss for
alternative welfare function

using $\hat{\beta}_1$			using $\hat{\beta}_2$		
L	$P[L]$	$LP[L]$	L	$P[L]$	$LP[L]$
2	0.1	0.2	1	0.1	0.1
1	0.2	0.2	0	0.2	0
0	0.4	0	1	0.4	0.4
1	0.2	0.2	2	0.2	0.4
2	0.1	0.2	4	0.1	0.4
Σ	1	$EL = 0.8$	Σ	1	$EL = 1.3$

However, even with this symmetric loss function, it is still not true that an unbiased estimator is *always* preferred to a biased estimator. To illustrate, suppose the statisticians found a new estimator $\hat{\beta}_3$ of β with distribution as given in table 6.7. Here we have $E(\hat{\beta}_3 - \beta) = 0.03$, that is $E\hat{\beta}_3 = \beta + 0.03$, and so $\hat{\beta}_3$ is a (positively) biased estimator of β. Thus (using our second welfare loss function) we can find the distribution of the loss, L, using $\hat{\beta}_3$. This is given in table 6.8, which shows that the expected loss using $\hat{\beta}_3$ is less than the expected loss using $\hat{\beta}_1$, even though $\hat{\beta}_3$ is biased while $\hat{\beta}_1$ is unbiased. The reason for this result should be apparent: although $\hat{\beta}_3$ is biased its variance is smaller, and the smaller variance outweighs the bias to give a smaller expected loss. (You should verify that var $\hat{\beta}_1 = 0.003$ and var $\hat{\beta}_3 = 0.0016$.) We illustrate the distributions of the three estimators in figure 6.1.

Table 6.7   Distribution of a third
estimator of β

$\hat{\beta}_3 - \beta$	$P[\hat{\beta}_3 - \beta]$	$(\hat{\beta}_3 - \beta)P[\hat{\beta}_3 - \beta]$
0	0.6	0
0.05	0.2	0.01
0.10	0.2	0.02
Σ	1	0.03

Table 6.8   Distribution
of welfare loss using $\hat{\beta}_3$

L	$P[L]$	$LP[L]$
0	0.6	0
1	0.2	0.2
2	0.2	0.4
Σ	1	$EL = 0.6$

We have shown in this section that the 'best' estimator to use in any particular application depends crucially on the losses incurred in misestimation. We can also see that an estimator that may be best in one particular application may not be best in another application. In general, therefore, the statistician is unable to state unambiguously that a particular estimator is always the best.

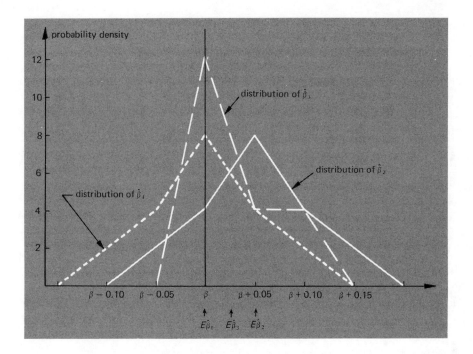

Figure 6.1   Distributions of the three estimators of β

6.3 Loss Functions in General

We continue the discussion in a more general form, using our notation θ for the parameter to be estimated, and $\hat{\theta}$ for an estimator of θ. In general the loss, L, incurred in mis-estimation depends on θ and on $\hat{\theta}$; that is,

$$L = L(\theta, \hat{\theta})$$

The particular form that the function L takes will of course depend on the particular application we are considering. In many cases, however, we can simplify the function and say that the loss depends solely on the mis-estimation, namely $\hat{\theta} - \theta$; that is,

$$L = L_1(\hat{\theta} - \theta).$$

(Both loss functions in our problem above were of this form.)

Whatever the form of L or L_1, it is reasonable to assume that no loss is incurred if $\hat{\theta} = \theta$; that is, $L(\theta, \theta) = 0$ and $L_1(0) = 0$. For $\hat{\theta} \neq \theta$ then the loss is non-negative.

One particular form L_1 may take is $L_1(\theta - \theta) = \lambda \mid \theta - \theta \mid$ $(\lambda > 0)$; that is, the loss is some constant multiple of the *magnitude* of the mis-estimation. In many applications, however, the loss becomes proportionately worse as the magnitude of the mis-estimation increases. One form of L_1 which would take this into account is the quadratic form, namely

$$L_1(\hat{\theta} - \theta) = \lambda(\hat{\theta} - \theta)^2 \quad (\lambda > 0) \tag{6.3}$$

that is, the loss is some constant multiple of the square of the mis-estimation.

However, both these last two forms consider the loss as being symmetric: the loss incurred from an underestimation of a certain magnitude is equal to the loss incurred from an overestimate of the same magnitude. Our first social welfare function was of a non-symmetric type, while the second was symmetric.

Whatever the form of our loss function, our objective is the same: to choose that estimator which has the smallest expected loss. Thus, if we have a choice between two estimators $\hat{\theta}_1$ and $\hat{\theta}_2$, and if $E[L(\theta, \hat{\theta}_1)] < E[L(\theta, \hat{\theta}_2)]$ then we prefer estimator $\hat{\theta}_1$ as it has a smaller expected loss.

In order to derive some desirable properties of estimators we need to use a particular form of the loss function. In many cases, the quadratic form (6.3) has been found applicable, and we will continue the discussion in terms of this particular form. We must, however, keep in mind that this may not be appropriate in all applications.

Assume, therefore, that the loss, L, is given by:

$$L = \lambda(\hat{\theta} - \theta)^2 \quad\quad (\lambda > 0). \tag{6.3}$$

The expected loss, using the estimator $\hat{\theta}$, is then:

$$EL = \lambda E(\hat{\theta} - \theta)^2$$
$$= \lambda E[(\hat{\theta} - E\hat{\theta}) + (E\hat{\theta} - \theta)]^2$$

where $E\hat{\theta}$ is of course the mean of $\hat{\theta}$, and will equal θ only if $\hat{\theta}$ is an unbiased estimator of θ. Thus

$$EL = \lambda[E(\hat{\theta} - E\hat{\theta})^2 + 2E\{(\hat{\theta} - E\hat{\theta})(E\hat{\theta} - \theta)\} + E(E\hat{\theta} - \theta)^2].$$

Now θ and $E\hat{\theta}$ are constants, and so

$$EL = \lambda[\text{var } \hat{\theta} + 2(E\hat{\theta} - \theta)E(\hat{\theta} - E\hat{\theta}) + (E\hat{\theta} - \theta)^2]$$

but $E(\hat{\theta} - E\hat{\theta}) = 0$, and $E\hat{\theta} - \theta$ (which is zero if $\hat{\theta}$ is an unbiased estimator of θ, but non-zero if it is biased) is the *bias* of the estimator of θ. Thus,

$$EL = \lambda\{\text{var } \hat{\theta} + (\text{bias})^2\}. \tag{6.4}$$

If $\hat{\theta}$ is an unbiased estimator of θ, then the expected loss is proportional to the variance of $\hat{\theta}$. So if we have two estimators, $\hat{\theta}_1$ and $\hat{\theta}_2$, of θ and if they are both unbiased, then

$$\text{using } \hat{\theta}_1 \quad\quad EL = \lambda \text{ var } \hat{\theta}_1$$
$$\text{using } \hat{\theta}_2 \quad\quad EL = \lambda \text{ var } \hat{\theta}_2.$$

If follows, therefore, that of all unbiased estimators the one with the smallest variance incurs the smallest expected loss, and is therefore preferred. We call the estimator with the smallest variance of all unbiased estimators the most *efficient*, or the *best unbiased estimator*. (Note carefully that all we have shown is that it is 'best' for the particular case of a quadratic loss function.)

Reverting to (6.4) we can see why we may prefer a biased to an unbiased estimator. Let $\hat{\theta}_1$ be an unbiased estimator of θ and $\hat{\theta}_2$ a biased estimator. Then $\hat{\theta}_2$ is preferred to $\hat{\theta}_1$ if $\lambda\{\text{var }\hat{\theta}_2 + (\text{bias of }\hat{\theta}_2)^2\} < \lambda \text{ var }\hat{\theta}_1$ that is, if $\text{var }\hat{\theta}_1 - \text{var }\hat{\theta}_2 > (\text{bias of }\hat{\theta}_2)^2$. So if $\hat{\theta}_2$ has a sufficiently smaller variance than $\hat{\theta}_1$, this might outweigh the bias to give a smaller expected loss. It follows, of course, that if two estimators have the same variance, but one is biased and the other not, then we prefer the unbiased one.

Note that a small variance by itself is not of much use. Consider, for example, the estimator $\hat{\theta}$ of θ, given by $\hat{\theta} = 2.3$. Then $\text{var }\hat{\theta} = 0$, as $\hat{\theta}$ is a constant. Would this be a very good estimator?

6.4 Properties of $\overline{X}$ as an Estimator of μ

In the light of the above discussion, we now look more closely at one of the estimators we have used, namely $\overline{X}$ as an estimator of μ. The sample information we have consists of n observations $X_1, X_2, \ldots X_n$. We know that for all i, $EX_i = \mu$ and $\text{var } X_i = \sigma^2$. We investigate whether or not $\overline{X}$ is the best way of using our sample information to estimate μ. In particular, we investigate whether other estimators, for example the sample median, or

$$\hat{\mu}_1 = \frac{X_1 + 2X_2 + 3X_3 + \ldots + nX_n}{1 + 2 + 3 + \ldots + n}$$

or

$$\hat{\mu}_2 = \frac{X_1^2 + X_2^2 + \ldots + X_n^2}{n}$$

etc. might not be better estimators of μ than is $\overline{X}$.

First, we note that $\overline{X}$ is a linear function of $X_1, \ldots, X_n$. It seems reasonable that a good estimator of μ must be a linear function of $X_1, \ldots, X_n$. To appreciate this, consider $\hat{\mu}_2$ as defined above; $E\hat{\mu}_2$ will contain terms in σ^2. Only a linear function of $X_1, \ldots, X_n$ will in general give an expected value containing only terms in μ. Therefore it seems reasonable to investigate only estimators of μ that are linear functions of the sample observations.

Consider the following most general linear estimator:

$$\hat{\mu} = a_1 X_1 + a_2 X_2 + \ldots + a_n X_n = \sum_1^n a_i X_i. \qquad (6.5)$$

What conditions must the a_i satisfy to make $\hat{\mu}$ the best estimator of μ? We have already discussed at length that this question has no answer unless a loss function is specified. So let us answer the more restricted question: what conditions must the a_i satisfy to make $\hat{\mu}$ the best *unbiased* estimator of μ? (By 'best' here we mean the unbiased estimator with the smallest variance.) For μ to be unbiased, we need $E\hat{\mu} = \mu$. Now:

$$E\hat{\mu} = E\left(\sum_1^n a_i X_i\right)$$

$$= \sum_1^n a_i E X_i$$

$$= \sum_1^n a_i \mu$$

Thus

$$E\hat{\mu} = \mu \sum_1^n a_i. \qquad (6.6)$$

Thus, for $\hat{\mu}$ to be unbiased we must have $\sum_1^n a_i = 1$. Given this restriction on the a_i how do we choose the a_i to make $\hat{\mu}$ the best unbiased estimator in the sense of smallest variance? Consider the variance of $\hat{\mu}$:

$$\text{var } \hat{\mu} = \text{var } \left(\sum_1^n a_i X_i\right)$$

$$= \sum_1^n \text{var } (a_i X_i)$$

assuming, as usual, that $X_1, X_2, \ldots, X_n$ are independent (using (3.46)). Thus:

$$\text{var } \hat{\mu} = \sum_1^n a_i^2 \text{ var } X_i = \sum_1^n a_i^2 \sigma^2$$

that is

$$\text{var } \hat{\mu} = \sigma^2 \sum_1^n a_i^2. \qquad (6.7)$$

So our problem is to choose the a_i to minimise $\sum_1^n a_i^2$ subject to $\sum_1^n a_i = 1$. It is useful to notice at this stage that if $a_i = 1/n$ (for all i) then

$$\hat{\mu} \equiv \bar{X} \quad \text{(see (6.5))}$$

In order to compare $\hat{\mu}$ with $\bar{X}$ let us write

$$a_i = \frac{1}{n} + d_i \quad (i = 1, 2, \ldots, n) \quad \text{where the } d_i \text{ are a set of constants.}$$

The condition for unbiasedness $(\Sigma_1^n a_i = 1)$ becomes

$$\sum_1^n d_i = 0 \quad \left(\text{since } \sum_1^n \frac{1}{n} = 1 \right)$$

Now from (6.7) the variance of $\hat{\mu}$ is given by

$$\text{var } \hat{\mu} = \sigma^2 \sum_1^n \left(\frac{1}{n} + d_i \right)^2$$

that is

$$\text{var } \hat{\mu} = \frac{\sigma^2}{n} + \frac{2\sigma^2}{n} \sum_1^n d_i + \sigma^2 \sum_1^n d_i^2$$

but we have already shown that $\Sigma_1^n d_i$ must be zero for $\hat{\mu}$ to be unbiased; we also know that $\sigma^2/n = \text{var } \bar{X}$ (from (5.5)). Thus:

$$\text{var } \hat{\mu} = \text{var } \bar{X} + \sigma^2 \sum_1^n d_i^2$$

This immediately shows (since $\sigma^2 \Sigma_1^n d_i^2$ must necessarily be greater than or equal to zero) that

$$\text{var } \hat{\mu} \geqslant \text{var } \bar{X}$$

and

$$\text{var } \hat{\mu} = \text{var } \bar{X} \quad \text{only if} \quad \sum_1^n d_i^2 = 0 \text{ (excluding the trivial case of } \sigma^2 |= 0),$$

that is, only if

$$d_i = 0 \quad \text{for all } i.$$

Further, if $d_i = 0$, then $a_i = 1/n \; (i = 1, 2, \ldots, n)$ and so

$$\hat{\mu} = \bar{X}$$

Thus $\bar{X}$ is indeed the best linear unbiased estimator of μ; this is our justification for using it so extensively.

Instead of asking the question: 'What is the best linear unbiased estimator of μ?', let us ask: 'What linear estimator minimises the expected loss when the loss function is quadratic, as in (6.3)?'. Equation (6.4) shows that the expected loss is minimised using that estimator with the smallest variance plus squared bias. Using our general linear estimator (6.5) and the result in (6.6) we see that the bias for any a_i $(i =$

$1, 2, \ldots, n)$ is:

$$\text{bias} = (E\hat{\mu} - \mu) = \mu \sum_{1}^{n} a_i - \mu$$

$$= \mu \left(\sum_{1}^{n} a_i - 1 \right)$$

Using (6.4) the expected loss is proportional to $(\text{var } \hat{\mu} + (\text{bias})^2)$ that is,

$$EL = \lambda \left\{ \sigma^2 \sum_{1}^{n} a_i^2 + \mu^2 \left(\sum_{1}^{n} a_i - 1 \right)^2 \right\} \quad \text{(using (6.7))}.$$

Choosing the a_i to minimise this involves solving $\partial EL/\partial a_i = 0$ $(i = 1, 2, \ldots, n)$,

$$\frac{\partial}{\partial a_i} EL = 0 \quad \text{gives} \quad 2a_i \sigma^2 + 2 \left(\sum_{1}^{n} a_j - 1 \right) \mu^2 = 0 \quad (i = 1, 2, \ldots, n).$$

We have, therefore, n linear equations in the n unknowns $a_1, a_2, \ldots, a_n$. These we can easily solve, but there is no point in doing so as we can see that the values of $a_1, a_2, \ldots, a_n$ which minimise EL will depend on μ and σ^2. If we do not know μ and σ^2 then we cannot find the a_i which minimise the expected loss. If we do know μ and σ^2, then we do not need to estimate them!

Thus when we are estimating μ, we cannot form the estimator $\hat{\mu}$ which minimises the expectation of a quadratic loss function. We thus use $\bar{X}$, which although the best of all linear unbiased estimators, may not necessarily be the estimator which minimises the expected loss.

6.5 Illustration of Properties of Estimators of μ

We illustrate these results with a simple numerical example. Suppose the population consists of just four members which take X values, 6, 12, 18 and 24; the population is thus uniformly distributed with mean $\mu = 15$ and variance $\sigma^2 = 45$.

Suppose we take samples of size $n = 3$, and for each sample compute the value of some possible estimators of the population mean. For example, we will examine the properties of the following four estimators:

$$\hat{\mu}_1 = \bar{X} = \frac{X_1 + X_2 + X_3}{3} \qquad\qquad \hat{\mu}_3 = \frac{2X_1 + X_2 + 3X_3}{6}$$

$$\hat{\mu}_2 = \text{sample median} \qquad\qquad \hat{\mu}_4 = 3X_1 - X_2 - X_3.$$

(You should construct some more examples yourself.)

Table 6.9   List of samples of size 3 from population

Sample number	X_1	X_2	X_3	$\hat{\mu}_1$	$\hat{\mu}_2$	$\hat{\mu}_3$	$\hat{\mu}_4$	Sample number	X_1	X_2	X_3	$\hat{\mu}_1$	$\hat{\mu}_2$	$\hat{\mu}_3$	$\hat{\mu}_4$
1	6	6	6	6	6	6	6	33	18	6	6	10	6	10	42
2	6	6	12	8	6	9	0	34	18	6	12	12	12	13	36
3	6	6	18	10	6	12	− 6	35	18	6	18	14	18	16	30
4	6	6	24	12	6	15	−12	36	18	6	24	16	18	19	24
5	6	12	6	8	6	7	0	37	18	12	6	12	12	11	36
6	6	12	12	10	12	10	− 6	38	18	12	12	14	12	14	30
7	6	12	18	12	12	13	−12	39	18	12	18	16	18	17	24
8	6	12	24	14	12	16	−18	40	18	12	24	18	18	20	18
9	6	18	6	10	6	8	− 6	41	18	18	6	14	18	12	30
10	6	18	12	12	12	11	−12	42	18	18	12	16	18	15	24
11	6	18	18	14	18	14	−18	43	18	18	18	18	18	18	18
12	6	18	24	16	18	17	−24	44	18	18	24	20	18	21	12
13	6	24	6	12	6	9	−12	45	18	24	6	16	18	13	24
14	6	24	12	14	12	12	−18	46	18	24	12	18	18	16	18
15	6	24	18	16	18	15	−24	47	18	24	18	20	18	19	12
16	6	24	24	18	24	18	−30	48	18	24	24	22	24	22	6
17	12	6	6	8	6	8	24	49	24	6	6	12	6	12	60
18	12	6	12	10	12	11	18	50	24	6	12	14	12	15	54
19	12	6	18	12	12	14	12	51	24	6	18	16	18	18	48
20	12	6	24	14	12	17	6	52	24	6	24	18	24	21	42
21	12	12	6	10	12	9	18	53	24	12	6	14	12	13	54
22	12	12	12	12	12	12	12	54	24	12	12	16	12	16	48
23	12	12	18	14	12	15	6	55	24	12	18	18	18	19	42
24	12	12	24	16	12	18	0	56	24	12	24	20	24	22	36
25	12	18	6	12	12	10	12	57	24	18	6	16	18	14	48
26	12	18	12	14	12	13	6	58	24	18	12	18	18	17	42
27	12	18	18	16	18	16	0	59	24	18	18	20	18	20	36
28	12	18	24	18	18	19	− 6	60	24	18	24	22	24	23	30
29	12	24	6	14	12	11	6	61	24	24	6	18	24	15	42
30	12	24	12	16	12	14	0	62	24	24	12	20	24	18	36
31	12	24	18	18	18	17	− 6	63	24	24	18	22	24	21	30
32	12	24	24	20	24	20	−12	64	24	24	24	24	24	24	24

If we sample with replacement, there are 64 possible samples, as listed in table 6.9. Each sample has probability 1/64. The distribution of X_1 is given in table 6.10, which shows that $EX_1 = 15 = \mu$ and var $X_1 = 45 = \sigma^2$, and that the 'shape' of the distribution of X_1 is exactly the same as the 'shape' of the population distribution. Thus, regarding X_1 as an estimator of μ, we see that it is an unbiased estimator, with variance 45. X_2 and X_3 have, of course, the same distribution as X_1.

The distribution of $\hat{\mu}_1 = \bar{X}$ is given in table 6.11. This shows (which, of course, we already knew) that $E\hat{\mu}_1 = 15 = \mu$, that is $\bar{X}$ is an unbiased estimator of μ, and that var $\hat{\mu}_1 = 15 = 45/3 = \sigma^2/n$.

The distribution of $\hat{\mu}_2$ (the sample median) is given in table 6.12. This shows that $E\hat{\mu}_2 = 15 = \mu$, and thus $\hat{\mu}_2$ is also an unbiased estimator of μ.

Table 6.10   Distribution of X_1

Sample numbers	Number of samples	X_1	$P[X_1]$ ($\times 4$)	$X_1 P[X_1]$ ($\times 4$)	$(X_1 - EX_1)^2 P[X_1]$ ($\times 4$)
1 to 16 inclusive	16	6	1	6	81
17 to 32 inclusive	16	12	1	12	9
33 to 48 inclusive	16	18	1	18	9
49 to 64 inclusive	16	24	1	24	81
Σ	64		4	60	180

$$EX_1 = \frac{60}{4} = 15 \qquad \text{var } X_1 = \frac{180}{4} = 45$$

Table 6.11   Distribution of $\hat{\mu}_1 = \bar{X}$

Sample numbers	Number of samples	$\hat{\mu}_1$	$P[\hat{\mu}_1]$ ($\times 64$)	$\hat{\mu}_1 P[\hat{\mu}_1]$ ($\times 64$)	$(\hat{\mu}_1 - E\hat{\mu}_1)^2$ $\times P[\hat{\mu}_1]$ ($\times 64$)
1	1	6	1	6	81
2, 5, 17	3	8	3	24	147
3, 6, 9, 18, 21, 33	6	10	6	60	150
4, 7, 10, 13, 19, 22, 25, 34, 37, 49	10	12	10	120	90
8, 11, 14, 20, 23, 26, 29, 35, 38, 41, 50, 53	12	14	12	168	12
12, 15, 24, 27, 30, 36, 39, 42, 45, 51, 54, 57	12	16	12	192	12
16, 28, 31, 40, 43, 46, 52, 55, 58, 61	10	18	10	180	90
32, 44, 47, 56, 59, 62	6	20	6	120	150
48, 60, 63	3	22	3	66	147
64	1	24	1	24	81
	64		64	960	960

$$E\hat{\mu}_1 = \frac{960}{64} = 15 \qquad \text{var } \hat{\mu}_1 = \frac{960}{64} = 15$$

This unbiasedness is a consequence of the population distribution being symmetrical; if the population was not symmetrically distributed, then the median would be a biased estimator of the population mean (see exercise 6.2). We also have that var $\hat{\mu}_2 = 63/2 > 15 = $ var $\hat{\mu}_1$, and thus $\hat{\mu}_2$ is a less efficient estimator than is $\hat{\mu}_1$.

Finally, the distributions of $\hat{\mu}_3$ and $\hat{\mu}_4$ are given in table 6.13. This shows that both $\hat{\mu}_3$ and $\hat{\mu}_4$ are unbiased estimators of μ (you should have been able to anticipate this) and that their variances are respectively 35/2 and 495.

Ranking our four estimators, all of which are unbiased, in order of efficiency, beginning with the most efficient, gives

$$\hat{\mu}_1, \hat{\mu}_3, \hat{\mu}_2, \hat{\mu}_4.$$

Table 6.12   Distribution of $\hat{\mu}_2$ (the sample median)

$\hat{\mu}_2$	$P[\hat{\mu}_2]$ (× 32)	$\hat{\mu}_2 P[\hat{\mu}_2]$ (× 32)	$(\hat{\mu}_2 - E\hat{\mu}_2)^2 P[\hat{\mu}_2]$ (× 32)
6	5	30	405
12	11	132	99
18	11	198	99
24	5	120	405
Σ	32	480	1,008

$$E\hat{\mu}_2 = \frac{480}{32} = 15 \quad \text{var } \hat{\mu}_2 = \frac{1008}{32} = \frac{63}{2}$$

Table 6.13   Distributions of $\hat{\mu}_3$ and $\hat{\mu}_4$

$\hat{\mu}_3$	$P[\hat{\mu}_3]$ (× 64)	$\hat{\mu}_3 P[\hat{\mu}_3]$ (× 64)	$(\hat{\mu}_3 - E\hat{\mu}_3)^2 P[\hat{\mu}_3]$ (× 64)	$\hat{\mu}_4$	$P[\hat{\mu}_4]$ (× 64)	$\hat{\mu}_4 P[\hat{\mu}_4]$ (× 64)	$(\hat{\mu}_4 - E\hat{\mu}_4)^2 P[\hat{\mu}_4]$ (× 64)
6	1	6	81	−30	1	−30	2,025
7	1	7	64	−24	2	−48	3,042
8	2	16	98	−18	3	−54	3,267
9	3	27	108	−12	5	−60	3,645
10	3	30	75	−6	5	−30	2,205
11	4	44	64	0	5	0	1,125
12	5	60	45	6	6	36	486
13	5	65	20	12	5	60	45
14	5	70	5	18	5	90	45
15	6	90	0	24	6	144	486
16	5	80	5	30	5	150	1,125
17	5	85	20	36	5	180	2,205
18	5	90	45	42	5	210	3,645
19	4	76	64	48	3	144	3,267
20	3	60	75	54	2	108	3,042
21	3	63	108	60	1	60	2,025
22	2	44	98				
23	1	23	64				
24	1	24	81				
Σ	64	960	1,120	Σ	64	960	31,680

$$E\hat{\mu}_3 = \frac{960}{64} = 15 \quad \text{var } \hat{\mu}_3 = \frac{1,120}{64} = \frac{35}{2} \qquad E\hat{\mu}_4 = \frac{960}{64} = 15 \quad \text{var } \hat{\mu}_4 = \frac{31,680}{64} = 495$$

We note that X_1 as an estimator of μ is more efficient than $\hat{\mu}_4$ but less efficient than $\hat{\mu}_1$, $\hat{\mu}_2$ and $\hat{\mu}_3$.

6.6   Properties of Estimators of σ^2

Let us now briefly examine the properties of s^2 and d^2 as estimators of σ^2. We know s^2 is an unbiased estimator of σ^2, but d^2 is biased. In fact the bias of d^2 is given by

$$Ed^2 - \sigma^2 = \left(\frac{n-1}{n}\right)\sigma^2 - \sigma^2 = -\frac{\sigma^2}{n}.$$

However, since $s^2 = (n/(n-1))d^2$, it obviously follows that

$$\text{var}(s^2) = \left(\frac{n}{n-1}\right)^2 \text{var}(d^2) > \text{var } d^2.$$

Thus, while d^2 is biased, it has a smaller variance than s^2, and if our loss function is quadratic we may prefer d^2 if the variance is sufficiently smaller to outweigh the bias.

If the population is normally distributed we can calculate the variances of s^2 and d^2 since we know that

$$(n-1)\frac{s^2}{\sigma^2} \equiv \frac{nd^2}{\sigma^2} \quad \text{is} \quad \chi^2_{n-1}.$$

Further, any book on mathematical statistics gives us the information that the variance of a variable with a chi-square distribution with k degrees of freedom is $2k$. Thus,

$$\text{var}\left[(n-1)\frac{s^2}{\sigma^2}\right] = \text{var}\left[\frac{nd^2}{\sigma^2}\right] = 2(n-1)$$

and so,

$$\text{var } s^2 = \frac{2(n-1)\sigma^4}{(n-1)^2} = \frac{2\sigma^4}{(n-1)} \quad \text{and} \quad \text{var } d^2 = \frac{2(n-1)\sigma^4}{n^2}.$$

Thus, from (6.4), the expected loss under a quadratic loss function using s^2 is

$$EL = \frac{\lambda 2\sigma^4}{(n-1)}$$

and using d^2 is

$$EL = \lambda \left\{\frac{2(n-1)\sigma^4}{n^2} + \frac{\sigma^4}{n^2}\right\} = \lambda \frac{(2n-1)\sigma^4}{n^2}.$$

You should be able to verify that the expected loss using d^2 is smaller than the expected loss using s^2. Hence, if our loss function is quadratic we should prefer d^2 to s^2 as our estimator of σ^2. Thus our use in this book of s^2 implies overriding concern with unbiasedness as a property of our estimator of σ^2.

6.7 Asymptotic Properties of Estimators

In some estimation problems we will be unable to find any unbiased estimator at all. This situation has not yet arisen. In the case of finding an unbiased estimator of σ^2 we were able to form s^2 from d^2 as we knew the magnitude of the bias of d^2 as an estimator of σ^2: the proportionate bias was a known number. In some cases the bias or proportionate bias depends on the unknown parameter, and we cannot form an unbiased estimator. Consider, for example, an estimator $\hat{\theta}$ which has expectation $E\hat{\theta}_1 = \theta + \theta^2$. There is no way of forming $\hat{\theta}_2$ from $\hat{\theta}_1$, so that $E\hat{\theta}_2 = \theta$. In some other cases, we will be unable to find the expected value of our estimator.

In these cases we have to resort to other ways of deciding which estimators are 'best'. One way of doing so is to investigate the properties of estimators when the sample size, n, increases without limit. If an estimator is biased, but the bias decreases to zero as the sample size goes to infinity, then we say the estimator is *asymptotically unbiased*. Thus, if we have two estimators, both of which are biased, but one of which is asymptotically unbiased while the other is not, then it would appear reasonable to prefer the asymptotically unbiased estimator. Formally, an estimator $\hat{\theta}$ is an asymptotically unbiased estimator of θ if

$$\underset{n \to \infty}{\text{Lt}} \ E\hat{\theta} = \theta$$

or

$$E\hat{\theta} \to \theta \quad \text{as} \quad n \to \infty.$$

We see, therefore, that d^2, although a biased estimator of σ^2, is asymptotically unbiased, since

$$Ed^2 = \left(\frac{n-1}{n}\right)\sigma^2 = \left(1 - \frac{1}{n}\right)\sigma^2 \to \sigma^2 \quad \text{as} \quad n \to \infty.$$

A further desirable property is that the variance of our estimator tends to zero as the sample size tends to infinity. If an estimator is asymptotically unbiased and its variance tends to zero, then we say that the estimator is *consistent*.

Formally, an estimator $\hat{\theta}$ is a consistent estimator of θ if

$$\underset{n \to \infty}{\text{Lt}} \ E\hat{\theta} = \theta \quad and \quad \underset{n \to \infty}{\text{Lt}} \ \text{var} \ \hat{\theta} = 0$$

or

$$E\hat{\theta} \to \theta \quad and \quad \text{var} \ \hat{\theta} \to 0 \quad \text{as} \quad n \to \infty.$$

We see therefore that $\bar{X}$ is a consistent estimator of μ since $E\bar{X} = \mu$ and var $\bar{X} = (\sigma^2/n) \to 0$ as $n \to \infty$. However X_1 (which is an unbiased estimator of

μ and therefore asymptotically unbiased) is *not* a consistent estimator of μ, since var $X_1 = \sigma^2 \nrightarrow 0$ as $n \to \infty$. Further, both d^2 and s^2 are consistent estimators of σ^2, since (see section 6.6)

$$Ed^2 = \left(1 - \frac{1}{n}\right)\sigma^2 \to \sigma^2 \quad \text{as} \quad n \to \infty$$

$$\text{var } d^2 = \frac{\sigma^4}{n^2} 2(n-1) = \sigma^4 2\left(\frac{1}{n} - \frac{1}{n^2}\right) \to 0 \quad \text{as} \quad n \to \infty$$

and

$$Es^2 = \sigma^2 ; \quad \text{var } s^2 = \frac{2\sigma^4}{(n-1)} \to 0 \quad \text{as} \quad n \to \infty$$

(p' is a consistent estimator of p, see exercise 6.4). Intuitively, a consistent estimator $\hat{\theta}$ is one whose distribution 'collapses' on to θ as $n \to \infty$. We illustrate this in figure 6.2.

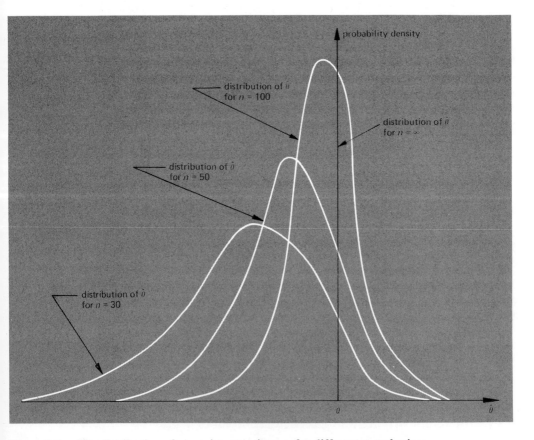

Figure 6.2　The distribution of a consistent estimator for different sample sizes

We can also intuitively generalise our ideas on efficiency to asymptotic efficiency. Loosely, we can say that the asymptotically most efficient estimator is that asymptotically unbiased estimator which has the smallest asymptotic variance of all asymptotically unbiased estimators. (If the estimators being compared are both consistent, then it is that estimator whose variance tends to zero fastest that is the asymptotically most efficient. Consider, for example, $\hat{\theta}_1$ and $\hat{\theta}_2$ which are such that

$$E\hat{\theta}_1 = \left(\frac{n-1}{n}\right)\theta \qquad E\hat{\theta}_2 = \left(\frac{n+1}{n}\right)\theta$$

$$\text{var } \hat{\theta}_1 = \frac{1}{n} \qquad \text{var } \hat{\theta}_2 = \frac{1}{n^2}$$

then the variance of $\hat{\theta}_2$ tends to zero faster, and $\hat{\theta}_2$ is thus the more asymptotically efficient of the two.)

6.8 Summary

We have shown in this chapter that the 'best' estimator to use in any application depends upon the consequences of mis-estimation. In particular, we may sometimes prefer a biased estimator to an unbiased estimator. We showed that $\bar{X}$ is the best linear unbiased estimator of μ; that is, of all linear unbiased estimators of μ, $\bar{X}$ has the smallest variance. We found that it was impossible to derive an estimator of μ that minimised the expected loss under a quadratic loss function without knowledge of μ and σ^2. Also we showed that s^2, although an unbiased estimator of σ^2, has a larger expected loss under a quadratic loss function than d^2, a biased estimator of σ^2. Finally, we considered ways of comparing estimators as the sample size increased without limit, and found that p', $\bar{X}$, s^2 and d^2 were all consistent estimators of p, μ, σ^2 and σ^2 respectively; that is, they tend in probability to the parameter they are estimating.

6.9 Exercises

6.1. Refer to section 6.2. Construct a loss-of-welfare function for which $\hat{\beta}_1$ would be preferred to both $\hat{\beta}_2$ and $\hat{\beta}_3$.

6.2. Given the following populations:

(i) 6   12   15   18
(ii) 9   12   15   18

(a) Determine for each population whether the sample mean and/or the sample median are unbiased estimators of the population mean. (You are actually asked to calculate the result; take samples of size 3 with replacement.)

(b) Calculate for each population which of the two estimators is the most efficient one, if relevant.

(c) Suppose the loss function was quadratic; determine which estimator would be preferred in each case.

(d) Would your answers be different if the samples were of size 2?

6.3. Suppose we have collected a random sample $X_1, X_2, \ldots, X_n$ from a large population which is rectangularly distributed with mean μ and variance σ^2. Show that $\bar{X}$ is both an unbiased and a consistent estimator of μ. What can you say about the statistic X_1 as an estimator of μ?

6.4. Is the sample proportion p' an unbiased and/or a consistent estimator of the population proportion p?

6.5. Construct an example for which we would prefer an estimator that equalled the parameter only 1 per cent of the time to an estimator that equalled the parameter 99 per cent of the time.

6.6. Discuss briefly the following statements (they are not necessarily all true). Illustrate your answers where applicable with specific examples.

(a) An estimator with a smaller variance than an alternative estimator is always preferred to it.

(b) We may sometimes prefer a biased estimator to an unbiased estimator with the same variance.

(c) An estimator with a zero variance is always optimal.

(d) Knowledge that an estimator is consistent is cold comfort if we are only able to obtain small samples.

(e) An estimator that is consistent is always to be preferred to one that is not consistent.

7

Elementary Regression Analysis

7.1 Introduction

We have covered in the first six chapters the statistical tools necessary for the remainder of the book. We now turn our attention to the major purpose of this book, that of applying these tools to the analysis of economic relationships. Virtually all the conclusions of economic theory are presented in the form of relationships between variables. The conventional theory of consumer behaviour, for example, leads to the conclusion that aggregate consumption expenditure is related to aggregate income and the rate of interest. In demand theory and the theory of the firm we are led to postulate relationships between the quantity demand of a good and its price, and between the quantity supplied of a good and its price respectively. We could give many more such examples.

One role of statistics is to test whether the conclusions of the theories are supported by the evidence. In respect of consumer theory we wish to test whether U.S. aggregate consumption has been related to U.S. aggregate income and the rate of interest. In the areas of demand theory and the theory of the firm we wish to test whether the quantity demanded of (say), potatoes, and the quantity supplied have been (in a particular market, town or country) related to the price of potatoes. If the empirical evidence and the conclusions of the theories are contradictory, then we will have to conclude that either the evidence is wrong or the theories are wrong. If the latter, then the theory will have to be modified in the light of the evidence.

The second role of statistics, once we have established that a meaningful relationship exists, is to determine the form and the strength of the relationship. In particular, if we have found that U.S. aggregate consumption is related to U.S. aggregate income, we want to know the form of the relationship and the magnitude of the parameters; we want to know whether the relationship is linear, and if so, what is the value of the marginal propensity to consume. We have seen in chapter 6 that the correct estimation of the marginal propensity to consume is of crucial importance for optimal government policy. We also want to know the numerical values of price elasticities of demand and supply of goods, so that optimal pricing and taxation decisions can be taken by firms and governments respectively.

In the following section we will investigate a particular formulation of an economic theory and consider how we might test its validity and find estimates of its parameters. We have deliberately chosen an over-simplified example to illustrate the main statistical problems; later we will show how to generalise it to make it more realistic.

7.2 Illustration of Regression

Let us consider the determinants of household consumption expenditure. Household consumption expenditure depends on many factors: household income, the household's assets, the availability of credit to the household, the size (and composition) of the household, the age of the head of the household, its tastes, and so on. Many texts, in order to analyse the determination of aggregate income, single out for particular attention the influence of household income on household consumption expenditure. By concentrating attention on the importance of income on expenditure considerable insight can be gained into the functioning of the economic system.

Let us start, therefore, by considering the theoretical proposition that household consumption expenditure depends on household income. Note carefully that we are not saying that expenditure depends *solely* on income, but that income is *one* determinant of expenditure; and that the isolation of this one factor provides us with a part of a simplified theory of the way the economy works.

We could qualify our theoretical proposition, as many texts do, by stating that household consumption expenditure depends on household income, *ceteris paribus*. This *ceteris paribus* assumption is a useful theoretical tool, as it enables us conceptually to hold constant the other determinants of household expenditure. However, even at a theoretical level we have to be careful about the use of this *ceteris paribus* assumption; we have to be sure that it is meaningful to talk about holding (for example) 'tastes' constant while we study the impact of varying income levels on expenditure. If 'tastes' change as income changes, which they may well do, then even at a theoretical level we could not meaningfully talk about *ceteris paribus*. Other examples spring to mind: can we hold credit availability, or age, constant as income varies?

We return from this digression on the use of *ceteris paribus* assumptions in theoretical work to restate our theoretical proposition in the original unqualified form: household consumption expenditure depends on household income. (We investigate the implications of this unqualified statement in the subsequent discussion; and, in particular, we discuss whether it is able to lead to a fair representation of the qualified theoretical proposition.) We will also go one stage further, and state our proposition in a form which often appears in textbooks:

'Household consumption expenditure depends *linearly* on household income.' (7.1)

That this is a stronger theoretical proposition is obvious, as we have now specified not only the variables in the relationship, but also the form of the relationship. The textbooks usually go further still, and specify that the intercept in this linear relationship is positive, and that the slope (the marginal propensity to consume) lies between zero and one. We see how we might test this linear relationship using actual data on household income and expenditure, and how we might find estimates of the intercept and the slope (enabling us to test the textbook *a priori* restrictions).

We now discuss what we might find if we looked at actual household consumption and income for all households in the U.S. Denote household consumption expenditure by Y and household income by X, both measured in dollars per day. Suppose we could find, at a particular point in time, all households with an income $X_1 = 40$ (dollars per day).* It is obvious that not all the households with an income of $40 will have the same consumption expenditure. We could calculate the (arithmetic) mean consumption expenditure for all these households with an income of $40. Some of the households would spend more than the mean, some less. Whether particular households spend more or less than the mean would depend on their other characteristics, such as size (composition), age, assets and so on.

We could also calculate the variance of household consumption expenditure for all these households with an income of $40. The more important household income was as a determinant of household expenditure, the smaller we would expect this variance to be. We illustrate these ideas with a simple numerical example; the actual numbers chosen are not meant to be realistic, but are kept simple to illustrate the main concepts. Suppose that there are 100 households in the total population with an income of $40 per day. Suppose we find that, of these 100 households, 4 spend $33 a day, 10 spend $34 a day, 21 spend $35, 30 spend $36, 21 spend $37, 10 spend $38 and the remaining 4 spend $39. We give this information in the frequency distribution of table 7.1.

Denote by Y_1 household consumption expenditure for households with income $X_1 = 40$. As we can usually (see below) interpret the frequency

---

*At this stage in our analysis, we want to focus attention on the necessary statistical concepts. Thus we will ignore problems raised by the theories of consumption of Duesenberry, Friedman, Modigliani and Brumberg, and interpret the income and expenditure figures as being the average (or expected) daily income and expenditure respectively. We are thus ignoring problems of fluctuating daily income and expenditure. We will return to this point later.

Table 7.1   Distribution of consumption of households with income of $40

Household consumption expenditure ($ per day)	Frequency f	Relative frequency f/m
33	4	0.04
34	10	0.10
35	21	0.21
36	30	0.30
37	21	0.21
38	10	0.10
39	4	0.04
Σ	$m = 100$	1.0

Table 7.2   Probability distribution of Y_1

Y_1	$P[Y_1]$	$Y_1 P[Y_1]$	$\{Y_1 - E(Y_1 \mid X_1)\}^2 P[Y_1]$
33	0.04	1.32	0.36
34	0.10	3.40	0.40
35	0.21	7.35	0.21
36	0.30	10.80	0
37	0.21	7.77	0.21
38	0.10	3.80	0.40
39	0.04	1.56	0.36
Σ	1.0	36.00	1.94

distribution of a variable as the probability distribution of the variable, and as it will be useful for expository purposes later, we will write the distribution of Y_1 in probability form in table 7.2 and from that calculate the mean and variance of Y_1.

We discuss in more detail the writing of the frequency distribution of Y_1 as a probability distribution. Of the 100 households, 4 spend $33, so the relative frequency of $Y_1 = 33$ is $4/100 = 0.04$. Hence if we pick one household at random out of the 100, the probability of this one household spending $Y_1 = 33$ is 0.04. Similarly the probability that one household picked at random spends $Y_1 = 38$ is 0.1. The mean household consumption of all households with an income $X_1 = 40$ is the mean of the above distribution, namely $36. We denote this by $E(Y_1 \mid X_1)$ to emphasise the fact that this is the mean household expenditure of those households with income X_1. Thus,

$$E(Y_1 \mid X_1) = E(Y_1 \mid 40) = 36.$$

The fourth column of table 7.2 gives us a measure of the spread of consumption expenditures for all these households, namely the variance. Denote this by var $(Y_1 \mid X_1)$. Thus,

$$\text{var } (Y_1 \mid X_1) = 1.94.$$

The distribution of Y_1 is illustrated in figure 7.1.

Suppose we now completely repeated the above conceptual experiment for all households with an income of $20 a day. We would again find that different households spend differing amounts, some more than average, some less. Suppose (for illustrative and conceptual simplicity) that there are again 100 households in the total population with an

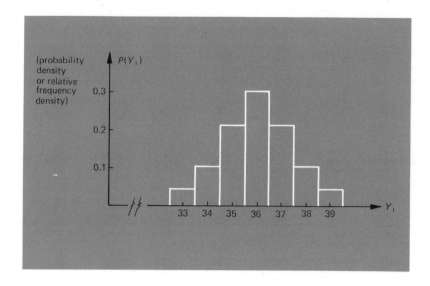

Fig. 7.1   Distribution of Y_1

income $X_2 = 20$ (dollars per day). Suppose we find that of these 100 households, 4 spend \$19 a day, 10 spend \$20, 21 spend \$21, 30 spend \$22, 21 spend \$23, 10 spend \$24 and the remaining 4 spend \$25. We put this information in the form of a probability distribution in table 7.3.

Table 7.3   Probability distribution of Y_2

Y_2	$P[Y_2]$	$Y_2 P[Y_2]$	$\{Y_2 - E(Y_2 \mid X_2)\}^2 P[Y_2]$
19	0.04	0.76	0.36
20	0.10	2.00	0.40
21	0.21	4.41	0.21
22	0.30	6.60	0
23	0.21	4.83	0.21
24	0.10	2.40	0.40
25	0.04	1.00	0.36
Σ	1.0	22.0	1.94

Table 7.3 shows that,

$$E(Y_2 \mid X_2) = E(Y_2 \mid 20) = 22$$
$$\text{var}(Y_2 \mid X_2) = \text{var}(Y_2 \mid 20) = 1.94.$$

Hence for all households with an income of \$20 per day, the mean expenditure is \$22, and the variance of expenditure is 1.94 (\$^{2}); this dis-saving is of course typical of low-income households. The distribution of Y_2 is illustrated in figure 7.2.

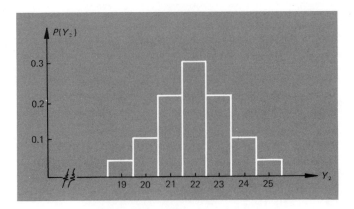

Fig. 7.2   Distribution of Y_2

You will notice that we have deliberately chosen the numbers so that the variances of Y_1 and Y_2 are the same, and so that the 'shape' of the distribution of Y_1 and the 'shape' of the distribution of Y_2 are the same. We have done this to illustrate the nature of the assumptions we will be making in subsequent sections. The only way the distributions of Y_1 and Y_2 differ is in their means. To bring out these points more clearly we introduce some new notation. We define a variable which measures, for each household, how much more (or less) that household spends above (or below) the mean expenditure of all households with the same income. Call this variable u_i for households with income X_i. Thus u_i is defined by:

$$u_i = Y_i - E(Y_i \mid X_i). \tag{7.2}$$

We consider again all those households with income $X_1 = 40$. The mean expenditure of these households was $E(Y_1 \mid X_1) = 36$. So the 4 households spending \$33 spend \$3 less than the mean, so for them $u_1 = -3$. Similarly the 10 households spending \$34 spend \$2 less than the mean, so their $u_1 = -2$. As we have the distribution of Y_1, we can continue in this way to calculate the distribution of u_1, as given in table 7.4. Thus the mean of u_1, denoted by $E(u_1 \mid X_1)$, or more simply, and without ambiguity, by Eu_1, is:

$$E(u_1 \mid X_1) = Eu_1 = 0.$$

You should have anticipated this result, as it follows immediately from the definition of u_1 (cf. (7.2)).

The variance of u_1, denoted by var $(u_1 \mid X_1)$, or more simply, and without ambiguity, by var u_1, is 1.94. Again you should have anticipated this result, as Y_1 and u_1 differ only by a constant (viz. $E(Y_1 \mid X_1)$) and must therefore have the same variance.

Table 7.4	Probability distribution of u_1		
u_1	$P[u_1]$	$u_1 P[u_1]$	$u_1{}^2 P[u_1]$
−3	0.04	−0.12	0.36
−2	0.10	−0.20	0.40
−1	0.21	−0.21	0.21
0	0.30	0	0
1	0.21	0.21	0.21
2	0.10	0.20	0.40
3	0.12	0.12	0.36
Σ	1.0	0.0	1.94

Table 7.5	Probability distribution of u_2
u_2	$P[u_2]$
−3	0.04
−2	0.10
−1	0.21
0	0.30
1	0.21
2	0.10
3	0.04
Σ	1.0

Consider now the households with an income of $(X_2 =)$ \$20. Of them, the 4 households spending \$19 spend \$3 below the mean (for their income group) and so their $u_2 = -3$. Continuing in this way we find the distribution of u_2, which is given in table 7.5.

As can be seen immediately, the distribution of u_2 is exactly the same as the distribution of u_1. Thus,

$$Eu_2 = 0 \qquad \mathrm{var}\, u_2 = 1.94$$

We have deliberately chosen the numbers in this illustrative example so that u_1 and u_2 have the same distribution. This is done to show exactly what is meant by the assumptions we will be making later. Whether in fact u_1 and u_2 do have the same distribution (and thus whether the assumptions we will be making later are valid) is essentially an empirical question which we will have to investigate in due course. This we leave to chapter 8.

We now return from this study of the u's, which are *deviations* from the mean, to look at the means themselves. We have:

$$E(Y_2 \mid X_2) = E(Y_2 \mid 20) = 22$$

and

$$E(Y_1 \mid X_1) = E(Y_1 \mid 40) = 36$$

Thus, as income increases from \$20 to \$40, mean household consumption increases from \$22 to \$36. An increase in income of \$20 leads to an increase (on average) in expenditure of \$14, giving a marginal propensity to consume of $14/20 = 0.7$. This, then, is how we must interpret our theoretical proposition (7.1) that 'household consumption expenditure depends . . . on household income'. It is saying that, as income changes, so does mean expenditure; that is, $E(Y_i \mid X_i)$ depends on X_i. If we now put back the missing word, 'linearly', we can write our theoretical proposition

(7.1) as:

$$E(Y_i \mid X_i) = \alpha + \beta X_i. \tag{7.3}$$

In terms of our numerical example we have $E(Y_2 \mid 40) = 36$ and $E(Y_1 \mid 20) = 22$. Thus:

$$36 = \alpha + 40\beta$$
$$22 = \alpha + 20\beta.$$

Solving we get $\alpha = 8$ and $\beta = 0.7$; that is,

$$E(Y_i \mid X_i) = 8 + 0.7X_i \tag{7.4}$$

Finally, let us consider what would happen if we took all households with income $X_3 = 30$. If (7.4) correctly describes the relationship between mean expenditure and income over the whole income range, and in particular for $X_3 = 30$, then Y_3, expenditure of households with an income of $30, will have mean, given by (7.4),

$$E(Y_3 \mid X_3) = 8 + 0.7(30) = 29.$$

If, further, the 'shape' of the distribution of Y_3 is the same as the 'shape' of the distributions of Y_1 and Y_2, and if the variance of Y_3 equals the variances of Y_1 and Y_2, then the distribution would be as given in table 7.6. You should verify that $E(Y_3 \mid X_3) = 29$ and var $(Y_3 \mid X_3) = 1.94$.

Table 7.6   Distribution of Y_3

Y_3	$P[Y_3]$
26	0.04
27	0.10
28	0.21
29	0.30
30	0.21
31	0.10
32	0.04
Σ	1.0

Figure 7.3 illustrates diagrammatically the hypothetical distributions of Y_1, Y_2 and Y_3 that we have introduced. In figure 7.3 each dot represents the consumption–income combination for one or more households. The number alongside each dot refers to the number of households with that particular consumption–income combination. We could of course take all households with $X_4 = 10$, $X_5 = 25$, etc. The same arguments apply to these households. You should note carefully that there will be a linear

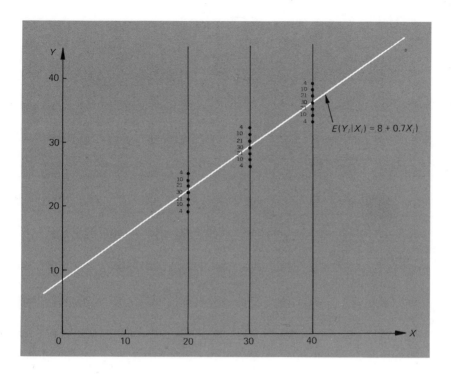

Fig. 7.3   Three consumption distributions

relationship between $E(Y_i \mid X_i)$ and X_i only if the proposition (7.1) is, in fact, correct.

Let us now combine equations (7.2) and (7.3). From (7.2)

$$E(Y_i \mid X_i) = Y_i - u_i.$$

Substituting in (7.3) for $E(Y_i \mid X_i)$ we get

$$Y_i - u_i = \alpha + \beta X_i$$

that is,

$$Y_i = \alpha + \beta X_i + u_i. \tag{7.5}$$

The interpretation of this equation is straightforward. For a household with an income X_i its consumption is given by the mean for its income group (namely $\alpha + \beta X_i$) plus a term (u_i) showing how far from the average that particular household's consumption is.

Our numerical example has now served its purpose. It has enabled us to illustrate the type of data we would expect to find if the theoretical proposition that household consumption expenditure depends linearly on household income was in fact correct. Our numerical example also illustrated what is meant by the statement that each u_i has the same distribution, and, in particular, that each u_i has the same variance. When we come to *estimate* the relationship between consumption and income, we will, initially, be making the assumption that each u_i has the same distribution (and the same variance). These assumptions will, of course, be subject to test, and we will discuss later how our analysis should be modified if the assumptions do not appear to hold in practice.

We now move on from a specific numerical example where all the parameters $(\alpha, \beta, \sigma^2)$ are known to a more general formulation. Suppose we wish to test the economic theory that household consumption expenditure depends linearly on household income and we wish to find out the numerical values of α and β in the U.S. Our first step is to repeat the argument above, and write the economic theory as meaning that $E(Y_i \mid X_i) = \alpha + \beta X_i$ (for all i). As before we can rewrite this as saying:

$$Y_i = \alpha + \beta X_i + u_i$$

where

$$Eu_i = 0 \quad \text{(for all } i\text{).}$$

Our second step is to make the simplifying assumption that var $u_i = \sigma^2$ (for all i).

Now it is fairly obvious that in order to collect data on household consumption and income for every household in the U.S. we are going to need a considerable amount of time and money. What perhaps is less obvious is that it may be unnecessary to do this in order to get good estimates of the parameters. It may be cheaper overall to collect a sample of households. The larger the sample the more accurate we would expect our results to be; but the larger the sample the higher are the costs of collection and processing. The optimal size for the sample would be that which gave the optimal trade-off between the costs of sampling, and the costs of getting and using inaccurate estimates. So here, as before, the size of the sample would depend on the use to which we are going to put the results of our analysis.

Suppose then we collect a random sample of n households, and for each of the households find their daily income and consumption expenditure. For the ith household, denote these by X_i and Y_i respectively. In general the income of the n households in the sample will be different, that is $X_i \neq X_j$ for $i \neq j$. For simplicity, let us number the households in such a way that income rises with i, that is $X_1 < X_2 < X_3 < \ldots < X_n$.

Let us concentrate attention first on the household with income X_1. We can consider this household to be a sample of one from all the households in the population with an income equal to X_1. We know from our theoretical model (if it is correct) that the mean consumption expenditure of all households with income X_1 is, of course, $E(Y_1 \mid X_1) = \alpha + \beta X_1$. The actual consumption expenditure of the household in our sample with income X_1 is denoted by Y_1. Does this sampled household spend more or less than the mean for its income group? That is, is

$$Y_1 > E(Y_1 \mid X_1) \quad \text{or} \quad Y_1 < E(Y_1 \mid X_1)?$$

or is

$$Y_1 = E(Y_1 \mid X_1)?$$

We rephrase the same question in terms of the deviation term u_1. For this particular household its u_1 is given by

$$u_1 = Y_1 - E(Y_1 \mid X_1)$$

Therefore

$$u_1 \geqslant 0 \quad \text{if} \quad Y_1 \geqslant E(Y_1 \mid X_1)$$

and

$$u_1 \leqslant 0 \quad \text{if} \quad Y_1 \leqslant E(Y_1 \mid X_1).$$

Thus our question as to whether this sampled household spends more or less than the mean for its income group can be rephrased by asking whether its u_1 is positive or negative. The obvious answer is that, if we do not know α or β and therefore do not know $E(Y_1 \mid X_1)$, we do not know whether this particular household has a positive or a negative u_1. We illustrate this point in figure 7.4.

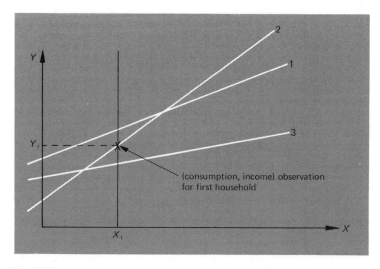

Fig. 7.4

We do not know α and β, so we do not know the position of the line $E(Y_i \mid X_i) = \alpha + \beta X_i$ in figure 7.4. In particular, we do not know where this line intersects the vertical line at X_1. Thus we do not know whether this household's Y_1 is less than $E(Y_1 \mid X_1)$ (that is, its u_1 is negative), equal to $E(Y_1 \mid X_1)$ (that is, its $u_1 = 0$), or greater than $E(Y_1 \mid X_1)$ (that is, its u_1 is positive) since we do not know whether the line $E(Y_i \mid X_i) = \alpha + \beta X_i$ is line 1, line 2 or line 3 (or any other line for that matter). All we know about this particular household's Y_1 is that it is somewhere in the distribution of all Y_1's, and that its u_1 is somewhere in the distribution of all u_1's.

We can apply a similar argument to the household in our sample with income X_2. Is its consumption, Y_2, above, below, or equal to the average consumption $E(Y_2 \mid X_2)$? That is, is its u_2 positive, negative or zero? Again we do not know as we do not know the values of α and β. Exactly the same argument applies to $u_3, u_4, \ldots, u_n$.

However, let us consider the following *conceptual* experiment: suppose we took a second sample of n households, and sampled so that the lowest income household in the second sample had the same income (X_1) as the lowest income household in the first sample; the second lowest income household in the second sample had the same income (X_2) as the second lowest income household in the first sample; ..., the highest income household in the second sample had the same income (X_n) as the highest income household in the first sample. Thus the incomes of the n households in the first sample are $X_1, X_2, \ldots, X_n$ and the incomes of the n households in the second sample are also $X_1, X_2, \ldots, X_n$.

Now, we introduce some new (but temporary) notation. Denote by $Y_i^{(1)}$ the consumption expenditure of the ith household (the one with income X_i) in the first sample, and by $Y_i^{(2)}$ the consumption expenditure of the ith household (also with income X_i) in the second sample. Thus (for example) $Y_1^{(1)}$ and $Y_1^{(2)}$ are two observations from the distribution Y_1 which has mean $E(Y_1 \mid X_1)$ and variance var $(Y_1 \mid X_1)$.

Suppose we took a third sample of n households in a similar way, that is, so that their incomes are $X_1, X_2, \ldots, X_n$ and their respective consumption expenditures are $Y_1^{(3)}, Y_2^{(3)}, \ldots, Y_n^{(3)}$. Suppose we took a fourth, a fifth and so on, all sampled so that the n households in each sample had income $X_1, X_2, \ldots, X_n$. Table 7.7 represents the results of this *conceptual* repeated sampling.

If we consider the first column of table 7.7 we see that we would have a series of observations on Y_1, namely $Y_1^{(1)}, Y_1^{(2)}, Y_1^{(3)}, \ldots, Y_1^{(m)}, \ldots$. (The number of possible different samples depends on the size of the population, the number of households in each income group, and the size n of each sample.)

Table 7.7   Results of *conceptual* repeated sampling

(table entries are household consumption figures)

Household number	1	2	3	...	i	...	n
Household income	X_1	X_2	X_3	...	X_i	...	X_n
First sample	$Y_1^{(1)}$	$Y_2^{(1)}$	$Y_3^{(1)}$	...	$Y_i^{(1)}$	...	$Y_n^{(1)}$
Second sample	$Y_1^{(2)}$	$Y_2^{(2)}$	$Y_3^{(2)}$	...	$Y_i^{(2)}$	...	$Y_n^{(2)}$
Third sample	$Y_1^{(3)}$	$Y_2^{(3)}$	$Y_3^{(3)}$	...	$Y_i^{(3)}$	...	$Y_n^{(3)}$
.	.	.	.		.		.
.	.	.	.		.		.
.	.	.	.		.		.
mth sample	$Y_1^{(m)}$	$Y_2^{(m)}$	$Y_3^{(m)}$	...	$Y_i^{(m)}$	...	$Y_n^{(m)}$
.	.	.	.		.		.
.	.	.	.		.		.
.	.	.	.		.		.

We know that the average value of $Y_1^{(1)}$, $Y_1^{(2)}$, $Y_1^{(3)}$, ..., $Y_1^{(m)}$, ..., is of course $E(Y_1 \mid X_1)$ and the variance of these values is var $(Y_1 \mid X_1)$. In general we know that the average value of $Y_i^{(1)}$, $Y_i^{(2)}$, $Y_i^{(3)}$, ..., $Y_i^{(m)}$, ..., is $E(Y_i \mid X_i)$ with variance var $(Y_i \mid X_i)$. Thus the value of $Y_i^{(1)}$ is one observation on a random variable which has mean $E(Y_i \mid X_i)$ and var $(Y_i \mid X_i)$.

We convert table 7.7 into one containing the deviation terms u. We can do this since $u_i^{(m)} = Y_i^{(m)} - E(Y_i \mid X_i)$ by definition. This is given in table 7.8.

The ith column of table 7.8 shows that we *would have* a series of realisations $u_i^{(1)}$, $u_i^{(2)}$, $u_i^{(3)}$, ..., $u_i^{(m)}$, ..., of u_i. The average value of these is $Eu_i = 0$ and the variance is var $u_i = \sigma^2$.

Thus, *before we take our sample*, u_i is a random variable with mean 0 and variance σ^2. For our future analysis we will require a further assumption about the sampling method, namely that the value of u_1 we get does not influence the value of u_2 that we get; and, in general, that the value of u_i that we get does not influence the value of u_j $(j \neq i)$ that we get. In other words u_i and u_j must be *independent random variables* $(j \neq i)$. In table 7.8 the realisations $u_i^{(1)}$, $u_i^{(2)}$, $u_i^{(3)}$, ..., $u_i^{(m)}$, ..., must have zero covariance with $u_j^{(1)}$, $u_j^{(2)}$, $u_j^{(3)}$, ..., $u_j^{(m)}$, ..., (for all $i, j; i \neq j$).

An example of when this assumption may not be met is if our sample consists of all the households in a particular street. If the households in

Table 7.8　　Results of *conceptual* repeated sampling

(table entries are household consumption figures expressed as deviations from mean expenditure for that income group)

Household number	1	2	3	... i	... n
Household income	X_1	X_2	X_3	... X_i	... X_n
First sample	$u_1^{(1)}$	$u_2^{(1)}$	$u_3^{(1)}$	... $u_i^{(1)}$	... $u_n^{(1)}$
Second sample	$u_1^{(2)}$	$u_2^{(2)}$	$u_3^{(2)}$	... $u_i^{(2)}$	... $u_n^{(2)}$
Third sample	$u_1^{(3)}$	$u_2^{(3)}$	$u_3^{(3)}$	... $u_i^{(3)}$	... $u_n^{(3)}$
.	.	.	.	.	.
.	.	.	.	.	.
.	.	.	.	.	.
mth sample	$u_1^{(m)}$	$u_2^{(m)}$	$u_3^{(m)}$	... $u_i^{(m)}$	... $u_n^{(m)}$
.	.	.	.	.	.
.	.	.	.	.	.
.	.	.	.	.	.

the street indulged in 'keeping up with the Joneses' we may well expect that, if one household spends above average for its income group, then the neighbouring household will also spend above average for *its* income group. Our realisation on the u's would therefore be dependent.

Our model so far can be summarised by:

$$
\left.
\begin{aligned}
Y_i &= \alpha + \beta X_i + u_i \\
Eu_i &= 0 && i = 1, 2, \ldots, n && \text{(A1)} \\
Eu_i^2 &= \sigma^2 && i = 1, 2, \ldots, n && \text{(A2)} \\
Eu_i u_j &= 0 && i, j = 1, 2, \ldots, n,\ i \neq j && \text{(A3)} \\
X_i &\text{ is fixed} && i = 1, 2, \ldots, n && \text{(A4)}
\end{aligned}
\right\}
\quad (7.6)
$$

The four assumptions (A1)–(A4) can be interpreted as follows:

(A1) means that, for each income, the average deviation is zero; that is, our sampling procedure is such that the expected value of Y_i is $E(Y_i \mid X_i) = \alpha + \beta X_i$.

(A2) means that the variance of each Y_i is the same, that is the variance of each Y_i about its mean is the same for all i.

(A3) means that the sampling method is such that the value we get for u_i does not influence the value we get for u_j. (cov $(u_i - u_j)$ $= E(u_i - Eu_i)(u_j - Eu_j) = Eu_i u_j$ since $Eu_i = Eu_j = 0$ from (A1).) Strictly speaking this only means that u_i, u_j have zero covariance. If, as we will later assume, each u_i is normal, then this implies independence.

(A4) means that (at least conceptually) we could collect many samples all of which have households with income X_1, X_2, ..., X_n. We will discuss the meaning of this assumption in more detail later, and consider cases where even conceptually we would be unable to fix the values of X_i ($i = 1, 2, ..., n$) in repeated samples.

In practice, of course, we only have *one* sample of n households, consisting of n income-consumption pairs:

$$(X_1, Y_1), (X_2, Y_2), ..., (X_n, Y_n).$$

which can be represented as n points in the graph of Y against X, as in figure 7.5.

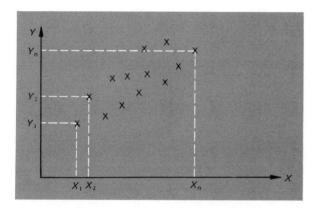

Fig. 7.5   Observations on income and consumption

In general, there is no reason to expect the points to lie on a straight line, as each (X_i, Y_i) pair is just one observation from many possible (X_i, Y_i) pairs.

We wish to use these observations to obtain estimates of α and β. As α and β define a straight line in the (X, Y) plane, it is apparent that using the observations to estimate α and β is equivalent to fitting a line to the observations such that the fitted line is an estimator of the line $E(Y_i \mid X_i) = \alpha + \beta X_i$. Calling this latter the *true line*, we want an *estimated line* which is an estimator of the true line. Denote the fitted, or estimated, line by:

$$\hat{Y}_i = \hat{\alpha} + \hat{\beta} X_i.$$

As the actual observations do not in general lie on a straight line, there will be a deviation between each observation and the point on the estimated line corresponding to that observation's value of X_i. We call the deviation of the ith observation from the estimated line the *estimated deviation* and denote it by e_i; that is

$$e_i = Y_i - \hat{Y}_i$$

Thus,

$$Y_i = \hat{\alpha} + \hat{\beta}X_i + e_i \quad (i = 1, 2, \ldots, n) \tag{7.7}$$

We illustrate this in figure 7.6.

Figure 7.7 shows in more detail the circled areas of Figure 7.6.

It is crucial to note that the u's, which from now on we will refer to as the *true deviations*, are the deviations of the observations from the *true* line (viz. $E(Y_i \mid X_i) = \alpha + \beta X_i$), while the e's, which we call the *estimated deviations*, are the deviations of the observations from the *estimated* line (viz. $\hat{Y}_i = \hat{\alpha} + \hat{\beta}X_i$)*. All deviations are measured in the vertical direction (that is, parallel to the Y-axis).

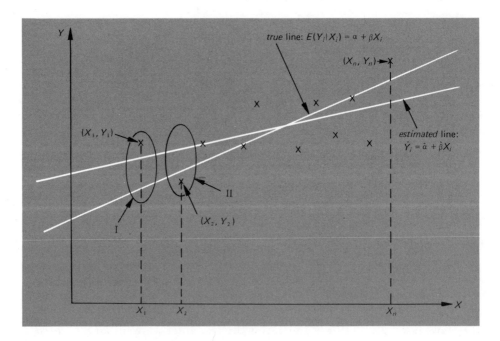

Fig. 7.6   The estimated line and the true line

*The terminology in other texts may differ from our terminology: the u's may elsewhere be called 'random errors' or 'stochastic disturbances', while the e's may be called 'residuals'. We prefer our terminology as it emphasises the fact that the u's measure *deviations* from the *true* line, while the e's measure *deviations* from the *estimated* line.

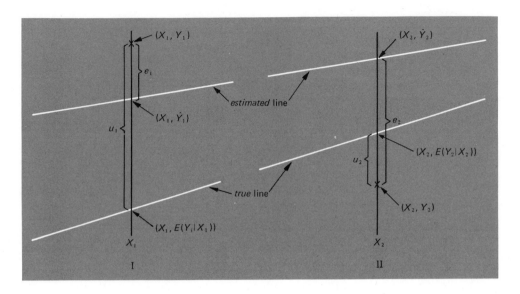

Fig. 7.7   Two details from figure 7.6

Now the (assumed) true relationship between Y and X is given by:

$$Y_i = \alpha + \beta X_i + u_i$$

and we have a sample of observations $(X_1, Y_1), (X_2, Y_2), \ldots, (X_n, Y_n)$ to which we wish to fit an estimated relationship

$$Y_i = \hat{\alpha} + \hat{\beta} X_i + e_i$$

in such a way that the estimated relationship is the 'best' estimator of the true relationship. We can rephrase this by saying that we wish to find estimators $\hat{\alpha}$ and $\hat{\beta}$ of α and β respectively, such that $\hat{\alpha}$ is the 'best' estimator of α and $\hat{\beta}$ the best estimator of β. We consider this problem in a general form in the next section.

7.3   Estimation of Parameters of Relationship

We continue the discussion in a general form. Suppose we wish to examine an economic theory which can be expressed in the form:

$$\left.\begin{array}{lll} Y_i = \alpha + \beta X_i + u_i & & \\ Eu_i = 0 & i = 1, 2, \ldots, n & \text{(A1)} \\ Eu_i^2 = \sigma^2 & i = 1, 2, \ldots, n & \text{(A2)} \\ Eu_i u_j = 0 & i, j = 1, 2, \ldots, n, i \neq j & \text{(A3)} \\ X_i \text{ is fixed} & i = 1, 2, \ldots, n & \text{(A4)} \end{array}\right\} \quad (7.8)$$

In our illustrative example, the variable Y was household consumption expenditure and X was household income. (7.8) could also represent the

Fig. 7.8   Scatter of observations

conclusion of demand theory with Y being quantity demanded and X the price. Obviously many economic relationships can be written in the general form (7.8).

We have a sample of n observations, $(X_1, Y_1), (X_2, Y_2), \ldots, (X_n, Y_n)$, and wish to use these to find an estimated relationship:

$$Y_i = \hat{\alpha} + \hat{\beta} X_i + e_i.$$

What is the 'best' method of fitting a line to the observations? We have a 'scatter' of observations as shown in figure 7.8. Obviously there are many ways of fitting a line to these points. We could fit the line by joining up the first and last observations. We could fit the line by dividing the data up into two halves, finding the average value of Y and X for each half, and joining the points representing the averages for each half. We could focus attention on the e's — the deviations between the observations and the fitted (or estimated) line (abbreviated to 'estimated deviations') — and fit the line so that $\Sigma_{i=1}^{n} e_i$ was zero. We could fit the line so that $\Sigma_{i=1}^{n} |e_i|$ was minimised. Which of these methods would be best? Our discussion of the last chapter showed that 'best' was only defined relative to the use to which the analysis was to be put and the consequences of mis-estimation. Let us therefore choose a particular method of fitting, and investigate the properties of the estimators of α and β using this fitting method. We will then see if other methods may be better for particular applications.

We will choose a method that we know provides estimators with interesting properties; then we will look at the problem from a more general point of view. (Compare this procedure with our original use of $\overline{X}$ as an estimator of μ in chapter 5, and our later investigation of alternative estimators of μ in chapter 6.)

Our fitting method is as follows: fit the line so that the sum of the squares of the estimated deviations is minimised. This is known as *least squares estimation*, or *least squares regression*. At this stage, this may seem a rather arbitrary method, particularly as the method places proportionally more emphasis on larger deviations than on smaller deviations; later we will see the rationale for such a method.

We fit the line, that is choose $\hat{\alpha}$ and $\hat{\beta}$, so that $\Sigma_1^n e_i^2$ is minimised. Denote $\Sigma_1^n e_i^2$ by S. Now,

$$Y_i = \hat{\alpha} + \hat{\beta}X_i + e_i$$

therefore

$$e_i = Y_i - \hat{\alpha} - \hat{\beta}X_i$$

and so

$$S \equiv \sum_1^n e_i^2 = \sum_1^n (Y_i - \hat{\alpha} - \hat{\beta}X_i)^2.$$

In order to minimise S with respect to $\hat{\alpha}$ and $\hat{\beta}$ we need to set the partial derivatives of S with respect to $\hat{\alpha}$ and $\hat{\beta}$ equal to zero. Now,

$$\frac{\partial S}{\partial \hat{\alpha}} = -2 \sum_1^n (Y_i - \hat{\alpha} - \hat{\beta}X_i)$$

and

$$\frac{\partial S}{\partial \hat{\beta}} = -2 \sum_1^n X_i(Y_i - \hat{\alpha} - \hat{\beta}X_i).$$

Thus

$$\frac{\partial S}{\partial \hat{\alpha}} = 0 \quad \text{gives} \quad \sum_1^n (Y_i - \hat{\alpha} - \hat{\beta}X_i) = 0 \qquad (7.9)$$

and

$$\frac{\partial S}{\partial \hat{\beta}} = 0 \quad \text{gives} \quad \sum_1^n X_i(Y_i - \hat{\alpha} - \hat{\beta}X_i) = 0. \qquad (7.10)$$

Taking each term in the summation in (7.9) separately, we get

$$\sum_1^n Y_i = n\hat{\alpha} + \hat{\beta}\sum_1^n X_i.$$

Dividing through by n, we get

$$\bar{Y} = \hat{\alpha} + \hat{\beta}\bar{X} \qquad (7.11)$$

where, of course,

$$\bar{X} = \frac{1}{n}\sum_1^n X_i \quad \text{and} \quad \bar{Y} = \frac{1}{n}\sum_1^n Y_i.$$

Now taking each term in the summation in (7.10) separately, we get

$$\sum_1^n Y_i X_i = \hat{\alpha} \sum_1^n X_i + \hat{\beta} \sum_1^n X_i^2.$$ (7.12)

Equations (7.11) and (7.12) are equations linear in $\hat{\alpha}$ and $\hat{\beta}$ and can therefore be solved to give $\hat{\alpha}$, $\hat{\beta}$ in terms of the observations. These equations are known as the *normal equations* (the word normal *is* connected with normal distributions, but the reason for this is outside the scope of this book; see (14) page 25).

We solve (7.11) and (7.12) for $\hat{\alpha}$ and $\hat{\beta}$: (7.12) can be written:

$$\sum_1^n Y_i X_i = \hat{\alpha} n \bar{X} + \hat{\beta} \sum_1^n X_i^2.$$ (7.13)

Multiplying (7.11) throughout by $n\bar{X}$ gives:

$$n\bar{Y}\bar{X} = \hat{\alpha} n\bar{X} + \hat{\beta} n\bar{X}^2$$ (7.14)

Subtracting (7.14) from (7.13) gives:

$$\sum_1^n Y_i X_i - n\bar{Y}\bar{X} = \hat{\beta} \left(\sum_1^n X_i^2 - n\bar{X}^2 \right).$$ (7.15)

This can be simplified by defining x_i, y_i, as before, by

$$\begin{aligned} x_i &= X_i - \bar{X} \\ y_i &= Y_i - \bar{Y} \end{aligned} \qquad i = 1, 2, \ldots, n.$$

Thus x_i is the deviation of X_i from the mean of the xs. We already know (from chapter 2) that

$$\sum_1^n x_i = 0$$ (7.16)

and

$$\sum_1^n y_i = 0$$ (7.17)

and

$$\sum_1^n x_i^2 = \sum_1^n X_i^2 - n\bar{X}^2.$$ (7.18)

Further, it can easily be shown that

$$\sum_1^n y_i x_i = \sum_1^n Y_i X_i - n\bar{Y}\bar{X}$$ (7.19)

(note that (7.18) is just a particular case of (7.19)).

Using (7.18) and (7.19) in (7.15) gives:

$$\sum_1^n y_i x_i = \hat{\beta} \sum_1^n x_i^2.$$

Hence

$$\hat{\beta} = \sum_1^n y_i x_i \Big/ \sum_1^n x_i^2 \tag{7.20}$$

and from (7.11)

$$\hat{\alpha} = \bar{Y} - \hat{\beta}\bar{X}. \tag{7.21}$$

Thus, using the criterion of fitting the line to the observations so as to minimise the sum of squared estimated deviations gives us a fitted line

$$Y_i = \hat{\alpha} + \hat{\beta}X_i + e_i$$

where $\hat{\alpha}$ and $\hat{\beta}$ are given by (7.21) and (7.20).

We now investigate the properties of these least squares estimators $\hat{\alpha}$ and $\hat{\beta}$ as estimators of α and β. First we examine whether $\hat{\alpha}$ and $\hat{\beta}$ are unbiased estimators of α and β, that is whether they will be right *on average*. In order to investigate this let us express $\hat{\beta}$ in terms of β. We can see from (7.20) that $\hat{\beta}$ contains y_i. Now, from the *true* relationship (7.8)

$$Y_i = \alpha + \beta X_i + u_i \quad (i = 1, 2, \ldots, n) \tag{7.22}$$

and so

$$\sum_1^n Y_i = n\alpha + \beta \sum_1^n X_i + \sum_1^n u_i.$$

Thus

$$\bar{Y} = \alpha + \beta\bar{X} + \bar{u} \quad \text{where} \quad \bar{u} = \frac{1}{n}\sum u_i. \tag{7.23}$$

Subtracting (7.23) from (7.22) we get

$$Y_i - \bar{Y} = \beta(X_i - \bar{X}) + u_i - \bar{u} \quad (i = 1, 2, \ldots, n)$$

which, in terms of our deviations x_i, y_i, is

$$y_i = \beta x_i + u_i - \bar{u} \quad (i = 1, 2, \ldots, n). \tag{7.24}$$

By substituting for y_i in (7.20) from (7.24) we can express $\hat{\beta}$ in terms of β:

$$\hat{\beta} = \frac{\sum_1^n y_i x_i}{\sum_1^n x_i^2} = \frac{\sum_1^n (\beta x_i + u_i - \bar{u}) x_i}{\sum_1^n x_i^2}$$

$$= \frac{\beta\sum_1^n x_i^2 + \sum_1^n x_i u_i - \bar{u}\sum_1^n x_i}{\sum_1^n x_i^2}.$$

Thus

$$\hat{\beta} = \beta + \frac{\sum\limits_{1}^{n} x_i u_i}{\sum\limits_{1}^{n} x_i^2} \qquad \text{(since } \sum\limits_{1}^{n} x_i = 0 \text{ from (7.16)).} \tag{7.25}$$

So, in any particular sample, $\hat{\beta}$ will equal β only if the second term on the right hand side of (7.25) is zero; in general there is no reason why $\sum_{1}^{n} x_i u_i$ should be zero, so in general $\hat{\beta}$ may not equal β in any particular sample. ($\sum_{1}^{n} x_i^2$ will in general be positive and finite.)

In order to simplify the following algebra, let us define a new expression w_i by

$$w_i = \frac{x_i}{x_1^2 + x_2^2 + \ldots + x_n^2} = \frac{x_i}{\sum\limits_{j=1}^{n} x_j^2} \qquad (i = 1, 2, \ldots, n). \tag{7.26}$$

These w_i have two important properties. First,

$$\sum\limits_{1}^{n} w_i = \sum\limits_{1}^{n} \left(\frac{x_i}{\sum\limits_{1}^{n} x_j^2} \right) = \frac{\sum\limits_{1}^{n} x_i}{\sum\limits_{1}^{n} x_j^2} = 0 \qquad \text{(from (7.16))} \tag{7.27}$$

Secondly,

$$\sum\limits_{1}^{n} w_i x_i = \sum\limits_{1}^{n} \left(\frac{x_i}{\sum\limits_{1}^{n} x_j^2} \right) x_i = \frac{\sum\limits_{1}^{n} x_i^2}{\sum\limits_{1}^{n} x_j^2} = 1. \tag{7.28}$$

From (7.25) $\hat{\beta}$ can be expressed in terms of β by

$$\hat{\beta} = \beta + \sum\limits_{1}^{n} w_i u_i. \tag{7.29}$$

We now find the expected value of $\hat{\beta}$. From (7.29)

$$E\hat{\beta} = \beta + E\left(\sum\limits_{1}^{n} w_i u_i \right) = \beta + \sum\limits_{1}^{n} (E w_i u_i). \tag{7.30}$$

Now we have assumed, see (A4) of (7.8), that each X_i is fixed. Thus the value of X for the ith member of the sample is the same in every sample (for all i). It follows that $\overline{X}$ must be the same in every sample, and so each x_i ($= X_i - \overline{X}$) must be the same in every sample. It immediately follows that each w_i is the same in every sample (inspect the definition (7.26)). Thus it follows that

$$E(w_i u_i) = w_i E u_i \quad \text{for all } i$$

and so, from (7.30)

$$E\hat{\beta} = \beta + \sum_1^n w_i E u_i$$

But $Eu_i = 0$ (for all i), by assumption (A1) of (7.8), and so

$$E\hat{\beta} = \beta \tag{7.31}$$

that is, $\hat{\beta}$ is an unbiased estimator of β.

We now investigate whether $\hat{\alpha}$ is an unbiased estimator of α. From (7.21) we have

$$\hat{\alpha} = \overline{Y} - \hat{\beta}\overline{X}.$$

In order to express $\hat{\alpha}$ in terms of α, we substitute for $\overline{Y}$ from (7.23). Thus

$$\hat{\alpha} = (\alpha + \beta\overline{X} + \overline{u}) - \hat{\beta}\overline{X}$$

that is,

$$\hat{\alpha} = \alpha + (\beta - \hat{\beta})\overline{X} + \overline{u}. \tag{7.32}$$

Therefore $\hat{\alpha}$ equals α in any particular sample, only if $(\beta - \hat{\beta})\overline{X} + \overline{u}$ is zero. However, *on average*, we have

$$E\hat{\alpha} = \alpha + E\{(\beta - \hat{\beta})\overline{X}\} + E\overline{u}.$$

The second term of this contains $\overline{X}$ which is the same in every sample, and so

$$E\{(\beta - \hat{\beta})\overline{X}\} = \overline{X}E(\beta - \hat{\beta}) = \overline{X}(\beta - E\hat{\beta}) = 0 \quad \text{from (7.31)}.$$

The third term of the above expression for $E\hat{\alpha}$ is

$$E\overline{u} = E\left(\frac{u_1 + u_2 + \ldots + u_n}{n}\right) = \frac{1}{n}(Eu_1 + Eu_2 + \ldots + Eu_n) = 0$$

from assumption (A1) of (7.8). Hence,

$$E\hat{\alpha} = \alpha \tag{7.33}$$

that is, $\hat{\alpha}$ is an unbiased estimator of α.

So we have proved that *on average* $\hat{\alpha}$ equals α, and *on average* $\hat{\beta}$ equals β. However, in any particular sample $\hat{\alpha}$ may not equal α, nor $\hat{\beta}$ equal β. We now find how large is the variation of $\hat{\alpha}$ around α, and how large is the variation of $\hat{\beta}$ around β; that is, we find the variances of $\hat{\alpha}$ and $\hat{\beta}$. Now,

$$\begin{aligned}
\text{var } \hat{\beta} &= E(\hat{\beta} - E\hat{\beta})^2 \\
&= E(\hat{\beta} - \beta)^2 \quad \text{since} \quad E\hat{\beta} = \beta \text{ from (7.31)} \\
&= E\left\{\sum_1^n w_i u_i\right\}^2 \qquad \text{from (7.29)}
\end{aligned}$$

Thus

$$\text{var } \hat{\beta} = E(w_1 u_1 + w_2 u_2 + \ldots + w_n u_n)^2$$
$$\text{var } \hat{\beta} = E(w_1^2 u_1^2 + w_2^2 u_2^2 + \ldots + w_n^2 u_n^2 + 2w_1 w_2 u_1 u_2$$
$$+ 2w_1 w_3 u_1 u_3 + \ldots + 2w_{n-1} w_n u_{n-1} u_n). \tag{7.34}$$

There are two types of terms in (7.34): terms in u_i^2 and terms in $u_i u_j$ ($i \neq j$). Let us take the first type first.

Consider

$$E(w_i^2 u_i^2) = w_i^2 E u_i^2 \quad \left\{ \begin{array}{l} \text{since } w_i^2 \text{ is the same in every sample as} \\ \text{deduced from (A4) of (7.8)} \end{array} \right.$$
$$= w_i^2 \sigma^2 \qquad \text{using assumption (A2) of (7.8)}$$

Now consider (for $i \neq j$)

$$E(w_i w_j u_i u_j) = w_i w_j E u_i u_j \quad \text{(from (A4) of (7.8) each } w_i \text{ is constant)}$$
$$= 0 \qquad \text{(from (A3) of (7.8))}.$$

Hence (7.34) becomes:

$$\text{var } \hat{\beta} = w_1^2 \sigma^2 + w_2^2 \sigma^2 + \ldots + w_n^2 \sigma^2$$

$$= \sigma^2 \sum_1^n w_i^2$$

$$= \sigma^2 \sum_1^n \left(\frac{x_i}{\sum_1^n x_j^2} \right)^2 = \sigma^2 \frac{\sum_1^n x_i^2}{\left(\sum_1^n x_j^2 \right)^2} = \frac{\sigma^2}{\sum_1^n x_i^2}$$

that is,

$$\text{var } \hat{\beta} = \frac{\sigma^2}{\sum_1^n x_i^2}. \tag{7.35}$$

Finally,

$$\text{var } \hat{\alpha} = E(\hat{\alpha} - E\hat{\alpha})^2 = E(\hat{\alpha} - \alpha)^2 \quad \text{from (7.33)}$$
$$= E\{(\beta - \hat{\beta})\bar{X} + \bar{u}\}^2 \qquad \text{from (7.32)}$$
$$= E\{(\hat{\beta} - \beta)^2 \bar{X}^2 - 2(\hat{\beta} - \beta)\bar{X}\bar{u} + \bar{u}^2\}.$$

Thus,

$$\text{var } \hat{\alpha} = \bar{X}^2 E(\hat{\beta} - \beta)^2 - 2\bar{X}E(\hat{\beta} - \beta)\bar{u} + E\bar{u}^2 \quad \text{(using (A4))}. \tag{7.36}$$

The first term of (7.36) is $\bar{X}^2 \text{var } \hat{\beta}$, which we know. The second term is

$-2\bar{X}$ times $E(\hat{\beta} - \beta)\bar{u}$ where

$$E(\hat{\beta} - \beta)\bar{u} = E\left(\sum_1^n w_i u_i\right)\bar{u} \quad \text{from (7.29)}$$

$$= E(w_1 u_1 + w_2 u_2 + \ldots + w_n u_n)\left(\frac{u_1 + u_2 + \ldots + u_n}{n}\right)$$

$$= \frac{1}{n}(w_1 \sigma^2 + w_2 \sigma^2 + \ldots + w_n \sigma^2) \quad \text{using assumptions}$$
$$\text{(A2) (A3) (A4)}$$

$$= \frac{1}{n}\sigma^2(w_1 + w_2 + \ldots + w_n).$$

Thus, using (7.27)

$$E(\hat{\beta} - \beta)\bar{u} = 0. \tag{7.37}$$

The third term in (7.36) is

$$E\bar{u}^2 = E\left(\frac{u_1 + u_2 + \ldots + u_n}{n}\right)^2$$

$$= E\left(\frac{u_1^2 + u_2^2 + \ldots + u_n^2 + 2u_1 u_2 + \ldots + 2u_{n-1} u_n}{n^2}\right)$$

$$= \frac{\sigma^2 + \sigma^2 + \ldots + \sigma^2}{n^2} \quad \text{using (A2) (A3)}$$

$$= \frac{n\sigma^2}{n^2}$$

$$E\bar{u}^2 = \frac{\sigma^2}{n}. \tag{7.38}$$

You should recognise this result.

Thus, from (7.36), using (7.35), (7.37) and (7.38), we have

$$\text{var } \hat{\alpha} = \bar{X}^2 \frac{\sigma^2}{\sum_1^n x_i^2} + \frac{\sigma^2}{n}$$

$$= \frac{\sigma^2}{n\sum_1^n x_i^2}\left(n\bar{X}^2 + \sum_1^n x_i^2\right).$$

Thus, from (7.18),

$$\text{var } \hat{\alpha} = \frac{\sigma^2 \sum_1^n X_i^2}{n\sum_1^n x_i^2} \tag{7.39}$$

Collecting our results together we have

$$
\left.\begin{array}{ll}
E\hat{\alpha} = \alpha & \text{var } \hat{\alpha} = \dfrac{\sigma^2 \sum\limits_{1}^{n} X_i^2}{n \sum\limits_{1}^{n} x_i^2} \\[3em]
E\hat{\beta} = \beta & \text{var } \hat{\beta} = \dfrac{\sigma^2}{\sum\limits_{1}^{n} x_i^2} .
\end{array}\right\} \tag{7.40}
$$

The last few pages have seen a lot of algebraic manipulations which tend to obscure the statistical analysis. In order to illustrate what we have done so far, we consider a simple example.

7.4   Illustration of Least Squares Estimation

Consider the following hypothetical example, where Y and X can be interpreted, for expository purposes, as household consumption and income respectively. (The numbers are chosen for simplicity, not for realism.) Suppose n, the sample size, is 3, and the three fixed X values X_1, X_2, X_3 are 2, 4 and 6 respectively. Suppose that of all households in the population with income X_1 $(= 2)$ ¼ of them spend $Y_1 = 2$, ½ spend $Y_1 = 3$ and the remaining ¼ spend $Y_1 = 4$. The distribution of expenditure for all households with income 2 is therefore as given in table 7.9.

Table 7.9   Distribution of Y_1

Y_1	Relative frequency
2	¼
3	½
4	¼
Σ	1

Table 7.10   Distribution of u_1

u_1	$P[u_1]$	$u_1 P[u_1]$	$u_1^2 P[u_1]$
−1	¼	−¼	¼
0	½	0	0
1	¼	¼	¼
Σ	1	$Eu_1 = 0$	var $u_1 = \frac{1}{2}$

The mean expenditure $E(Y_1 \mid X_1) = 3$. Defining $u_1 = Y_1 - E(Y_1 \mid X_1)$ as before, we see that those households with expenditure 2 spend 1 less than the mean for their income group, and so ¼ of those with income X_1 have $u_1 = -1$. Similarly ½ have $u_1 = 0$ and ¼ have $u_1 = 1$. As we are going to take a sample of one from these households we can interpret the frequency distribution of u_1 as a probability distribution; this is given in table 7.10.

Now, suppose that of all households in the population with income X_2 $(= 4)$, ¼ spend 3, ½ spend 4 and the remaining ¼ spend 5. Thus

Table 7.11 Distribution of u_2		
u_2	$P[u_2]$	
−1	¼	$Eu_2 = 0$
0	½	var $u_2 = \frac{1}{2}$
1	¼	
Σ	1	

Table 7.12 Distribution of u_3		
u_3	$P[u_3]$	
−1	¼	$Eu_3 = 0$
0	½	var $u_3 = \frac{1}{2}$
1	¼	
Σ	1	

Table 7.13 Distribution of u_i $(i = 1, 2, 3)$		
u_i	$P[u_i]$	
−1	¼	$Eu_i = 0$
0	½	var $u_i = \frac{1}{2}$
1	¼	
Σ	1	

$E(Y_2 \mid X_2) = 4$ and defining u_2 in the usual way, we find the distribution of u_2, as given in table 7.11.

Finally, suppose that of all households in the population with income X_3 (= 6), ¼ spend 4, ½ spend 5, and ¼ spend 6. Thus $E(Y_3 \mid X_3) = 5$, and u_3 has the distribution given in table 7.12.

Figure 7.9 illustrates the distribution of the population by income and consumption.

The line joining the means is

$$E(Y_i \mid X_i) = 2 + 0.5X_i$$

that is, $E(Y_i \mid X_i) = \alpha + \beta X_i$, where $\alpha = 2$, $\beta = 0.5$. The relationship between consumption and income can thus be written

$$Y_i = \alpha + \beta X_i + u_i$$

where $\alpha = 2$, $\beta = 0.5$, and each u_i has the distribution as given in table 7.13.

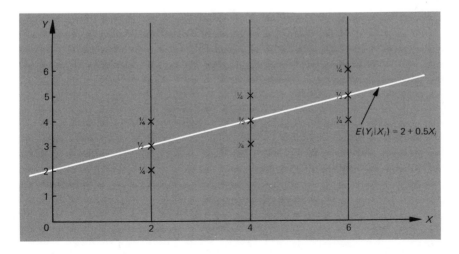

Fig. 7.9   Population income–consumption scatter

Now suppose we took samples of size 3; one household with an income of X_1, one with an income of X_2 and one with an income of X_3. One possible sample we may get is: (2,2), (4,5), (6,5); that is,

$$Y_1 = 2 \qquad Y_2 = 5 \qquad Y_3 = 5$$
$$X_1 = 2 \qquad X_2 = 4 \qquad X_3 = 6.$$

The scatter of these sample observations is illustrated in figure 7.10.

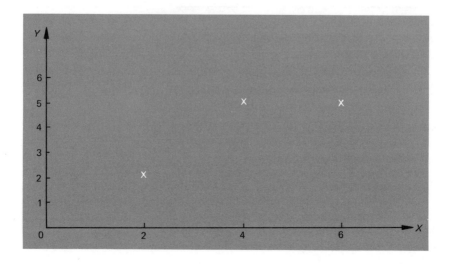

Fig. 7.10   Scatter of sample observations

Our least squares fitted line would be

$$Y_i = \hat{\alpha} + \hat{\beta}X_i + e_i$$

where

$$\hat{\beta} = \frac{\sum\limits_{1}^{3} y_i x_i}{\sum\limits_{1}^{3} x_i^2} \qquad \hat{\alpha} = \bar{Y} - \hat{\beta}\bar{X} \quad \text{from (7.20) and (7.21)}$$

The necessary calculations are given in table 7.14. Thus,

$$\hat{\beta} = \frac{6}{8} = \frac{3}{4} \qquad \hat{\alpha} = 4 - \left(\frac{3}{4}\right)4 = 1.$$

The (least squares) fitted line is thus,

$$Y_i = 1 + \frac{3}{4}X_i + e_i.$$

The estimated deviations* are given by

$$e_i = Y_i - 1 - \frac{3}{4}X_i.$$

Thus,

$$e_1 = -\tfrac{1}{2} \quad e_2 = 1 \quad e_3 = -\tfrac{1}{2}.$$

Note that

$$\sum_1^3 e_i = e_1 + e_2 + e_3 = 0.$$

This is, in fact, a result that holds generally for all least squares deviations; we will prove this result in due course. Note also that

$$\sum_1^3 e_i X_i = e_1 X_1 + e_2 X_2 + e_3 X_3 = 0.$$

Again this is a general result, which we will prove later.

Table 7.14   Calculation of least squares coefficients

i	X_i	Y_i	$x_i = X_i - \bar{X}$	$y_i = Y_i - \bar{Y}$	x_i^2	$y_i x_i$
1	2	2	−2	−2	4	4
2	4	5	0	1	0	0
3	6	5	2	1	4	2
Σ	12	12	0	0	8	6

$$\bar{X} = 4 \quad \bar{Y} = 4$$

Our fitted line is illustrated in figure 7.11.

So, if we obtained the sample $(2,2)$, $(4,5)$, $(6,5)$ our least squares fitted line would be $Y_i = 1 + \tfrac{3}{4}X_i + e_i$. This sample has $Y_1 = 2$, $Y_2 = 5$ and $Y_3 = 5$; that is $u_1 = -1$, $u_2 = 1$, $u_3 = 0$. What is the probability of obtaining such a sample? The probability of u_1 equalling -1 is $\tfrac{1}{4}$, the probability of u_2 being 1 is also $\tfrac{1}{4}$ and the probability of u_3 being zero is $\tfrac{1}{2}$. Thus, as long as these are independent (that is, the value u_1 takes does not change the probability of u_2 taking certain values, etc.) then the probability of getting this particular sample is

$$P[u_1 = -1]P[u_2 = 1]P[u_3 = 0] = \frac{1}{4} \cdot \frac{1}{4} \cdot \frac{1}{2} = \frac{2}{64}.$$

*Remember that 'estimated deviations' is just an abbreviation for 'deviations from the estimated line'.

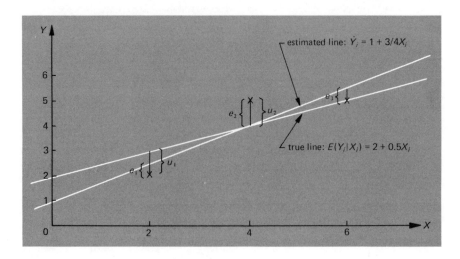

Fig. 7.11   Least squares fitted line

Table 7.15   List of all possible samples

Sample number	Probability (× 64)	u_1	u_2	u_3	$\bar{u}$ (× 3)	Y_1	Y_2	Y_3	$\hat{\alpha}$ (× 3)	$\hat{\beta}$ (× 4)
1	1	−1	−1	−1	−3	2	3	4	3	2
2	2	−1	−1	0	−2	2	3	5	1	3
3	1	−1	−1	1	−1	2	3	6	−1	4
4	2	−1	0	−1	−2	2	4	4	4	2
5	4	−1	0	0	−1	2	4	5	2	3
6	2	−1	0	1	0	2	4	6	0	4
7	1	−1	1	−1	−1	2	5	4	5	2
8	2	−1	1	0	0	2	5	5	3	3
9	1	−1	1	1	1	2	5	6	1	4
10	2	0	−1	−1	−2	3	3	4	7	1
11	4	0	−1	0	−1	3	3	5	5	2
12	2	0	−1	1	0	3	3	6	3	3
13	4	0	0	−1	−1	3	4	4	8	1
14	8	0	0	0	0	3	4	5	6	2
15	4	0	0	1	1	3	4	6	4	3
16	2	0	1	−1	0	3	5	4	9	1
17	4	0	1	0	1	3	5	5	7	2
18	2	0	1	1	2	3	5	6	5	3
19	1	1	−1	−1	−1	4	3	4	11	0
20	2	1	−1	0	0	4	3	5	9	1
21	1	1	−1	1	1	4	3	6	7	2
22	2	1	0	−1	0	4	4	4	12	0
23	4	1	0	0	1	4	4	5	10	1
24	2	1	0	1	2	4	4	6	8	2
25	1	1	1	−1	1	4	5	4	13	0
26	2	1	1	0	2	4	5	5	11	1
27	1	1	1	1	3	4	5	6	9	2

Now, as each u_i can take three different values, there are obviously 27 different possible samples that we may obtain. For each sample we can compute the appropriate probability. For each sample we compute the least squares estimates, and we thus obtain table 7.15 (the particular sample we illustrated was number 8).

(Note, for *all* samples

$$X_1 = 2, \quad X_2 = 4, \quad X_3 = 6, \quad \bar{X} = 4,$$
$$x_1 = -2, \quad x_2 = 0, \quad x_3 = 2.)$$

As a quick check, we find the distribution of u_2. This is given in table 7.16.

Table 7.16   Distribution of u_2

Sample numbers	u_2	$P[u_2]$	
1, 2, 3, 10, 11, 12, 19, 20, 21	-1	¼	$Eu_2 = 0$
4, 5, 6, 13, 14, 15, 22, 23, 24	0	½	var $u_2 = ½ = \sigma^2$
7, 8, 9, 16, 17, 18, 25, 26, 27	1	¼	
	Σ	1	

Table 7.17   Distribution of $\bar{u}$

Sample numbers	$\bar{u}$ (× 3)	$P[\bar{u}]$ (× 64)	$\bar{u}^2 P[\bar{u}]$ (× 576)
1	-3	1	9
2, 4, 10	-2	6	24
3, 5, 7, 11, 13, 19	-1	15	15
6, 8, 12, 14, 16, 20, 22	0	20	0
9, 15, 17, 21, 23, 25	1	15	15
18, 24, 26	2	6	24
27	3	1	9
Σ		64	96

The distribution of $\bar{u}$ is given in table 7.17. This shows that

$$E\bar{u} = 0$$

and

$$\text{var } \bar{u} = \frac{96}{576} = \frac{1}{6} \quad \left(= \frac{½}{3} = \frac{\sigma^2}{n}; \text{ cf. } (7.38) \right).$$

Table 7.18   Distribution of $\hat{\beta}$

Sample numbers	$\hat{\beta}$ (× 4)	$P[\hat{\beta}]$ (× 64)	$\hat{\beta}P[\hat{\beta}]$ (× 256)	$(\hat{\beta} - E\hat{\beta})^2 P[\hat{\beta}]$ (× 1024)
19, 22, 25	0	4	0	16
10, 13, 16, 20, 23, 26	1	16	16	16
1, 4, 7, 11, 14, 17, 21, 24, 27	2	24	48	0
2, 5, 8, 12, 15, 18	3	16	48	16
3, 6, 9	4	4	16	16
Σ		64	128	64

The distribution of $\hat{\beta}$ is given in table 7.18. Thus

$$E\hat{\beta} = \frac{128}{256} = 0.5 = \beta \quad (\text{cf. } (7.31))$$

and

$$\text{var } \hat{\beta} = \frac{64}{1024} = \frac{1}{16} = \frac{\frac{1}{2}}{8} = \frac{\sigma^2}{\sum\limits_{1}^{n} x_i^2} \quad (\text{cf. } (7.35)).$$

Finally, the distribution of $\hat{\alpha}$ is given in table 7.19. Thus

$$E\hat{\alpha} = \frac{384}{192} = 2 = \alpha \quad (\text{cf. } (7.33))$$

Table 7.19   Distribution of $\hat{\alpha}$

Sample numbers	$\hat{\alpha}$ (× 3)	$P[\hat{\alpha}]$ (× 64)	$\hat{\alpha}P[\hat{\alpha}]$ (× 192)	$(\hat{\alpha} - E\hat{\alpha})^2 P[\hat{\alpha}]$ (× 576)
3	−1	1	−1	49
6	0	2	0	72
2, 9,	1	3	3	75
5	2	4	8	64
1, 8, 12	3	5	15	45
4, 15	4	6	24	24
7, 11, 18	5	7	35	7
14	6	8	48	0
10, 17, 21	7	7	49	7
13, 24	8	6	48	24
16, 20, 27	9	5	45	45
23	10	4	40	64
19, 26	11	3	33	75
22	12	2	24	72
25	13	1	13	49
Σ		64	384	672

and

$$\text{var } \hat{\alpha} = \frac{672}{576} = \frac{7}{6} = \frac{\frac{1}{2} \times 56}{3 \times 8} = \frac{\sigma^2 \sum_{1}^{3} X_i^2}{n \sum_{1}^{n} x_i^2} \quad (\text{cf. } (7.39)).$$

We draw, in figure 7.12, the least squares fitted line for each of the 27 possible samples.

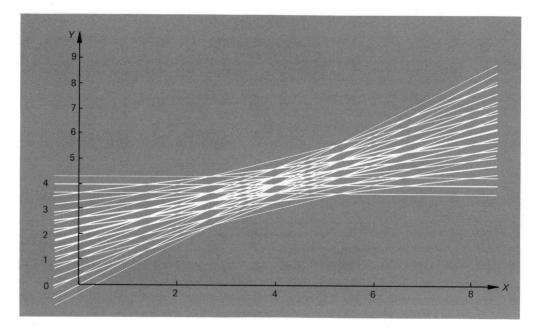

Fig. 7.12   Distribution of least squares fitted lines

7.5   Properties of Least Squares Estimators

We have already shown that least squares estimators are unbiased. We will now investigate whether of all unbiased estimators they are the 'best' in the sense of having minimum variance. We note first that the least squares estimators are 'linear' in that they are linear combinations of the observations $Y_1, Y_2, \ldots, Y_n$. They are not linear in the X_i, but these are presumed fixed from sample to sample, and the only parts of $\hat{\alpha}$ and $\hat{\beta}$ that vary from sample to sample are the Y_i. Again it seems reasonable to restrict ourselves to consideration of linear estimators of α and β, since if our estimators were non-linear in the Y_i then the expected values of our estimators would contain terms in α^2, $\alpha\beta$, β^2, σ^2 and so on. Therefore let us investigate whether of all linear unbiased estimators, the least squares estimators of α and β are the 'best' in the sense of minimum variance. Our procedure follows closely that of section 6.4.

Consider then the most general linear estimator of β.

$$\beta^* = a_1 Y_1 + a_2 Y_2 + \ldots + a_n Y_n = \sum_1^n a_i Y_i \tag{7.41}$$

where the a_i are constant over samples.

Express β^* in terms of β by replacing Y_i in terms of β from (7.22)

$$\beta^* = \sum_1^n a_i(\alpha + \beta X_i + u_i) = \alpha \sum_1^n a_i + \beta \sum_1^n a_i X_i + \sum_1^n a_i u_i \tag{7.42}$$

Hence

$$E\beta^* = \alpha \sum_1^n a_i + \beta \sum_1^n a_i X_i + E\left(\sum_1^n a_i u_i\right)$$

(since a_i and X_i are constant over samples, for all i). Now,

$$E\left(\sum_1^n a_i u_i\right) = \sum_1^n a_i E u_i \quad \text{since each } a_i \text{ is constant over samples}$$

$$= 0 \quad\quad\quad \text{from assumption (A1)}$$

Hence

$$E\beta^* = \alpha \sum_1^n a_i + \beta \sum_1^n a_i X_i$$

Thus, for β^* to be an unbiased estimator of β, for all α and β, we require

$$\sum_1^n a_i = 0 \quad \sum_1^n a_i X_i = 1$$

Taken together, these two conditions imply that

$$\sum_1^n a_i x_i = 1 \tag{7.43}$$

since

$$\sum_1^n a_i x_i = \sum_1^n a_i(X_i - \bar{X}) = \sum_1^n a_i X_i - \bar{X} \sum_1^n a_i$$

If we impose these conditions (7.42) becomes

$$\beta^* = \beta + \sum_1^n a_i u_i \tag{7.44}$$

Thus the variance of β^* is:

$$\text{var } \beta^* = E(\beta^* - E\beta^*)^2$$

$$= E(\beta^* - \beta)^2 \qquad \text{since we have constrained } \beta^* \text{ to be unbiased}$$

$$= E\left\{\sum_1^n a_i u_i\right\}^2 \qquad \text{from (7.44)}$$

$$= E\left(a_1 u_1 + a_2 u_2 + \ldots + a_n u_n\right)^2$$

$$= E\left(\begin{array}{l} a_1{}^2 u_1{}^2 + a_2{}^2 u_2{}^2 + \ldots + a_n{}^2 u_n{}^2 \\ + 2a_1 a_2 u_1 u_2 + 2a_1 a_3 u_1 u_3 + \ldots + 2a_{n-1} a_n u_{n-1} u_n \end{array}\right)$$

$$= a_1{}^2 \sigma^2 + a_2{}^2 \sigma^2 + \ldots + a_n{}^2 \sigma^2 \qquad \begin{array}{l}\text{using assumptions (A2) and} \\ \text{(A3)}\end{array}$$

Thus

$$\text{var } \beta^* = \sigma^2 \sum_1^n a_i{}^2 \tag{7.45}$$

It is useful to note at this stage that if $a_i = w_i$ (for all i) then $\beta^* \equiv \beta$ since

$$\sum_1^n w_i Y_i = \sum_1^n w_i(y_i + \bar{Y}) = \sum_1^n w_i y_i + \bar{Y}\sum_1^n w_i$$

$$= \sum_1^n w_i y_i \qquad \text{(using (7.27))}$$

$$= \frac{\sum_1^n x_i y_i}{\sum_1^n x_i{}^2} \qquad \text{(from the definition of } w_i \text{ (7.26))}$$

In order to compare β^* with $\hat{\beta}$, let us write

$$a_i = w_i + d_i \qquad (i = 1, 2, \ldots, n) \tag{7.46}$$

where the d_i are a set of constants.
Thus, from (7.45), the variance of β^* is given by

$$\text{var } \beta^* = \sigma^2 \sum_1^n (w_i + d_i)^2$$

that is,

$$\text{var } \beta^* = \sigma^2 \sum_1^n w_i{}^2 + 2\dot{\sigma}^2 \sum_1^n w_i d_i + \sigma^2 \sum_1^n d_i{}^2 \tag{7.47}$$

Now, our condition that β^* be unbiased (7.43) implies that

$$\sum_1^n w_i d_i = 0$$

since, from (7.43) and (7.46) we require

$$\sum_1^n (w_i + d_i)x_i = 0$$

But

$$\sum_1^n w_i x_i = 1, \text{ from (7.28), and so (7.43) implies that}$$

$$\sum_1^n w_i d_i = \frac{\sum_1^n d_i x_i}{\sum_1^n x_j^2} = 0$$

Further,

$$\sigma^2 \sum_1^n w_i^2 = \text{var } \hat{\beta} \quad \text{(from (7.35))}.$$

Thus (7.47) becomes

$$\text{var } \beta^* = \text{var } \hat{\beta} + \sigma^2 \sum_1^n d_i^2$$

This shows immediately (since $\sigma^2 \sum_1^n d_i^2 \geqslant 0$) that

$$\text{var } \beta^* \geqslant \text{var } \hat{\beta}$$

and

$$\text{var } \beta^* = \text{var } \hat{\beta} \quad \text{only if} \quad \sum_1^n d_i^2 = 0$$

(excluding the trivial case of $\sigma^2 = 0$), that is, only if

$$d_i = 0 \quad (i = 1, 2, \dots, n)$$

Further if $d_i = 0$, then $a_i = w_i$ $(i = 1, 2, \dots, n)$ and so

$$\beta^* = \hat{\beta}$$

Thus, the linear unbiased estimator with the smallest variance is indeed the least squares estimator. Hence the best (in the sense of minimum variance) linear unbiased estimator of β is the least squares estimator $\hat{\beta}$. This is our justification for using the least squares estimation method.

We can also show that $\hat{\alpha}$, the least squares estimator of α, is the best linear unbiased estimator of α.

If we look for an estimator of β which would be optimal under a quadratic loss function, we come across the same problems as we did when trying to find such an estimator for μ. We find that the equations to solve for the a_i which make $\beta^* = \sum_1^n a_i Y_i$ the best estimator under a quadratic loss function contain terms in both β and σ^2, and are in general unsolvable unless we know β and σ^2.

7.6 Illustration of Properties of Estimators of α and β

To illustrate the result that the least squares estimators of α and β are the best linear unbiased estimators, we will find the properties of some alternative linear unbiased estimators. We will use the example of section 7.4 for illustrative purposes.

Consider the following alternative fitting methods (you should try some others yourself):

(1) the least squares method;

$$\hat{\alpha}_1 = \hat{\alpha} = \bar{Y} - \hat{\beta}\bar{X}; \quad \hat{\beta}_1 = \hat{\beta} = \frac{\Sigma x_i y_i}{\Sigma x_i^2}$$

(2) join the first and last observations:

$$\hat{\alpha}_2 = \frac{Y_1 X_3 - Y_3 X_1}{X_3 - X_1} \qquad \hat{\beta}_2 = \frac{Y_3 - Y_1}{X_3 - X_1}$$

(3) join the first observation and the mean of the last two observations:

$$\hat{\alpha}_3 = \frac{Y_1 X_2 + Y_1 X_3 - Y_2 X_1 - Y_3 X_1}{X_2 + X_3 - 2X_1} \qquad \hat{\beta}_3 = \frac{Y_2 + Y_3 - 2Y_1}{X_2 + X_3 - 2X_1}$$

(4) join the second and third observations

$$\hat{\alpha}_4 = \frac{Y_2 X_3 - Y_3 X_2}{X_3 - X_2} \qquad \hat{\beta}_4 = \frac{Y_3 - Y_2}{X_3 - X_2}.$$

To illustrate what the fitted line would look like in each case consider the previously illustrated sample (sample number 8):

$$(2,2), (4,5), (6,5)$$

that is,

$$(X_1, Y_1), (X_2, Y_2), (X_3, Y_3).$$

We have already shown

$$\hat{\alpha}_1 = 1 \qquad \hat{\beta}_1 = \frac{3}{4}.$$

Further, using the above formulae

$$\hat{\alpha}_2 = \frac{1}{2} \qquad \hat{\beta}_2 = \frac{3}{4}$$
$$\hat{\alpha}_3 = 0 \qquad \hat{\beta}_3 = 1$$
$$\hat{\alpha}_4 = 5 \qquad \hat{\beta}_4 = 0$$

These four fitted lines are shown in figure 7.13. The line

$\hat{Y}_i = \hat{\alpha}_1 + \hat{\beta}_1 X_i$ is the line fitted by least squares;

$\hat{Y}_i = \hat{\alpha}_2 + \hat{\beta}_2 X_i$ joins the first and last observations;

$\hat{Y}_i = \hat{\alpha}_3 + \hat{\beta}_3 X_i$ goes through the first observation and the mean of

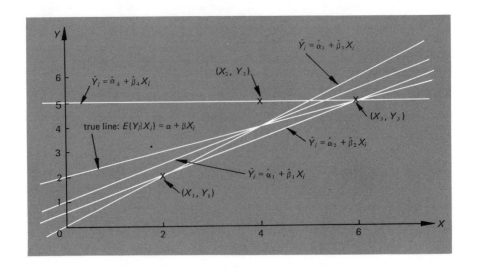

Fig. 7.13　Alternative fitting methods

the last two $\{(Y_2 + Y_3)/2 = 5, (X_2 + X_3)/2 = 5\}$;

$\hat{Y}_i = \hat{\alpha}_4 + \hat{\beta}_4 X_i$ joins the second and third observations.

We now fit lines by each of these four methods to all 27 possible samples. Table 7.20 lists these 27 samples. (Again note for *all* samples

$$X_1 = 2, \quad X_2 = 4, \quad X_3 = 6, \quad \bar{X} = 4$$
$$x_1 = -2, \quad x_2 = 0, \quad x_3 = 2.)$$

The distributions of $\hat{\alpha}_1$ and $\hat{\beta}_1$ we have already found, and give

$$E\hat{\alpha}_1 = 2 \quad (= \alpha) \qquad \mathrm{var}\, \hat{\alpha}_1 = 7/6 \qquad (= \sigma^2 \Sigma X_i^2 / n \Sigma x_i^2)$$
$$E\hat{\beta}_1 = \tfrac{1}{2} \quad (= \beta) \qquad \mathrm{var}\, \hat{\beta}_1 = 1/16 \quad (= \sigma^2 / \Sigma x_i^2).$$

The distribution of $\hat{\beta}_2$ (in this example) is exactly the same as that of $\hat{\beta}_1$. (This is no coincidence, as

$$\hat{\beta}_1 = \frac{\Sigma x_i y_i}{\Sigma x_i^2} = \frac{-2y_1 + 0y_2 + 2y_3}{(-2)^2 + (0)^2 + (2)^2} = \frac{y_3 - y_1}{4}$$
$$= \frac{(Y_3 - \bar{Y}) - (Y_1 - \bar{Y})}{X_3 - X_1} = \frac{Y_3 - Y_1}{X_3 - X_1} = \hat{\beta}_2 .\Bigg)$$

The distribution of $\hat{\alpha}_2$ is given in table 7.21. This shows that

$$E\hat{\alpha}_2 = 2 \quad (= \alpha) \qquad \mathrm{var}\, \hat{\alpha}_2 = 5/4$$

and, from above

$$E\hat{\beta}_2 = \tfrac{1}{2} \quad (= \beta) \qquad \mathrm{var}\, \hat{\beta}_2 = 1/16.$$

Table 7.20   List of all possible samples

Sample number	Probability of sample (× 64)	$\hat{\alpha}_1$ (× 3)	$\hat{\alpha}_2$ (× 2)	$\hat{\alpha}_3$ (× 3)	$\hat{\alpha}_4$ (× 1)	$\hat{\beta}_1$ (× 4)	$\hat{\beta}_2$ (× 4)	$\hat{\beta}_3$ (× 6)	$\hat{\beta}_4$ (× 2)
1	1	3	2	3	1	2	2	3	1
2	2	1	1	2	1	3	3	4	2
3	1	−1	0	1	−3	4	4	5	3
4	2	4	2	2	4	2	2	4	0
5	4	2	1	1	2	3	3	5	1
6	2	0	0	0	0	4	4	6	2
7	1	5	2	1	7	2	2	5	−1
8	2	3	1	0	5	3	3	6	0
9	1	1	0	−1	3	4	4	7	1
10	2	7	5	8	1	1	1	1	1
11	4	5	4	7	−1	2	2	2	2
12	2	3	3	6	−3	3	3	3	3
13	4	8	5	7	4	1	1	2	0
14	8	6	4	6	2	2	2	3	1
15	4	4	3	5	0	3	3	4	2
16	2	9	5	6	7	1	1	3	−1
17	4	7	4	5	5	2	2	4	0
18	2	5	3	4	3	3	3	5	1
19	1	11	8	13	1	0	0	−1	1
20	2	9	7	12	−1	1	1	0	2
21	1	7	6	11	−3	2	2	1	3
22	2	12	8	12	4	0	0	0	0
23	4	10	7	11	2	1	1	1	1
24	2	8	6	10	0	2	2	2	2
25	1	13	8	11	7	0	0	1	−1
26	2	11	7	10	5	1	1	2	0
27	1	9	6	9	3	2	2	3	1

Table 7.21   Distribution of $\hat{\alpha}_2$

Sample numbers	$\hat{\alpha}_2$ (× 2)	$P[\hat{\alpha}_2]$ (× 64)	$\hat{\alpha}_2 P[\hat{\alpha}_2]$ (× 32)	$(\hat{\alpha}_2 - E\hat{\alpha}_2)^2 P[\hat{\alpha}_2]$ (× 64)
3, 6, 9	0	4	0	16
2, 5, 8	1	8	2	18
1, 4, 7	2	4	2	4
12, 15, 18	3	8	6	2
11, 14, 17	4	16	16	0
10, 13, 16	5	8	10	2
21, 24, 27	6	4	6	4
20, 23, 26	7	8	14	18
19, 22, 25	8	4	8	16
Σ		64	64	80

$$E\hat{\alpha}_2 = 64/32 = 2 \quad \text{var } \hat{\alpha}_2 = 80/64 = 5/4$$

Table 7.22   Distributions of $\hat{\alpha}_3$ and $\hat{\beta}_3$

Sample numbers	$\hat{\alpha}_3$ (× 3)	$P[\hat{\alpha}_3]$ (× 64)	Sample numbers	$\hat{\beta}_3$ (× 6)	$P[\hat{\beta}_3]$ (× 64)
9	−1	1	19	−1	1
6, 8	0	4	20, 22	0	4
3, 5, 7	1	6	10, 21, 23, 25	1	8
2, 4	2	4	11, 13, 24, 26	2	12
1	3	1	1, 12, 14, 16, 27	3	14
18	4	2	2, 4, 15, 17	4	12
15, 17	5	8	3, 5, 7, 18	5	8
12, 14, 16	6	12	6, 8	6	4
11, 13	7	8	9	7	1
10	8	2			
27	9	1		Σ	64
24, 26	10	4			
21, 23, 25	11	6			
20, 22	12	4			
19	13	1			

The distribution of $\hat{\alpha}_3$ and $\hat{\beta}_3$ are given in table 7.22, and you should verify from this that

$$E\hat{\alpha}_3 = 2 \quad (= \alpha) \qquad \text{var } \hat{\alpha}_3 = 3/2$$
$$E\hat{\beta}_3 = \tfrac{1}{2} \quad (= \beta) \qquad \text{var } \hat{\beta}_3 = 1/12.$$

By deriving the probability distribution of $\hat{\alpha}_4$ and $\hat{\beta}_4$ you should be able to verify that

$$E\hat{\alpha}_4 = 2 \quad (= \alpha) \qquad \text{var } \hat{\alpha}_4 = 13/2$$
$$E\hat{\beta}_4 = \tfrac{1}{2} \quad (= \beta) \qquad \text{var } \hat{\beta}_4 = \tfrac{1}{4}.$$

So $\hat{\alpha}_1, \hat{\alpha}_2, \hat{\alpha}_3, \hat{\alpha}_4$, are all unbiased estimators of α and

$$\text{var } \hat{\alpha}_1 \; (= 7/6) < \text{var } \hat{\alpha}_2 \; (= 5/4) < \text{var } \hat{\alpha}_3 \; (= 3/2) < \text{var } \hat{\alpha}_4 \; (= 13/2);$$

that is, the least squares estimator of α is the most efficient.

Also $\hat{\beta}_1, \hat{\beta}_2, \hat{\beta}_3, \hat{\beta}_4$, are all unbiased estimators of β and

$$\text{var } \hat{\beta}_1 \; (= 1/16) = \text{var } \hat{\beta}_2 \; (= 1/16) < \text{var } \hat{\beta}_3 \; (= 1/12) < \text{var } \hat{\beta}_4 \; (= 1/4)$$

that is, the least squares estimator of β is the most efficient, (as is, in this case, $\hat{\beta}_2$).

Figures 7.14, 7.15 and 7.16 illustrate the distributions of the fitted lines using each fitting method. The distribution for the least-squares method (method 1) has already been illustrated in figure 7.12. These figures show visually what we have already proved algebraically, namely, that the lines fitted by the least-squares method have a smaller dispersion around the true line than do lines fitted by other methods.

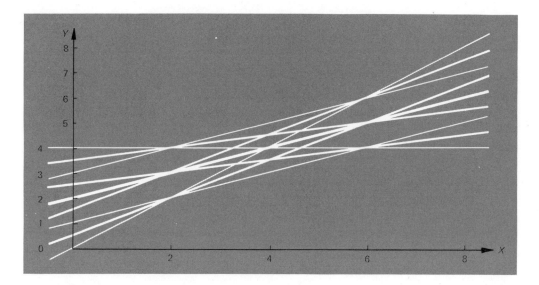

Fig. 7.14 Distribution of fitted lines using method 2

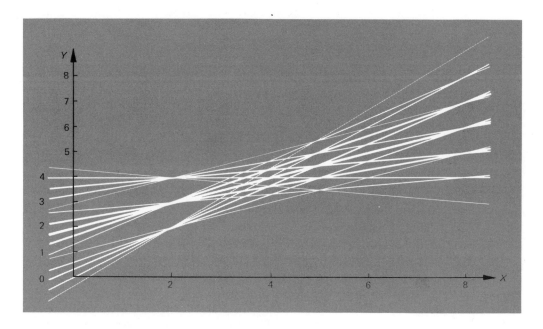

Fig. 7.15 Distribution of fitted lines using method 3

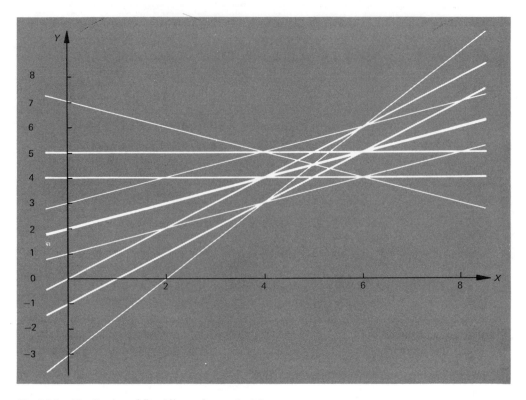

Fig. 7.16   Distribution of fitted lines using method 4

7.7 Distribution of Least Squares Estimators

We have now shown that the least squares estimators of α and β are unbiased; we have found their variances; and we have shown that of all linear unbiased estimators of α and β the least squares estimators have the smallest variance.

Note very carefully that all these results rely crucially on the four assumptions, (A1)–(A4), that we made earlier. In particular we used (A1) and (A4) to prove unbiasedness; then we used (A2) and (A3) to derive the variances, and all four were used to show they were best linear unbiased. Thus if any of these assumptions do not hold we will have to modify the results that we have obtained so far. For the time being, we will continue to assume that they hold, but later we will investigate how to *test* whether they hold or not and, if not, how we must modify our analysis.

This section will consider how to use our estimates to form confidence intervals for, and test hypotheses about, the relevant population parameters. We have seen in chapters 4 and 5 that, in order to do this, we need to know not only the mean and variance of our estimators, but also the (shape of the) distribution of our estimators.

Now, we see from (7.29) that

$$\hat{\beta} = \beta + \sum_1^n w_i u_i.$$

Thus the distribution of $\hat{\beta}$ depends on the distribution of the u_i.

In order to find the distribution of $\hat{\beta}$ we need to make some assumption about the distribution of each u_i. Now u_i contains all the factors that influence Y_i other than X_i; in general we can suppose that these are a large number of factors contained in u_i, and we can resort to the Central Limit Theorem to assert that u_i has a normal distribution. We therefore use assumption (A5) for the moment, an assumption which again we need to subject to test at a later point:

u_i is normal $\qquad (i = 1, 2, \ldots, n)$. $\qquad$ (A5)

We can combine (A1), (A2) and (A5) to give

u_i is $N(0, \sigma^2)$ $\qquad (i = 1, 2, \ldots, n)$.

(A3) of course says that u_i, u_j are independent for all $i, j; i \neq j$. With this assumption, $\hat{\beta}$ is a linear combination of independent, normally distributed random variables, and thus

$\hat{\beta}$ is normally distributed. $\qquad\qquad\qquad$ (7.48)

Further, from (7.32) we see that $\hat{\alpha} = \alpha + (\beta - \hat{\beta})\overline{X} + \bar{u}$ Therefore, if each u_i is normal, then so is $\bar{u}$ (see chapter 5), and as we have already shown that $\hat{\beta}$ is normal, then $\hat{\alpha}$ is a linear combination of independent* normal random variables, and thus

$\hat{\alpha}$ is normally distributed. $\qquad\qquad\qquad$ (7.49)

Finally, combining (7.40) with (7.48) and (7.49) we get:

$$\hat{\beta} \text{ is } N\left(\beta, \frac{\sigma^2}{\sum_1^n x_i^2}\right) \qquad \hat{\alpha} \text{ is } N\left(\alpha, \frac{\sigma^2 \sum_1^n X_i^2}{n \sum_1^n x_i^2}\right) \qquad (7.50)$$

or

$$\frac{\hat{\beta} - \beta}{\sigma / \sqrt{\sum_1^n x_i^2}} \text{ is } N(0, 1) \quad \text{and} \quad \frac{\hat{\alpha} - \alpha}{\sigma \sqrt{(\Sigma X_i^2 / n \Sigma x_i^2)}} \text{ is } N(0, 1). \qquad (7.51)$$

We focus attention on forming confidence intervals for, and testing hypotheses about, β. Similar procedures are, of course, followed for α.

---

*We will show later that $\hat{\beta}$ and $\bar{u}$ are independent.

(7.51) now gives us information in a recognisable form. Thus, to form a 95 per cent confidence interval for β based on sample information we proceed as usual:

Since

$$\frac{\hat{\beta} - \beta}{\sigma/\sqrt{\Sigma x_i{}^2}} \quad \text{is } N(0, 1) \quad \text{(from (7.51))}$$

then

$$P\left[-1.96 \leqslant \frac{\hat{\beta} - \beta}{\sigma/\sqrt{\Sigma x_i{}^2}} \leqslant 1.96\right] = 0.95 \text{ from the unit normal area tables.}$$

Rearranging this in the usual way, we get:

$$P[\hat{\beta} - 1.96\, \sigma/\sqrt{\Sigma x_i{}^2} \leqslant \beta \leqslant \hat{\beta} + 1.96\, \sigma/\sqrt{\Sigma x_i{}^2}] = 0.95$$

Thus a 95 per cent confidence interval for β, based on a sample estimate $\hat{\beta}$, is:

$$(\hat{\beta} - 1.96\, \sigma/\sqrt{\Sigma x_i{}^2}, \quad \hat{\beta} + 1.96\, \sigma/\sqrt{\Sigma x_i{}^2}). \tag{7.52}$$

Note that we have our usual problem: we cannot form this interval unless we know σ^2. We will consider ways of estimating σ^2 in the next section. We proceed for the moment assuming we know σ^2. Secondly, let us test a hypothesis about β. The usual null hypothesis of interest in economics is:

$$H_0 : \beta = 0$$

This H_0 of course hypothesises that the relationship between Y_i and X_i is given by

$$Y_i = \alpha + u_i \qquad (i = 1, 2, \ldots, n)$$

that is, X does not influence Y. If our sample rejects this null hypothesis then we can be reasonably confident that a relationship between X and Y does exist.

The alternative hypothesis depends upon the *a priori* economic theory indicating the likely sign of the coefficient. If economic theory indicates that X should influence Y positively, then our alternative hypothesis would be of the form:

$$H_1 : \beta > 0.$$

If *a priori* theory indicates a negative relationship, then our alternative would be

$$H_1 : \beta < 0.$$

If economic theory is uncertain about the sign of β we could hypothesise the alternative $H_1 : \beta \neq 0$.

Consider (for illustrative purposes) testing

$$H_0 : \beta = 0$$

against

$$H_1 : \beta > 0.$$

Now we know that $(\hat{\beta} - \beta)/(\sigma/\sqrt{\Sigma x_i^2})$ is $N(0,1)$ (from 7.51); therefore if H_0 is correct, that is $\beta = 0$, then $\hat{\beta}/(\sigma/\sqrt{\Sigma x_i^2})$ is $N(0,1)$. We illustrate in figure 7.17. If our observed value of $\hat{\beta}/(\sigma/\sqrt{\Sigma x_i^2})$ is 'large' we will prefer H_1 to H_0. Choose a critical value C^* to determine an acceptable probability of a Type I error. Suppose we take a 5 per cent significance level, that is choose C^* to make the probability of a Type I error equal to 0.05, then $C^* = 1.64$ (from unit normal area tables) and our decision rule is

$$\left. \begin{array}{ll} \text{if } \dfrac{\hat{\beta}}{\sigma/\sqrt{\Sigma x_i^2}} > C^* & \text{accept } H_1 \\[2em] \text{if } \dfrac{\hat{\beta}}{\sigma/\sqrt{\Sigma x_i^2}} < C^* & \text{accept } H_0. \end{array} \right\} \qquad (7.53)$$

(Note that the Type II error is not defined.)

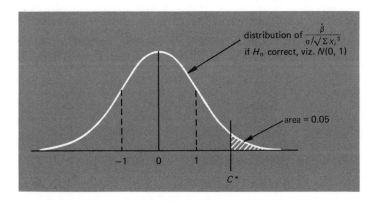

Fig. 7.17   Testing a hypothesis about β

To illustrate, consider the first sample obtained in our sample example in section 7.4. It was

$$Y_1 = 2 \qquad Y_2 = 5 \qquad Y_3 = 5$$
$$X_1 = 2 \qquad X_2 = 4 \qquad X_3 = 6.$$

These gave $\hat{\alpha} = 1$, $\hat{\beta} = \frac{3}{4}$. Now

$$\sum_{1}^{n} x_i{}^2 = (-2)^2 + (0)^2 + (2)^2 = 8$$

and σ^2, which we are assuming known, was $\frac{1}{2}$. (Note that in this example the distribution of the u_i was *not* normal: we are using this example only for illustrative purposes.)

Using (7.52) our 95 per cent confidence interval for β, based on our sample estimate $\hat{\beta} = \frac{3}{4}$, is

$$\left(\frac{3}{4} - 1.96 \, \frac{1/\sqrt{2}}{\sqrt{8}}, \ \frac{3}{4} + 1.96 \, \frac{1/\sqrt{2}}{\sqrt{8}} \right)$$

that is, $(0.75 - 0.49, 0.75 + 0.49)$; that is, $(0.26, 1.24)$. Notice how wide this 95 per cent confidence interval is (a consequence of the very small sample size) but note also that in this case it *does* contain the true value of $\beta (= 0.5)$.

Let us now see if our sample $\hat{\beta} = \frac{3}{4}$ leads us to reject the null hypothesis $H_0 : \beta = 0$ against the alternative that $H_1 : \beta > 0$. Our critical value $C^* = 1.64$ as argued before for a 5 per cent significance level. Our observed value of $\hat{\beta}/(\sigma/\sqrt{\sum x_i{}^2})$ is $0.75/\frac{1}{4} = 3$ which is greater than C^* and so the sample evidence would lead us to reject H_0 in favour of H_1 at the 5 per cent level.

Throughout this section we have assumed that we know σ^2. This is unlikely in practice. In the next section we consider how we may estimate σ^2 when it is unknown.

7.8   Estimation of σ^2

σ^2 is of course the variance of the true deviations. We do not know the values of the true deviations, but we do have a set of estimated deviations. It would seem intuitively reasonable to use the variance of the estimated deviations as an estimator of the variance of the true deviations. Let us investigate the properties of such an estimator.

The estimated deviations are $e_1, e_2, \ldots, e_n$ and so their variance is

$$\frac{1}{n} \sum_{1}^{n} (e_i - \bar{e})^2.$$

However

$$\bar{e} = 0 \quad \text{since } e_i = Y_i - \hat{\alpha} - \hat{\beta} X_i$$

that is,

$$\bar{e} = \bar{Y} - \hat{\alpha} - \hat{\beta}\bar{X}$$

but

$$\hat{\alpha} = \bar{Y} - \hat{\beta}\bar{X} \qquad \text{(from (7.21))}$$

and so

$$\bar{e} = 0. \tag{7.54}$$

Thus the variance of the estimated deviations is

$$d^2 \equiv \frac{1}{n}\sum_1^n e_i^2.$$

We now investigate whether d^2 is an unbiased estimator of σ^2; that is, whether $Ed^2 = \sigma^2$.

To investigate this, note that $e_i = Y_i - \hat{\alpha} - \hat{\beta}X_i$ but

$$\hat{\alpha} = \bar{Y} - \hat{\beta}\bar{X} \qquad \text{(from (7.21))}.$$

Thus

$$e_i = Y_i - \bar{Y} - \hat{\beta}(X_i - \bar{X})$$

that is,

$$e_i = y_i - \hat{\beta}x_i.$$

Substituting for y_i in terms of β and u from (7.24),

$$e_i = \beta x_i + u_i - \bar{u} - \hat{\beta}x_i$$

that is,

$$e_i = (\beta - \hat{\beta})x_i + u_i - \bar{u}. \tag{7.55}$$

Thus,

$$d^2 = \frac{1}{n}\sum_1^n e_i^2 = \frac{1}{n}\sum_1^n \left\{ -(\hat{\beta} - \beta)x_i + u_i - \bar{u} \right\}^2$$

and so

$$d^2 = \frac{1}{n}\left\{ (\hat{\beta} - \beta)^2 \sum_1^n x_i^2 + \sum_1^n u_i^2 + n\bar{u}^2 - 2(\hat{\beta} - \beta)\sum_1^n x_i u_i \right. \\ \left. + 2\bar{u}(\hat{\beta} - \beta)\sum_1^n x_i - 2\bar{u}\sum_1^n u_i. \right\} \tag{7.56}$$

The fifth term in (7.56) is zero, since $\sum_1^n x_i = 0$. The final term is $-2\bar{u}n\bar{u} = -2n\bar{u}^2$, which combines with the third term to give $-n\bar{u}^2$.

Let us use an algebraic 'trick' on the fourth term: Note from (7.25) that

$$(\hat{\beta} - \beta) = \sum_{1}^{n} x_i u_i \Big/ \sum_{1}^{n} x_i^2 .$$

That is,

$$\sum_{1}^{n} x_i u_i = (\hat{\beta} - \beta) \sum_{1}^{n} x_i^2$$

hence the fourth term in (7.56) is

$$-2(\hat{\beta} - \beta)(\hat{\beta} - \beta) \sum_{1}^{n} x_i^2$$

which combines with the first term to give

$$-(\hat{\beta} - \beta)^2 \sum_{1}^{n} x_i^2 .$$

Hence (7.56) becomes:

$$d^2 = \frac{1}{n} \left\{ -(\hat{\beta} - \beta)^2 \sum_{1}^{n} x_i^2 + \sum_{1}^{n} u_i^2 - n\bar{u}^2 \right\} . \tag{7.57}$$

We take expected values, noting that:

$$E(\hat{\beta} - \beta)^2 = \text{var } \hat{\beta} = \sigma^2 \Big/ \sum_{1}^{n} x_i^2 \quad \text{from (7.35)}$$

$$E u_i^2 = \sigma^2 \quad \text{assumption (A2)}$$

and

$$E \bar{u}^2 = \sigma^2/n. \quad \text{from (7.38)}$$

Thus,

$$E d^2 = \frac{1}{n} \left\{ -\frac{\sigma^2}{\sum\limits_{1}^{n} x_i^2} \sum_{1}^{n} x_i^2 + n\sigma^2 - \sigma^2 \right\}$$

and so,

$$E d^2 = \left(\frac{(n-2)}{n} \right) \sigma^2 . \tag{7.58}$$

Hence on average d^2 underestimates σ^2.

Note that, even though d^2 is a biased estimator of σ^2, it is asymptotically unbiased, since $(n - 2)/n \to 1$ as $n \to \infty$.

In the spirit of our previous approach, let us define an estimator s^2 of σ^2 by:

$$s^2 = \left(\frac{n}{n-2} \right) d^2$$

then

$$Es^2 = \left(\frac{n}{n-2}\right)Ed^2 = \left(\frac{n}{n-2}\right)\left(\frac{n-2}{n}\right)\sigma^2 = \sigma^2$$

That is,

$Es^2 = \sigma^2$, and thus s^2 is an unbiased estimator of σ^2. (7.59)

Now

$$s^2 = \left(\frac{n}{n-2}\right)d^2 = \left(\frac{n}{n-2}\right)\frac{1}{n}\sum_1^n e_i^2 = \left(\frac{1}{n-2}\right)\sum_1^n e_i^2 .$$

Hence

$$s^2 = \frac{1}{n-2}\sum_1^n e_i^2 \text{ is an unbiased estimator of } \sigma^2 .$$

This is a modification of the variance of the estimated deviations.

To illustrate these results, consider the values of d^2 and s^2 for each of the 27 samples in the illustrative example of section 7.4. In sample 8, we had

$e_1 = -\frac{1}{2}$

$e_2 = 1$

$e_3 = -\frac{1}{2}$.

Thus (in sample 8)

$$\sum_1^n e_i^2 = 3/2$$

and so

$$d^2 = \frac{1}{n}\sum_1^n e_i^2 = \frac{1}{2} \qquad s^2 = \frac{1}{n-2}\sum_1^n e_i^2 = \frac{3}{2} \quad (n = 3).$$

We have computed d^2 and s^2 for each of the 27 samples, obtaining table 7.23.

Thus the distribution of d^2 and s^2 are as given in table 7.24, and from this we see that

$$Ed^2 = 192/1152 = 1/6 = \left(\frac{3-2}{3}\right)\frac{1}{2} = \left(\frac{n-2}{n}\right)\sigma^2$$

and

$$Es^2 = \frac{1}{2} = \sigma^2 .$$

Table 7.23   Values of d^2 and s^2 in the 27 samples

Sample number	Probability (x 64)	d^2 (x 18)	s^2 (x 6)	Sample number	Probability (x 64)	d^2 (x 18)	s^2 (x 6)
1	1	0	0	15	4	1	1
2	2	1	1	16	2	9	9
3	1	4	4	17	4	4	4
4	2	4	4	18	2	1	1
5	4	1	1	19	1	4	4
6	2	0	0	20	2	9	9
7	1	16	16	21	1	16	16
8	2	9	9	22	2	0	0
9	1	4	4	23	4	1	1
10	2	1	1	24	2	4	4
11	4	4	4	25	1	4	4
12	2	9	9	26	2	1	1
13	4	1	1	27	1	0	0
14	8	0	0				

Table 7.24   Distributions of d^2 and s^2

d^2 (x 18)	$P[d^2]$ (x 64)	$d^2 P[d^2]$ (x 1152)	s^2 (x 6)	$P[s^2]$ (x 14)	$s^2 P[s^2]$ (x 384)
0	14	0	0	14	0
1	24	24	1	24	24
4	16	64	4	16	64
9	8	72	9	8	72
16	2	32	16	2	32
Σ	64	192	Σ	64	192

We have now proved and illustrated that s^2 $(= \Sigma_1^n e_i^2 /(n-2))$ is an unbiased estimator of σ^2. Thus, if n is large, we can replace σ by its estimate s when forming confidence intervals for, and testing hypotheses about, α and β.

However, if n is small, s^2 in a particular sample may be a long way from σ^2; and we would thus introduce a large error into our confidence intervals for α and β. (We can see this very clearly in our example above; since we know $\sigma^2 = \frac{1}{2}$, and even though $Es^2 = \sigma^2$, s^2 itself varies from 0 to 8/3 over samples.)

Examination of the procedures used when we encountered this problem in chapter 5 suggests that we look at the distribution of s^2. We know s^2 is always positive so the obvious distribution to investigate is the chi-square distribution. From equation (7.57) and the fact that $(n-2)s^2 = \Sigma_1^n e_i^2 = nd^2$, we have

$$(n-2)s^2 = -(\hat{\beta} - \beta)^2 \sum_1^n x_i^2 + \sum_1^n u_i^2 - n\bar{u}^2.$$

Re-arranging, we get

$$(n - 2)s^2 + (\hat{\beta} - \beta)^2 \sum_1^n x_i^2 + n\bar{u}^2 = \sum_1^n u_i^2.$$

Dividing throughout by σ^2, we get

$$(n - 2)\frac{s^2}{\sigma^2} + \left(\frac{\hat{\beta} - \beta}{\sigma/\sqrt{\Sigma x_i^2}}\right)^2 + \left(\frac{\bar{u}}{\sigma/\sqrt{n}}\right)^2 = \left(\frac{u_1}{\sigma}\right)^2 + \left(\frac{u_2}{\sigma}\right)^2 + \ldots + \left(\frac{u_n}{\sigma}\right)^2.$$

$$(7.60)$$

The right-hand side of (7.60) consists of the sum of n independent unit normals squared (by assumptions (A1) (A2) (A3) and (A5)). Hence the right-hand side of (7.60) has by definition a chi-square distribution with n degrees of freedom.

The second term on the left-hand side of (7.60) is a unit normal squared, (since $\hat{\beta}$ is $N(\beta, \sigma^2/\Sigma x_i^2)$); that is, it has a chi-square distribution with one degree of freedom.

The third term on the left-hand side of (7.60) is a unit normal squared (since $\bar{u}$ is $N(0, \sigma^2/n)$); that is, it has a chi-square distribution with one degree of freedom.

Hence if the three terms on the left-hand side are independent it follows that the first term on the left-hand side, namely $(n - 2)s^2/\sigma^2$, has a chi-square distribution with $(n - 2)$ degrees of freedom (from (5.35)).

Now, we have already shown (see (7.37)) that $E(\hat{\beta} - \beta)\bar{u} = 0$ and as $\hat{\beta} - \beta$ and $\bar{u}$ are both normal, it follows that they are independent. (See the appendix.) Also, we know that

$$e_i = (\beta - \hat{\beta})x_i + u_i - \bar{u} \text{ from } (7.55) \tag{7.61}$$

thus

$$Ee_i\bar{u} = Ex_i(\beta - \hat{\beta})\bar{u} + Eu_i\bar{u} - E\bar{u}^2.$$

The first term on the right-hand side we have already shown to be zero. Thus

$$Ee_i\bar{u} = \frac{\sigma^2}{n} - \frac{\sigma^2}{n} = 0.$$

Therefore e_i, $\bar{u}$ have zero covariance, and as they are both normal, it follows that they are independent (for all i) (see the appendix). Hence

$$e_i^2, \bar{u}^2 \text{ are independent (for all } i)$$

and so,

$$\Sigma e_i^2, \bar{u}^2 \text{ are independent; that is } s^2 \text{ and } \bar{u}^2 \text{ are independent.}$$

Finally, from (7.61),

$$Ee_i(\beta - \hat{\beta}) = x_i \operatorname{var} \hat{\beta} + E(\beta - \hat{\beta})(u_i - \bar{u})$$

$$= \frac{x_i\sigma^2}{\Sigma x_i^2} - \frac{x_i\sigma^2}{\Sigma x_i^2} = 0.$$

Hence e_i, $(\beta - \hat{\beta})$ have zero covariance, and as they are both normal it follows (see appendix) that they are independent. Thus s^2 and $(\hat{\beta} - \beta)^2$ are independent. Thus the three terms on the left-hand side of (7.54) are independent, and our result follows that:

$$(n - 2)\frac{s^2}{\sigma^2} \quad \text{has a chi-square distribution with} \atop (n - 2) \text{ degrees of freedom.} \tag{7.62}$$

This result enables us to form confidence intervals for, and test hypotheses about, σ^2 given a sample value of s^2. We will not go into any detail as the procedure was illustrated in detail in chapter 5.

7.9   Interval Estimation and Hypothesis Testing with Unknown σ^2

We are now in a position to find exact confidence intervals for α and β for the case when σ^2 is unknown and has to be estimated using s^2. We first concentrate on β.

We know $(\hat{\beta} - \beta)/(\sigma/\sqrt{\Sigma x_i^2})$ is $N(0, 1)$ from (7.50). We know $(n - 2)s^2/\sigma^2$ is a chi-square with $(n - 2)$ degrees of freedom from (7.62), where

$$s^2 = \frac{1}{n - 2}\sum_1^n e_i^2.$$

Also, we have already shown that s^2 and $(\hat{\beta} - \beta)$ are independent. Hence

$$\frac{(\hat{\beta} - \beta)/(\sigma/\sqrt{\Sigma x_i^2})}{\sqrt{\left(\frac{(n - 2)s^2/\sigma^2}{(n - 2)}\right)}}$$

has a t-distribution with $(n - 2)$ degrees of freedom (by the definition (5.41)). Simplifying this expression, it reduces to

$$\frac{\hat{\beta} - \beta}{s/\sqrt{\Sigma x_i^2}} \quad \text{has a } t\text{-distribution with } (n - 2) \text{ degrees of freedom.} \tag{7.63}$$

Similarly we can show that

$$\frac{\hat{\alpha} - \alpha}{s\sqrt{(\Sigma X_i^2/n\Sigma x_i^2)}} \quad \text{has a } t\text{-distribution with } (n - 2) \atop \text{degrees of freedom.} \tag{7.64}$$

Note carefully that these results rely crucially on assumption (A5) holding; if (A5) is not valid, then (7.62) (7.63) and (7.64) will not be valid.

Before we illustrate this large amount of theory with a numerical example, we have one more theoretical construct to examine.

7.10 Goodness of Fit

As should be apparent, we can 'fit' a line to any set of data by a variety of methods. The particular method we have focused attention on, because of its useful properties, is that of least squares. However, we have not discussed yet how well our fitted line fits the observations. What would be useful is a measure of how well our line fits the points, that is how close the points are to the fitted line. For each individual observation (X_i, Y_i) the distance from the line is given by the estimated deviation e_i. The overall variance of the observations about the line is thus given by the variance of these estimated deviations (the e_i), namely $d^2 = \Sigma_1^n e_i^2 / n$.

The further the observations are from the line, the larger will be the e_i, and the larger the variance d^2. If the line fits the points exactly, then each observation will lie on the line, and thus each e_i will be zero, and so will d^2. The worst that can happen is that X is found not to influence Y at all — that is, the fitted line is horizontal. In this case $\hat{\beta}$ will be zero, and therefore, from (7.21), $\hat{\alpha} = \bar{Y}$. Thus, using $e_i = Y_i - \hat{\alpha} - \hat{\beta}X_i$, we see that if the fitted line is horizontal then $e_i = Y_i - \bar{Y}$, that is $e_i = y_i$ for all i. Hence, in this case

$$d^2 = \frac{1}{n}\sum_1^n e_i^2 = \frac{1}{n}\Sigma y_i^2$$

and so d^2 is equal to the variance of the Ys. Notice that this is the worst that can happen, that is d^2 can never be greater than $\Sigma_1^n y_i^2 / n$, since

$$Y_i = \hat{\alpha} + \hat{\beta}X_i + e_i$$

and

$$\bar{Y} = \hat{\alpha} + \hat{\beta}\bar{X}. \quad \text{from (7.21)}$$

Thus

$$Y_i - \bar{Y} = \hat{\beta}(X_i - \bar{X}) + e_i$$

that is,

$$y_i = \hat{\beta}x_i + e_i \tag{7.65}$$

and so

$$\sum_1^n y_i{}^2 = \hat{\beta}^2 \sum_1^n x_i{}^2 + \sum_1^n e_i{}^2 + 2\hat{\beta}\sum_1^n x_i e_i \tag{7.66}$$

but

$$\sum_1^n x_i e_i = \sum_1^n x_i(y_i - \hat{\beta}x_i) \quad \text{from (7.65)}$$

$$= \sum_1^n y_i x_i - \hat{\beta}\sum_1^n x_i{}^2 .$$

Thus, using (7.20),

$$\sum_1^n x_i e_i = 0. \tag{7.67}$$

Substituting this in (7.66), we get

$$\sum_1^n y_i{}^2 = \hat{\beta}^2 \sum_1^n x_i{}^2 + \sum_1^n e_i{}^2 . \tag{7.68}$$

Dividing through by n and rearranging:

$$\frac{1}{n}\sum_1^n y_i{}^2 - d^2 = \hat{\beta}^2 \sum_1^n x_i{}^2 \geqslant 0$$

and so

$$d^2 \leqslant \frac{1}{n}\sum_1^n y_i{}^2$$

and the equality only holds if $\hat{\beta} = 0$ or $\sum_1^n x_i{}^2 = 0$ (this latter is only true if $X_1 = X_2 = \ldots = X_n$). d^2 is necessarily non-negative, and so:

$$0 \leqslant d^2 \leqslant \frac{1}{n}\sum_1^n y_i{}^2$$

or

$$0 \leqslant \frac{d^2}{\sum y_i{}^2/n} \leqslant 1 \tag{7.69}$$

Putting

$$d^2 = \sum_1^n e_i{}^2/n \quad \text{in (7.69) we get:}$$

$$0 \leqslant \sum_1^n e_i{}^2 / \sum_1^n y_i{}^2 \leqslant 1. \tag{7.70}$$

Now the smaller $\sum_1^n e_i{}^2$ is the better the fit, and the larger the worse the fit.

Let us define our 'goodness of fit' measure R^2 by:

$$R^2 = 1 - \sum_1^n e_i^2 / \sum_1^n y_i^2.$$
(7.71)

(7.70) immediately implies that $0 \leqslant R^2 \leqslant 1$, and $R^2 = 1$ only when $\Sigma_1^n e_i^2 = 0$, that is the fit is perfect, and $R^2 = 0$ when $\Sigma_1^n e_i^2 = \Sigma_1^n y_i^2$, that is, when no relationship has been found.

We must note carefully two things about R^2. First, it is a random variable, since both $\Sigma_1^n e_i^2$ and $\Sigma_1^n y_i^2$ are — that is, R^2 will vary from sample to sample. Secondly, it is a measure of the goodness of fit of the fitted line to the observations — it is *not*, repeat *not*, a measure of goodness of fit of the fitted line to the true line. These two points taken together imply the important point that a high R^2 does not *necessarily* mean that the fitted line is close to the true line. Even if the true relationship is that there is no relationship, we could get samples that gave a high R^2. There are tests to see whether R^2 is significantly different from zero, that is to see whether a significant relationship exists. We do not intend to describe such tests, for two reasons: first, these tests would necessitate the introduction of a new theoretical distribution (the F distribution), and we do not want to burden and confuse the reader; secondly, such tests on R^2 are unnecessary, as we already have a perfectly good test of whether a significant relationship exists between X and Y: that of testing whether $\hat{\beta}$ is significantly different from zero. In our simple model, a test as to whether R^2 is significantly different from zero is identical in its results to a test as to whether $\hat{\beta}$ is significantly different from zero.

To illustrate these remarks, we have computed the value of R^2 for each of the 27 samples in our illustrative example. To demonstrate the procedure, consider once again sample number 8. There we had

$$Y_1 = 2 \qquad Y_2 = 5 \qquad Y_3 = 5$$

and

$$e_1 = -\tfrac{1}{2} \qquad e_2 = 1 \qquad e_3 = -\tfrac{1}{2}.$$

Thus

$$\bar{Y} = 4 \quad \text{and} \quad y_1 = -2 \qquad y_2 = 1 \qquad y_3 = 1.$$

Hence

$$\sum_1^3 e_i^2 = 3/2 \quad \text{and} \quad \Sigma y_i^2 = 6.$$

Thus

$$R^2 = 1 - \left(\sum_1^n e_i^2 / \sum_1^n y_i^2 \right) = 1 - \frac{3/2}{6} = 1 - \frac{1}{4} = \frac{3}{4}.$$

Table 7.25   Values of R^2 in the 27 samples

Sample number	Probability (x 64)	R^2 (x 364)	Sample number	Probability (x 64)	R^2 (x 364)
1	1	364	15	4	351
2	2	351	16	2	91
3	1	336	17	4	273
4	2	273	18	2	351
5	4	351	19	1	0
6	2	364	20	2	91
7	1	156	21	1	156
8	2	273	22	2	*
9	1	336	23	4	273
10	2	273	24	2	273
11	4	273	25	1	0
12	2	273	26	2	273
13	4	273	27	1	364
14	8	364			

*R^2 in sample number 22 is indeterminate, since $\sum_1^3 y_i^2 = \sum_1^n e_i^2 = 0$.

The value of R^2 for each of the 27 samples is given in table 7.25.

Let us first look at those samples with $R^2 = 1$; these are sample numbers 1, 6, 14 and 27. The fitted lines in these four samples were:

$$\text{sample 1} : \hat{Y}_i = 1 + 0.5X_i \qquad \text{sample 14}: \hat{Y}_i = 2 + 0.5X_i$$

$$\text{sample 6} : \hat{Y}_i = \qquad X_i \qquad \text{sample 27}: \hat{Y}_i = 3 + 0.5X_i.$$

(Note also that sample 22 was a 'perfect fit' in that $e_i = 0$ all i, and the fitted line was $\hat{Y}_i = 4$.)

So of these four samples, all of which were perfect fits, only one (sample 14) was the true line. Further, if we look at samples 4 and 7 we see:

$$\text{sample 4}: \hat{Y}_i = 4/3 + 0.5X_i \quad R^2 = 273/364$$

$$\text{sample 7}: \hat{Y}_i = 5/3 + 0.5X_i \quad R^2 = 156/364$$

that is, sample 7 gives a fitted line closer to the true line than sample 4, yet its R^2 is lower than sample 4's R^2.

These rather extreme results are a consequence of the small sample size; but the point is made that a high R^2 does not necessarily indicate that the fitted line is close to the true line.

Our goodness of fit measure R^2, usually called the *coefficient of determination*, was defined in terms of $\Sigma_1^n e_i^2$ and $\Sigma_1^n y_i^2$. For computational simplicity, it is often found useful to express R^2 in terms of the original observations. We can do this as follows:

we have from (7.68) that

$$\sum_1^n e_i^2 = \sum_1^n y_i^2 - \hat{\beta}^2 \sum_1^n x_i^2 \qquad \text{but from (7.20)}$$

$$\hat{\beta} = \frac{\sum_1^n x_i y_i}{\sum_1^n x_i^2}.$$

Thus

$$\sum_1^n e_i^2 = \sum_1^n y_i^2 - \left(\frac{\sum_1^n x_i y_i}{\sum x_i^2} \right)^2 \sum_1^n x_i^2 = \sum_1^n y_i^2 - \frac{\left(\sum_1^n x_i y_i \right)^2}{\sum_1^n x_i^2}.$$

Hence

$$R^2 = 1 - \frac{\sum_1^n e_i^2}{\sum_1^n y_i^2} = 1 - 1 + \frac{\left(\sum_1^n x_i y_i \right)^2}{\sum_1^n x_i^2 \sum_1^n y_i^2}$$

that is,

$$R^2 = \frac{\left(\sum_1^n x_i y_i \right)^2}{\sum_1^n x_i^2 \sum_1^n y_i^2} \qquad (7.72)$$

hence R^2 is the square of r, the correlation coefficient. The results of this section prove the assertion made in chapter 2, that $-1 < r < 1$.

At this stage we should make the customary warning that regression analysis does not and cannot tell us anything about causation. Even if we find that there *is* a relationship between two variables Y and X, and that the coefficient of X is significantly different from zero, this does not mean that we can conclude that X causes Y. Nor does it necessarily mean that Y causes X; nor even that X and Y have a common cause. Finding a relationship between two variables indicates nothing about causality. We have to look at *a priori* theory to help us investigate direction of causality, and no amount of statistical theory will ever *prove* that one variable causes another.

7.11 Examples of Regression

We now look at two applications of the methods derived and investigated in this chapter. Both are concerned with the consumption-income relationship, but differ in that the first application uses non-section data, while the second uses one-series data. To add further variety (permissible since the consumption function is not a nation-specific concept) we use U.K. data in the first application and U.S. data in the second.

First, therefore, we consider the problem of unmasking and testing the relationship between household consumption expenditure and household income on a non-section basis. One source of data on these variables is the

Family Expenditure Survey, published annually by the U.K. Department of Employment. (Similar data for the U.S. is contained in the *Current Population Reports—Consumer Income* (series P-60) and *Consumer Buying Indicators* (series P-65) published by the U.S. Bureau of the Census). Unfortunately, the data given there are not for individual households, for reasons of confidentiality, but are averages over many households. For example in appendix 6A of the 1971 *Family Expenditure Survey*, we see that a sample of 622 households with income less than £10 per week was selected, with the average weekly income of these 622 households being £7.89. From table 1 of the same *Survey* we see that these 622 households spent on average £9.73 a week. Similarly there were 679 households sampled in the income range £10 to £15, with an average of £12.40, and average expenditure £14.14. We have not developed in this book the necessary theory for dealing with situations where the data are averaged. At this stage, however, we wish to illustrate regression as simply as possible, and thus intend to interpret the data as if they were obtained as follows: we will suppose that one household was sampled, and its average weekly income was £7.89 and its average weekly expenditure was £9.73. (Thus we are considering the 622 households as one; obviously this raises severe problems of interpretation, particularly in the estimation of the variance of household expenditure, but we propose to ignore these problems at this stage.)* Similarly, we will suppose the second household sampled had an income of £12.40 and an expenditure of £14.14. (Again we are considering the 679 households as one.) The *F.E.S.* gives data on household income and expenditure for twelve income groups. We will consider these figures as figures for twelve individual households. Thus from appendix 6A and table 1 we obtain the data given in table 7.26. The data are given graphically in figure 7.18.

We use these twelve observations to estimate the parameters of the (assumed) true relationship:

$$Y_i = \alpha + \beta X_i + u_i \qquad\qquad i = 1, 2, \ldots, n$$

where

$$Eu_i = 0 \qquad\qquad i = 1, 2, \ldots, n \quad \text{(A1)}$$

$$Eu_i^2 = \sigma^2 \qquad\qquad i = 1, 2, \ldots, n \quad \text{(A2)}$$

$$Eu_i u_j = 0 \qquad\qquad i \neq j, i, j = 1, 2, \ldots, n \quad \text{(A3)}$$

$$X_i \text{ is fixed} \qquad\qquad i = 1, 2, \ldots, n \quad \text{(A4)}$$

$$u_i \text{ is normal} \qquad\qquad i = 1, 2, \ldots, n \quad \text{(A5)}.$$

*As the survey covered mainly income and expenditure of the 622 households in a particular fortnight in 1971, we could interpret the data as being the averages for one household for 622 fortnightly periods.

Table 7.26 Cross-section income—consumption data

i : 'household number'	Y_i : weekly expenditure	X_i : weekly income
1	9.73	7.89
2	14.14	12.40
3	18.78	17.55
4	22.34	22.52
5	25.58	27.49
6	28.87	32.50
7	30.77	37.45
8	33.61	42.52
9	38.19	47.42
10	42.12	54.69
11	48.31	68.14
12	70.92	112.26

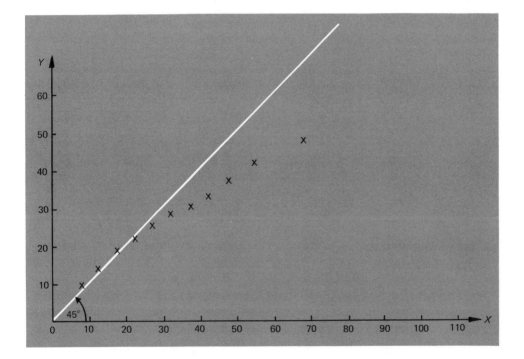

Fig. 7.18   Cross-section income-consumption data

(By studying the sampling method used by the *F.E.S.* you should get some idea of the likely validity of assumptions (A3) and (A4). Assumptions (A1), (A2) and (A5) do not depend solely for their validity on the sample procedure. However, you should consider whether these assumptions are likely to be valid in this particular application.)

We denote our estimated relationship, as usual, by:

$$Y_i = \hat{\alpha} + \hat{\beta}X_i + e_i \quad (i = 1, 2, \ldots, n).$$

We will estimate the relationship by least squares regression since we know that, if assumptions (A1) to (A4) hold, then least squares is the best linear unbiased estimation method.

From (7.20) and (7.21) our least squares estimators, $\hat{\alpha}$ and $\hat{\beta}$, of α and β are given by:

$$\hat{\beta} = \frac{\sum\limits_{1}^{n} y_i x_i}{\sum\limits_{1}^{n} x_i^2} \qquad \hat{\alpha} = \bar{Y} - \hat{\beta}\bar{X}.$$

Arranging our data in a form convenient for computing these estimates, we carry out the necessary computations in table 7.27.

Table 7.27   Calculation of least squares coefficients

i	Y_i	X_i	$y_i = Y_i - \bar{Y}$	$x_i = X_i - \bar{X}$	$y_i x_i$	x_i^2	y_i^2	X_i^2
1	9.73	7.89	−22.217	−32.346	718.6311	1,046.2637	493.5951	62.2521
2	14.14	12.40	−17.807	−27.836	495.6757	774.8429	317.0892	153.7600
3	18.78	17.55	−13.167	−22.686	298.7066	514.6546	173.3699	308.0025
4	22.34	22.52	−9.607	−17.716	170.1976	313.8567	92.2944	507.1504
5	25.58	27.49	−6.367	−12.746	81.1538	162.4605	40.5387	755.7001
6	28.87	32.50	−3.077	−7.736	23.8037	59.8457	9.4679	1,056.2500
7	30.77	37.45	−1.177	−2.786	3.2791	7.7618	1.3853	1,402.5025
8	33.61	42.52	1.663	2.284	3.7983	5.2167	2.7656	1,807.9504
9	38.19	47.42	6.243	7.184	44.8497	51.6099	38.9750	2,248.6564
10	42.12	54.69	10.173	14.454	147.0405	208.9181	103.4899	2,990.9961
11	48.31	68.14	16.363	27.904	456.5932	778.6332	267.7478	4,643.0596
12	70.92	112.26	38.973	72.024	2,806.9914	5,187.4566	1,518.8947	12,602.3076
Σ	383.36	482.83	−0.004*	−0.002*	5,250.7207	9,111.5204	3,059.6135	28,538.5877

$\bar{Y} = 31.947$ $\bar{X} = 40.236$

*should be zero, but slight rounding error is due to approximation of means.

Thus,

$$\bar{X} = 40.236 \qquad\qquad \bar{Y} = 31.947$$

$$\sum_{1}^{n} x_i^2 = 9111.5204 \qquad \sum_{1}^{n} y_i x_i = 5250.7207$$

and so

$$\hat{\beta} = \frac{\sum\limits_{1}^{n} y_i x_i}{\sum\limits_{1}^{n} x_i^2} = \frac{5250.7207}{9111.5204} = 0.57627$$

and

$$\hat{\alpha} = \bar{Y} - \hat{\beta}\bar{X} = 31.947 - (0.57627)(40.236) = 8.76.$$

Our fitted relationship is therefore:

$$Y_i = 8.76 + 0.5763X_i + e_i \qquad i = 1, 2, \ldots, n \qquad (7.73)$$

which is graphed in figure 7.19.

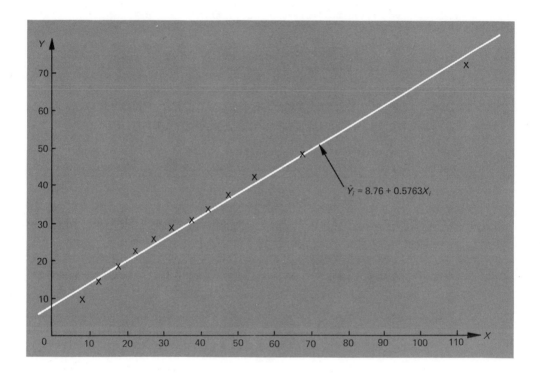

Fig. 7.19   Least squares line fitted to cross-section income—consumption data

The estimated deviations e_i can be calculated from (7.73) as:

$$e_i = Y_i - 8.76 - 0.5763X_i$$

or

$$e_i = Y_i - \hat{Y}_i \quad \text{where} \quad \hat{Y}_i = 8.76 + 0.5763X_i \qquad (\text{all } i).$$

This gives table 7.28. Hence

$$\sum_1^n e_i^2 = 33.7645$$

and so the coefficient of determination R^2 (our measure of goodness of fit) is:

$$R^2 = 1 - \frac{\sum\limits_1^n e_i^2}{\sum\limits_1^n y_i^2} = 1 - \frac{33.7645}{3059.6135} = 0.9890. \tag{7.74}$$

Note that we showed that R^2 can also be expressed as:

$$R^2 = \frac{\left(\sum\limits_1^n y_i x_i\right)^2}{\left(\sum\limits_1^n y_i^2\right)\left(\sum\limits_1^n x_i^2\right)} = \frac{(5250.7207)^2}{(3059.6135)(9111.5204)} = 0.9890$$

which, of course, is the same as that previously calculated.

Table 7.28   Calculation of estimated deviations

i	Y_i	X_i	$\hat{Y}_i = 8.76 + 0.5763X_i$	$e_i = Y_i - \hat{Y}_i$	$e_i X_i$	e_i^2
1	9.73	7.89	13.31	−3.58	−28.2462	12.8164
2	14.14	12.40	15.91	−1.77	−21.9480	3.1329
3	18.78	17.55	18.87	−0.09	−1.5795	0.0081
4	22.34	22.52	21.74	0.60	13.5120	0.3600
5	25.58	27.49	24.60	0.98	26.9402	0.9604
6	28.87	32.50	27.49	1.38	44.8500	1.9044
7	30.77	37.45	30.34	0.43	16.1035	0.1849
8	33.61	42.52	33.26	0.35	14.8820	0.1225
9	38.19	47.42	36.09	2.10	99.5820	4.4100
10	42.12	54.69	40.28	1.84	100.6296	3.3856
11	48.31	68.14	48.03	0.28	19.0792	0.0784
12	70.92	112.26	73.45	−2.53	−284.0178	6.4009
Σ	383.36	482.48	383.37*	−0.01**	−0.2130**	33.7645

*should equal ΣY_i, but slight inaccuracy is due to rounding error;
**should be zero, but slight inaccuracy is due to rounding error (see (7.54) and (7.66)).

We now estimate the variance of the true deviations, using our unbiased estimator s^2, given by:

$$s^2 = \frac{1}{n-2} \sum\limits_1^n e_i^2 = \frac{1}{10} 33.7645 = 3.37645.$$

Hence

$$s = 1.8375. \tag{7.75}$$

The variance of our estimators, $\hat{\alpha}$ and $\hat{\beta}$ are given by (from (7.40)):

$$\text{var } \hat{\alpha} = \sigma^2 \frac{\overset{n}{\underset{1}{\Sigma}} X_i^2}{n \overset{n}{\underset{1}{\Sigma}} x_i^2} = \frac{\sigma^2 \, 28538.5877}{12(9111.5204)} = 0.2610\sigma^2$$

$$\text{var } \hat{\beta} = \frac{\sigma^2}{\overset{n}{\underset{1}{\Sigma}} x_i^2} = \frac{\sigma^2}{9111.5204} = 0.0001098 \, \sigma^2.$$

As we do not know σ^2, we estimate it by s^2, and so get:

$$\text{estimated var } \hat{\alpha} = \frac{s^2 \overset{n}{\underset{1}{\Sigma}} X_i^2}{n \overset{n}{\underset{1}{\Sigma}} x_i^2} = (3.37645)(0.2610) = 0.8813$$

$$\text{estimated var } \hat{\beta} = \frac{s^2}{\overset{n}{\underset{1}{\Sigma}} x_i^2} = (3.37645)(0.0001098) = 0.0003707.$$

Hence the estimated standard deviations of $\hat{\alpha}$ and $\hat{\beta}$ are given by:

$$\left. \begin{array}{l} \text{estimated s.d.}(\hat{\alpha}) = \sqrt{\text{estimated var } \hat{\alpha}} = \sqrt{0.8813} \quad = 0.9388 \\ \text{estimated s.d.}(\hat{\beta}) = \sqrt{\text{estimated var } \hat{\beta}} = \sqrt{0.0003707} = 0.01925 \end{array} \right\} \quad (7.76)$$

Many texts and writers refer to the estimated standard deviation of the coefficients as the *'standard error'* of the coefficients. We do not like this term, as it may be misleading, and prefer to use the more cumbersome, but more accurate, description of *'estimated standard deviation'*.

Drawing together results (7.73), (7.74) and (7.76), we can express our estimated relationship between household consumption expenditure and household income by:

$$Y_i = 8.76 + 0.5763X_i + e_i \quad R^2 = 0.9890. \qquad (7.77)$$
$$(0.94) \quad (0.0192)$$

(The numbers in brackets under the estimated coefficients are their estimated standard deviations.)

The estimated coefficients $\hat{\alpha}$ and $\hat{\beta}$ are point estimates of the true coefficients α and β. To derive interval estimates for $\hat{\alpha}$ and $\hat{\beta}$ we proceed in the usual fashion.

From (7.63) we have that

$$\frac{\hat{\beta} - \beta}{s/\sqrt{\Sigma x_i^2}}$$

has a t-distribution with $(n - 2)$ degrees of freedom. Now, in this application, $n = 12$, and so, using the t-tables we can form, for example, a

95 per cent confidence interval for β as follows:

$$P\left[-2.228 \leqslant \frac{\hat{\beta} - \beta}{s/\sqrt{\Sigma x_i^2}} \leqslant 2.228\right] = 0.95$$

Rearranging gives:

$$P\left[\hat{\beta} - 2.228 \frac{s}{\sqrt{\Sigma x_i^2}} \leqslant \beta \leqslant \hat{\beta} + 2.228 \frac{s}{\sqrt{\Sigma x_i^2}}\right] = 0.95.$$

And so, our 95 per cent confidence interval for β is:

$$\left(\hat{\beta} - 2.228 \frac{s}{\sqrt{\Sigma x_i^2}} , \hat{\beta} + 2.228 \frac{s}{\sqrt{\Sigma x_i^2}}\right)$$

where $\hat{\beta}$ is our estimated coefficient, and $s/\sqrt{\Sigma x_i^2}$ is the estimated standard deviation of $\hat{\beta}$. From (7.73) and (7.76) we have $\hat{\beta} = 0.5763$ and $s/\sqrt{\Sigma x_i^2} = 0.01925$; thus our 95 per cent confidence interval for β is:

$$(0.5763 - (2.228)(0.01925), 0.5763 + (2.228)(0.01925))$$

that is,

$$(0.5334, 0.6192).$$

Hence, on the basis of our twelve observations, we can be 95 per cent confident that the marginal propensity to consume is covered by the interval from 0.5334 to 0.6192.

Similarly a 99 per cent confidence interval for β is given by:

$$(0.5763 - (3.169)(0.01925), 0.5763 + (3.169)(0.01925))$$

that is,

$$(0.5153, 0.6373).$$

Also from (7.64) we have the information necessary to provide interval estimates for α, namely that:

$$\frac{\hat{\alpha} - \alpha}{s\sqrt{(\Sigma X_i^2/n\Sigma x_i^2)}}$$ has a t-distribution with $(n - 2)$ degrees of freedom;

thus (with $n = 12$)

$$P\left[-2.228 \leqslant \frac{\hat{\alpha} - \alpha}{s\sqrt{(\Sigma X_i^2/n\Sigma x_i^2)}} \leqslant 2.228\right] = 0.95$$

and so our 95 per cent confidence interval for α is:

$$(\hat{\alpha} - 2.228\, s\sqrt{(\Sigma X_i^2/n\Sigma x_i^2)}, \hat{\alpha} + 2.228\, s\sqrt{(\Sigma X_i^2/n\Sigma x_i^2)})$$

where $\hat{\alpha}$ is our estimated intercept (= 8.76 from (7.73)) and

$$s\sqrt{(\Sigma X_i^2 / n\Sigma x_i^2)}$$

is its estimated standard deviation (= 0.9388 from (7.76)). Hence the 95 per cent confidence interval for α is:

$$(8.76 - (2.228)\,(0.9388),\ 8.76 + (2.228)\,(0.9388))$$

that is,

$$(6.67, 10.85).$$

Similarly, the 99 per cent confidence interval for α is (5.78, 11.74).

Let us use our results to test a hypothesis about the value of the true coefficient β. Let us posit the null hypothesis that no relationship exists, that is, that $\beta = 0$. We will test this against the alternative that there is a (positive) relationship. Our hypotheses are thus:

$$H_0 : \beta = 0$$

$$H_1 : \beta > 0.$$

(H_1 embodies our theoretical expectations of the value of β.) We know from (7.63) that

$$\frac{\hat{\beta} - \beta}{s/\sqrt{\Sigma x_i^2}}$$

has a t-distribution with $(n - 2)$ degrees of freedom. Thus, if H_0 is correct (i.e. $\beta = 0$), then

$$\frac{\hat{\beta}}{s/\sqrt{\Sigma x_i^2}}$$

has a t-distribution with $(n - 2)$ degrees of freedom. We illustrate in figure 7.20. If our observed value of $\hat{\beta}/(s/\sqrt{\Sigma x_i^2})$ is 'considerably' greater than zero we should reject H_0 in favour of H_1. Thus we choose a critical value C^* to achieve a desired significance level and accept H_0 if our observed value of $\hat{\beta}/(s/\sqrt{\Sigma x_i^2})$ is less than C^*, and reject H_0 if our observed value of $\hat{\beta}/(s/\sqrt{\Sigma x_i^2})$ is greater than C^*. For 5 per cent significance on a one-tailed test, $C^* = 1.812$ for a t-distribution with 10 degrees of freedom ($n = 12$). Our observed value of $\hat{\beta}/(s/\sqrt{\Sigma x_i^2})$ is the ratio of the estimated coefficient to its estimated standard deviation (this ratio is often referred to as the coefficient's t-ratio), and is, in this case, 0.5763/0.01925 = 29.9. This is obviously greater than C^*, and so we reject the null hypothesis that $\beta = 0$ in favour of the alternative that $\beta > 0$.

As we are often interested in whether the estimated coefficients are significantly different from zero, many texts and writers give the value of the t-ratio (the ratio of the estimated coefficient to its estimated standard deviation) in brackets under the estimated coefficient, rather than its

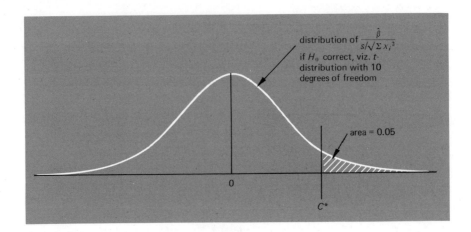

Fig. 7.20   Testing the hypothesis that $\beta = 0$

estimated standard deviation. In this form, our equation would be:

$$Y_i = 8.76 + 0.5763X_i + e_i \quad R^2 = 0.9890 \tag{7.78}$$
$$(9.3) \quad\quad (29.9)$$

(the numbers in brackets under the estimated coefficients are their t-ratios). The value of the t-ratio for $\hat{\alpha}$ shows that $\hat{\alpha}$ is also significantly different from zero.

Finally let us test to see if the estimated marginal propensity to consume is significantly less than 1. Our hypotheses in this case are:

$$H_0 : \beta = 1$$
$$H_1 : \beta < 1.$$

From (7.63) we know that, if H_0 is correct, then

$$\frac{\hat{\beta} - 1}{s/\sqrt{\Sigma x_i^2}}$$

has a t-distribution with $(n - 2) = 10$ degrees of freedom. Figure 7.21 illustrates this. If $(\hat{\beta} - 1)/(s/\sqrt{\Sigma x_i^2})$ is less than some critical value C^* we will reject H_0 in favour of H_1. Choosing a 1 per cent significance level, $C^* = -3.169$ and

$$\frac{\hat{\beta} - 1}{s/\sqrt{\Sigma x_i^2}} = \frac{0.5763 - 1}{0.01925} = -\frac{0.4237}{0.01925} = -22.0,$$

which is less than C^*, and so the evidence suggests that we reject $H_0 : \beta = 1$ at the 1 per cent level.

We now consider a second example of simple regression. We will again estimate a consumption–income relation, but instead of using cross-

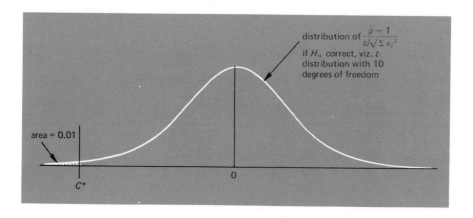

distribution of $\frac{\hat{\beta}-1}{s/\sqrt{\Sigma x_i^2}}$

if H_0 correct, viz. t-distribution with 10 degrees of freedom

area = 0.01

c^*

0

Fig. 7.21   Testing the hypothesis that $\beta = 1$

section data on households (i.e. a sample of households at one point in time), we will use time-series data on aggregate consumption and aggregate income for the United States.

As usual we write our relationship as:

$$Y_i = \alpha + \beta X_i + u_i \quad (i = 1, 2, \ldots, n) \tag{7.79}$$

where Y_i is aggregate U.S. Personal Consumption Expenditures in year i, and X_i is aggregate U.S. Gross National Product in year i, and we will make the usual assumptions about the nature of the deviation term:

$$Eu_i = 0 \qquad\qquad i = 1, 2, \ldots, n \quad \text{(A1)}$$
$$Eu_i^2 = \sigma^2 \qquad\qquad i = 1, 2, \ldots, n \quad \text{(A2)}$$
$$Eu_i u_j = 0 \qquad i \neq j, i, j = 1, 2, \ldots, n \quad \text{(A3)}$$
$$X_i \text{ is fixed} \qquad i = 1, 2, \ldots, n \quad \text{(A4)}$$
$$u_i \text{ is normal} \qquad i = 1, 2, \ldots, n \quad \text{(A5)}$$

Now, from the *Survey of Current Business*, January 1976, we obtain the data given in table 7.29.

At this stage it might be useful to interpret the nature of our postulated relationship (7.29) and of the assumptions (A1) to (A5). Consider, for example, $i = 21$ (i.e. 1966). In 1966, U.S. Gross National Product was $981 billion and Personal Consumption Expenditures was $586.1 billion. If GNP was ever $981 billion again, it would be highly unlikely that consumption would be exactly $586.1 billion again. If we had a whole series of years in which GNP was $981 billion we would expect a whole series of different values of consumption to be recorded, simply since GNP is not the sole determinant of consumption. Thus our observed value of $586.1

Table 7.29   Aggregate U.S. income-consumption data

Year	i	P.C.E.* Y_i	G.N.P* X_i	
1946	1	301.4	477.6	
1947	2	306.2	468.3	
1948	3	312.8	487.7	
1949	4	320.0	490.7	
1950	5	338.1	533.5	
1951	6	342.3	576.5	
1952	7	350.9	598.5	
1953	8	364.2	621.8	
1954	9	370.9	613.7	
1955	10	395.1	654.8	
1956	11	406.3	668.8	
1957	12	414.7	680.9	
1958	13	419.0	679.5	
1959	14	441.5	720.4	
1960	15	453.0	736.8	
1961	16	462.2	755.3	
1962	17	482.9	799.1	
1963	18	501.4	830.7	
1964	19	528.7	874.4	
1965	20	558.1	925.9	
1966	21	586.1	981.0	
1967	22	603.2	1,007.7	
1968	23	633.4	1,051.8	
1969	24	655.4	1,078.8	
1970	25	668.9	1,075.3	
1971	26	691.9	1,107.5	
1972	27	733.0	1,171.1	*P.C.E. : Personal Consumption Expenditures
1973	28	766.3	1,233.4	G.N.P. : Gross National Product.
1974	29	759.8	1,210.7	Both in billions of 1972 dollars

Source: *Survey of Current Business* (National Income Issue) Jan.
1976, Vol. 56, no. 1, Part II, U.S. Department of Commerce Bureau
of Economic Analysis; Table 1 (pp. 6–7) and Table 2.4 (pp. 48–49).

billion for consumption in 1966 is regarded as one observation from the
distribution of all possible consumption figures corresponding to a GNP
figure of $981 billion. Assumption (A1) thus means that the average value
of consumption corresponding to a GNP of X_i is given by $a + \beta X_i$. Assumption (A2) means that the spread of the distribution of consumption corresponding to a given value of GNP is the same irrespective of the value of
GNP. Assumption (A3) means that there is no connection between the
deviations for different years; in other words, if consumption is above
average (for the GNP prevailing) in one year, this does not influence
whether consumption will be above or below average (for the GNP prevailing) in other years. You may suspect that this is a very strong assumption, which is unlikely to hold in practice. However we will maintain it
for the moment, and investigate theoretical and empirical tests for its

validity in the next chapter. Assumption (A4) implies that, at least conceptually, we could observe another twenty-nine years' figures with the same GNP values (i.e. the X_i in the twenty-nine years 1946–74 inclusive). Obviously this might be difficult and rather unpopular; and you may wonder whether even conceptually it is a meaningful assumption. We will investigate it further in the next chapter. Assumption (A5) should be self-evident.

Let us now use our observations to calculate the least squares estimators, $\hat{\alpha}$ and $\hat{\beta}$, of α and β, respectively. The necessary calculations are given in table 7.30.

Table 7.30   Calculation of least squares coefficients

i	Y_i	X_i	$y_i = Y_i - \bar{Y}$	$x_i = X_i - \bar{X}$	$y_i x_i$	x_i^2	y_i^2	X_i^2
1	301.4	477.6	−187.1	−319.4	59,759.7	102,016.4	36,006.4	228,101.8
2	306.2	468.3	−182.3	−328.7	59,922.0	108,049.7	33,233.3	219,304.9
3	312.8	487.7	−175.7	−309.3	54,344.0	95,666.5	30,870.5	237.851.3
4	320.0	490.7	−168.5	−306.3	51.611.6	93,819.7	23,392.3	240,786.5
5	338.1	533.5	−150.4	−263.5	39,630.4	69,432.2	22,620.2	284.622.2
6	342.3	576.5	−146.2	−220.5	32,237.1	48,620.3	21,374.4	332,352.3
7	350.9	598.5	−137.6	−198.5	27,313.6	39,402.2	18,933.8	358.202.2
8	364.2	621.8	−124.3	−175.2	21,777.4	30,695.0	15,450.5	386.635.2
9	370.9	613.7	−117.6	−183.3	21,556.1	33,598.9	13,829.8	376.627.7
10	395.1	654.8	− 93.4	−142.2	13,281.5	20,220.8	8,723.6	428,763.0
11	406.3	668.8	− 82.2	−128.2	10,538.0	16,435.2	6,756.8	447,293.4
12	414.7	680.9	− 73.8	−116.1	8,568.2	13,479.2	5,446.4	463,624.8
13	419.0	679.5	− 69.5	−117.5	8,116.2	13,806.2	4,830.2	461,720.3
14	441.5	720.4	− 47.0	− 76.6	3,600.2	5,867.6	2,209.0	518,976.2
15	453.0	736.8	− 35.5	− 60.2	2,137.1	3,624.0	1,260.2	542,874.2
16	462.2	755.3	− 26.3	− 41.7	1,096.7	1,738.9	691.7	570,478.1
17	482.9	799.1	− 5.6	2.1	− 11.8	4.4	31.4	638,560.8
18	501.4	830.7	12.9	33.7	434.7	1,135.7	116.4	690,062.5
19	528.7	874.4	40.2	77.4	3,111.5	5,990.8	1,616.0	764,575.4
20	558.1	925.9	69.6	128.9	8,971.4	16,615.2	4,844.2	857,290.8
21	586.1	981.0	97.6	184.0	17,958.4	33,856.0	9,525.8	962,361.0
22	603.2	1,007.7	114.7	210.7	24,167.3	44,394.5	13,156.1	1,015,459.2
23	633.4	1,051.8	144.9	254.8	36,920.5	64,923.0	20,996.0	1,106,283.2
24	655.4	1,078.8	166.9	281.8	47,032.4	79,411.2	27,855.6	1,163,809.4
25	668.9	1,075.3	180.4	278.3	50,205.3	77,450.9	32,544.2	1,156,270.0
26	691.9	1,107.5	203.4	310.5	63,155.7	96,410.2	41,371.6	1,226,556.2
27	733.0	1,171.1	244.5	374.1	91,467.4	139,950.8	59,780.2	1,371,475.2
28	766.3	1,233.4	277.8	436.4	121,231.9	190,445.0	77,172.8	1,521,275.5
29	759.8	1,210.7	271.3	413.7	112,236.8	171,147.7	73,603.7	1,465,794.4
Σ	14,167.7	23,112.2	1.2*	− 0.8*	992,421.3	1,618,208.2	612,293.1	20,037,987.7

$$\bar{Y} = 488.5 \quad \bar{X} = 797.0 \quad \sum_1^n y_i x_i = 992,421.3 \quad \sum_1^n x_i^2 = 1,618,208.2$$

$$\hat{\beta} = \frac{\sum_1^n y_i x_i}{\sum_1^n x_i^2} = \frac{992,421.3}{1,618,208.2} = 0.61328$$

and

$$\hat{\alpha} = \bar{Y} - \hat{\beta}\bar{X} = 488.5 - (0.61328)(797.0) = -0.28$$

Our fitted relationship is therefore:

$$Y_i = -0.28 + 0.6133X_i + e_i \quad i = 1, 2, \ldots, n.$$

We illustrate in figure 7.22.

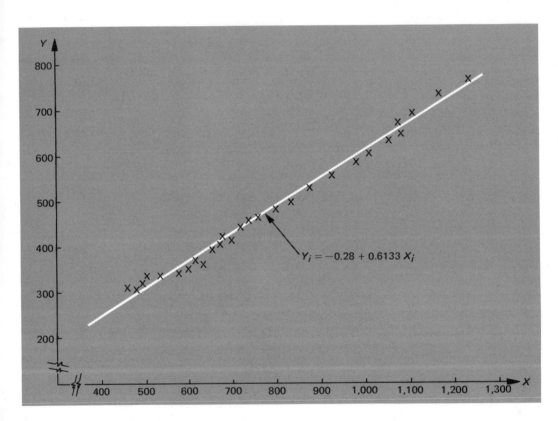

Fig. 7.22   Least squares line fitted to aggregate income—consumption data

Table 7.31 contains the calculations necessary to find the estimated deviations. Hence

$$R^2 = 1 - \frac{\Sigma e_i^2}{\Sigma y_i^2} = 1 - \frac{3,660.64}{612,293.1} = 0.994$$

$$s^2 = \frac{1}{n-2} \Sigma e_i^2 = \frac{1}{27}(3,660.64) = 135.57925$$

$$s = 11.64385$$

Table 7.31   Calculation of estimated deviations

i	Y_i	X_i	$\hat{Y}_i = 1.0223 + 0.60958X_i$	$e_i = Y_i - \hat{Y}_i$	e_iX_i	e_i^2
1	301.4	477.6	292.6	8.8	4,202.9	77.44
2	306.2	468.3	286.9	19.3	9,038.2	372.49
3	312.8	487.7	298.8	14.0	6,827.8	196.00
4	320.0	490.7	300.7	19.3	9,470.5	372.49
5	338.1	533.5	326.9	11.2	5,975.2	125.44
6	342.3	576.5	353.3	−11.0	−6,341.5	121.00
7	350.9	598.5	366.8	−15.9	−9,516.1	252.81
8	364.2	621.8	381.1	−16.9	−10,508.4	285.61
9	370.9	613.7	376.1	−5.2	−3,191.2	27.04
10	395.1	654.8	401.3	−6.2	−4,059.8	38.44
11	406.3	668.8	409.9	−3.6	−2,407.7	12.96
12	414.7	680.9	417.3	−2.6	−1,770.3	6.76
13	419.0	679.5	416.4	2.6	1,766.7	6.76
14	441.5	720.4	441.5	0.0	0.0	0.00
15	453.0	736.8	451.6	1.4	1,031.5	1.96
16	462.2	755.3	462.9	−0.7	−528.7	0.49
17	482.9	799.1	489.8	−6.9	−5,513.8	47.61
18	501.4	830.7	509.2	−7.8	−6,479.5	60.84
19	528.7	874.4	536.0	−7.3	−6,383.1	53.29
20	558.1	925.9	567.6	−9.5	−8,796.1	90.25
21	586.1	981.0	601.3	−15.2	−14,911.2	231.04
22	603.2	1,007.7	617.7	−14.5	−14,611.7	210.25
23	633.4	1,051.8	644.8	−11.4	−11,990.5	129.96
24	655.4	1,078.8	611.3	−5.9	−6,364.9	34.81
25	668.9	1,075.3	659.2	9.7	10,430.4	94.09
26	691.9	1,107.5	678.9	13.0	14,397.5	169.00
27	733.0	1,177.1	717.9	15.1	17,683.6	228.01
28	766.3	1,233.4	756.1	10.2	12,580.7	104.04
29	759.8	1,210.7	742.2	17.6	21,308.3	309.76
Σ	14,167.7	23,112.2	14,166.1	1.6*	1,338.8*	3,660.64

*slight rounding error.

Hence:
$$\text{estimated s.d.}(\hat{\alpha}) = s\sqrt{\frac{\Sigma X_i^2}{n\Sigma x_i^2}} = (11.64385)\sqrt{\frac{(20,037,987.7)}{29(1,618,208.2)}} = 7.609$$

$$\text{estimated s.d.}(\hat{\beta}) = s/\sqrt{\Sigma x_i^2} = \frac{11.643.85}{\sqrt{1,618,208.2}} = 0.009153$$

The t-ratios for α and β are (respectively) −0.04 and 67.00, and so our estimated relationship can be written:

$$Y_i = -0.28 + 0.6133 X_i + e_i \qquad R^2 = 0.994 \qquad (7.80)$$
$$(-0.04) \quad (67.00)$$

(the numbers in brackets under the estimated coefficients are their t-ratios). You should check for yourself that $\hat{\beta}$ is significantly different from zero at the 1 per cent level on a one-tailed test (the critical value for a t-distribution with 27 degrees of freedom is 2.473). Also we can show that $\hat{\beta}$ is significantly different from one. We can also derive in the usual way the 95 and 99 per cent confidence intervals for β; these are, respectively: (0.5945, 0.6321) and (0.5879, 0.6386).

Comparing our point estimates of the marginal propensity to consume, we see that the estimate obtained from the cross-section data (0.5763) is lower than that obtained from the time-series data (0.6133). This is a typical result which has been found by many researchers. As a result of these apparently conflicting results, several more advanced theories of consumers' behaviour were proposed. These theories include the permanent income hypothesis of Friedman (11), the life-cycle hypothesis Modigliani and Brumberg (20) and the relative income hypothesis of Duesenberry (7). Readers interested in the reconciliation of these two apparently different estimates of the marginal propensity to consume should consult these quoted references. Some of the explanations concern the statistical procedure used, and in particular question the validity of the assumptions we have made to obtain our estimates. We will be making some comments along these lines in the next chapter. Other explanations concern the theoretical formulation of the income—consumption relation, and the definitions and measurement of the appropriate concepts of income and consumption. We do not propose to look in detail at the economic arguments; also, some of the statistical reasons are outside the scope of this book. However it is useful to note that the 95 per cent confidence interval for the marginal propensity to consume obtained from the cross-section data and the 95 per cent confidence interval for the marginal propensity to consume obtained from the time-series data overlap. In fact it can be shown that the difference between the two point estimates is not significantly different from zero at the 5 per cent level.

7.12 Prediction

We now examine how we can use our estimated relationships to *predict* the values of the dependent variable, Y, corresponding to a given value of the independent variable, X. First, let us consider the problem of prediction in the situation where we know all the parameters of the true relationship exactly; that is, where we know α, β and σ^2. Our true relationship is:

$$Y_i = \alpha + \beta X_i + u_i \quad (i = 1, 2, \ldots, n).$$

Thus there is no unique value of Y corresponding to a particular value of X. There is a whole range of possible values of Y corresponding to a particular value of X. In terms of our household consumption—income example, even if we know α, β and σ^2, we would be unable to predict the actual consumption of a particular household with a given income. What we could do is to predict the average value of household consumption for all households with the given income. This average is of course given by

$$E(Y_i \mid X_i) = \alpha + \beta X_i.$$

So if we knew α and β we could find $E(Y_i \mid X_i)$ for any given X_i. However, if we knew σ^2 and the shape of the distribution of the u_i, we could form a confidence interval for the consumption of a particular household. Suppose we are interested in a household with income X_0. The average consumption of all households with income X_0 is $E(Y_0 \mid X_0) = \alpha + \beta X_0$. Now if the usual assumptions hold with respect to u_0, we know that u_0 is $N(0, \sigma^2)$. Hence, for example,

$$P\left[-1.96 \leqslant \frac{u_0}{\sigma} \leqslant 1.96\right] = 0.95 \tag{7.81}$$

but

$$Y_0 = \alpha + \beta X_0 + u_0, \text{ so } u_0 = Y_0 - \alpha - \beta X_0.$$

Hence the probability statement (7.81) can be written

$$P\left[-1.96 \leqslant \frac{Y_0 - \alpha - \beta X_0}{\sigma} \leqslant 1.96\right] = 0.95$$

which can be expressed as

$$P[\alpha + \beta X_0 - 1.96\,\sigma \leqslant Y_0 \leqslant \alpha + \beta X_0 + 1.96\,\sigma] = 0.95. \tag{7.82}$$

Thus if we know α, β and σ^2 we can form the 95 per cent confidence interval $(\alpha + \beta X_0 - 1.96\sigma, \; \alpha + \beta X_0 + 1.96\sigma)$ for the expenditure of a particular household. (7.82) of course says that 95 per cent of all households with income X_0 spend between 1.96 standard deviations less and 1.96 standard deviations more than the mean expenditure for all these households.

Thus if we know α, β and σ^2 we can predict the average value of Y corresponding to a given value of X exactly, and form confidence intervals for a particular value of Y corresponding to a given value of X.

However, if we do not know the true parameters, α, β and σ^2 exactly but only have estimates of them, then we will be unable to predict exactly even the average value of Y corresponding to a given value of X. However we can form an interval estimate of this average. We proceed as follows: we wish to estimate $E(Y_0 \mid X_0) = \alpha + \beta X_0$. We have estimates of α and β and an estimated relationship given by $Y_i = \hat{\alpha} + \hat{\beta} X_i + e_i$. The obvious estimator of $E(Y_0 \mid X_0) = \alpha + \beta X_0$ is of course $\hat{Y}_0 = \hat{\alpha} + \hat{\beta} X_0$. Notice that $\hat{Y}_0$ is an unbiased estimator of $E(Y_0 \mid X_0)$ since

$$E\hat{Y}_0 = E(\hat{\alpha} + \hat{\beta} X_0) = E\hat{\alpha} + (E\hat{\beta})X_0$$

$$= \alpha + \beta X_0 \quad \text{(since } \hat{\alpha} \text{ and } \hat{\beta} \text{ are unbiased estimators of } \alpha \text{ and } \beta\text{)}$$

$$= E(Y_0 \mid X_0).$$

Now, $\hat{Y}_0 = \hat{\alpha} + \hat{\beta}X_0$ and X_0 is constant; thus the variance of $\hat{Y}_0$ is given by (from exercise 3.24):

$$\text{var } \hat{Y}_0 = \text{var } \hat{\alpha} + 2 \text{ cov } (\hat{\alpha}, \hat{\beta})X_0 + \text{var } \hat{\beta}X_0^2 . \tag{7.83}$$

We know the variances of $\hat{\alpha}$ and $\hat{\beta}$, but as yet have not calculated their covariance. We do this now. From (7.32)

$$\hat{\alpha} - \alpha = - (\hat{\beta} - \beta)\bar{X} + \bar{u}.$$

Thus

$$(\hat{\alpha} - \alpha)(\hat{\beta} - \beta) = - (\hat{\beta} - \beta)^2 \bar{X} + (\hat{\beta} - \beta)\bar{u}.$$

Therefore

$$\begin{aligned}
\text{cov } (\hat{\alpha}, \hat{\beta}) &= E\{(\hat{\alpha} - \alpha)(\hat{\beta} - \beta)\} \quad \text{since} \quad E\hat{\alpha} = \alpha, \ E\hat{\beta} = \beta \\
&= -E(\hat{\beta} - \beta)^2 \bar{X} + E(\hat{\beta} - \beta)\bar{u}.
\end{aligned}$$

We have already shown ((7.37)) that $E(\hat{\beta} - \beta)\bar{u} = 0$; hence

$$\text{cov } (\hat{\alpha}, \hat{\beta}) = -\bar{X} \text{ var } \hat{\beta} = \frac{-\sigma^2 \bar{X}}{\Sigma x_i^2} . \tag{7.84}$$

Note the intuitive reasonableness of this result: if $\bar{X}$ is positive, then $\hat{\alpha}$ and $\hat{\beta}$ are inversely related; the higher $\hat{\beta}$ is, the lower $\hat{\alpha}$ is and vice versa.

Thus, from (7.84) and (7.40) we can express (7.83) as:

$$\text{var } \hat{Y}_0 = \sigma^2 \frac{\Sigma X_i^2}{n\Sigma x_i^2} - \frac{2\sigma^2 \bar{X}X_0}{\Sigma x_i^2} + \frac{\sigma^2 X_0^2}{\Sigma x_i^2}$$

$$= \frac{\sigma^2}{n\Sigma x_i^2} (\Sigma X_i^2 - 2n\bar{X}X_0 + nX_0^2).$$

But

$$\Sigma X_i^2 = \Sigma x_i^2 + n\bar{X}^2$$

that is,

$$\text{var } \hat{Y}_0 = \sigma^2 \left(\frac{1}{n} + \frac{(X_0 - \bar{X})^2}{\Sigma x_i^2} \right) . \tag{7.85}$$

Note that var $\hat{Y}_0$ increases with $(X_0 - \bar{X})^2$; that is, the further is the value of X_0 from $\bar{X}$, the larger is the variance of our estimator.

Finally we note that $\hat{Y}_0$ is a linear combination of normally distributed variables ($\hat{\alpha}$ and $\hat{\beta}$), and so $\hat{Y}_0$ is normal. Collecting results we have that

$$\hat{Y}_0 \text{ is } N(\alpha + \beta X_0, \text{ var } \hat{Y}_0)$$

where var $\hat{Y}_0$ is given by (7.85). Thus

$$(\hat{Y}_0 - \alpha - \beta X_0)/\sigma \sqrt{\left(\frac{1}{n} + \frac{(X_0 - \bar{X})^2}{\Sigma x_i^2}\right)} \quad \text{is } N(0, 1) \tag{7.86}$$

and so, as usual, we can show that

$$(Y_0 - \alpha - \beta X_0)/s \sqrt{\left(\frac{1}{n} + \frac{(X_0 - \bar{X})^2}{\Sigma x_i^2}\right)} \quad \begin{array}{l} \text{has a } t\text{-distribution with} \\ (n-2) \text{ degrees of freedom.} \end{array} \tag{7.87}$$

Hence we can form confidence intervals for, and test hypotheses about the value of, $\alpha + \beta X_0$ given sample values for $\hat{Y}_0$, s, etc.

What we have been considering is using $\hat{Y}_0 = \hat{\alpha} + \hat{\beta} X_0$ as a predictor of $E(Y_0 \mid X_0) = \alpha + \beta X_0$; i.e., we have been predicting the *average* value of Y corresponding to a given value of X. Consider now the problem of predicting a *particular* value of Y corresponding to a given value of X. That is, we wish to predict

$$Y_0 = \alpha + \beta X_0 + u_0.$$

The thing we want to predict, Y_0, is itself a random variable. Again the obvious predictor is $\hat{Y}_0 = \hat{\alpha} + \hat{\beta} X_0$.

Consider

$$E(\hat{Y}_0 - Y_0) = E(\hat{\alpha} + \hat{\beta} X_0 - \alpha - \beta X_0 - u_0)$$
$$= 0 \quad \text{since } E\hat{\alpha} = \alpha \quad E\hat{\beta} = \beta \quad Eu_0 = 0.$$

So in the sense that $E(\hat{Y}_0 - Y_0) = 0$, $\hat{Y}_0$ is an unbiased estimator of Y_0 even though Y_0 is itself a random variable. Consider the variance of $(\hat{Y}_0 - Y_0)$:

$$E(\hat{Y}_0 - Y_0)^2 = E\{(\hat{\alpha} - \alpha) + (\hat{\beta} - \beta)X_0 - u_0\}^2$$
$$= E\{(\hat{\alpha} - \alpha) + (\hat{\beta} - \beta)X_0\}^2 + Eu_0^2 \tag{7.88}$$

since $E(\hat{\alpha} - \alpha)u_0 = 0$ and $E(\hat{\beta} - \beta)u_0 = 0$ since u_0 is independent of the $u_1, u_2, \ldots, u_n$ which determine $\hat{\alpha}$ and $\hat{\beta}$. Hence, from (7.88)

$$E(\hat{Y}_0 - Y_0)^2 = \sigma^2 \left\{1 + \frac{1}{n} + \frac{(X_0 - \bar{X})^2}{\Sigma x_i^2}\right\} \quad \text{using (7.85).} \tag{7.89}$$

Further, $\hat{Y}_0 - Y_0$ is normal, since $\hat{Y}_0$ and Y_0 both are. Hence, collecting results

$$\hat{Y}_0 - Y_0 \text{ is } N \left\{0, \sigma^2 \left(1 + \frac{1}{n} + \frac{(X_0 - \bar{X})^2}{\Sigma x_i^2}\right)\right\}$$

and so

$$(\hat{Y}_0 - Y_0)/\sigma \sqrt{\left(1 + \frac{1}{n} + \frac{(X_0 - \bar{X})^2}{\Sigma x_i^2}\right)} \quad \text{is } N(0, 1) \tag{7.90}$$

and in the usual fashion

$$(\hat{Y}_0 - Y_0)/s \sqrt{\left(1 + \frac{1}{n} + \frac{(X_0 - \bar{X})^2}{\Sigma x_i^2}\right)} \qquad \text{has a } t\text{-distribution with } (n - 2) \text{ degrees of freedom.}$$

$$\tag{7.91}$$

You should note carefully the difference between using $\hat{Y}_0$ as an estimator of $E(Y_0 \mid X_0) = \alpha + \beta X_0$, and using $\hat{Y}_0$ as an estimator of $Y_0 = \alpha + \beta X_0 + u_0$. You should also compare carefully (7.87) and (7.91) and rationalise intuitively the difference.

Let us illustrate these results with reference to our cross-section consumption—income relationship estimated in section 7.11. The estimated relationship is, from (7.73):

$$Y_i = 8.76 + 0.5763 X_i + e_i.$$

Consider first the problem of predicting the *average* consumption for all households with an income $X_0 = £50$ per week. Our estimated value $\hat{Y}_0$ is given by:

$$\hat{Y}_0 = 8.76 + (0.5763)(50) = 8.76 + 28.81 = 37.57.$$

Further we had

$$
\begin{aligned}
s &= \quad 1.8375 \\
\bar{X} &= \quad 40.236 \\
\Sigma x_i^2 &= 9111.5204.
\end{aligned}
$$

Now from (7.87) we have (since $n = 12$)

$$P\left[-2.228 \leqslant (\hat{Y}_0 - \alpha - \beta X_0)/s \sqrt{\left(\frac{1}{n} + \frac{(X_0 - \bar{X})^2}{\Sigma x_i^2}\right)} \leqslant 2.228\right] = 0.95$$

i.e.

$$P\left[\hat{Y}_0 - 2.228\, s \sqrt{\left(\frac{1}{n} + \frac{(\bar{X}_0 - \bar{X})^2}{\Sigma x_i^2}\right)} \leqslant \alpha + \beta X_0 \leqslant \right.$$

$$\left. \hat{Y}_0 + 2.228\, s \sqrt{\left(\frac{1}{n} + \frac{(X_0 - \bar{X})^2}{\Sigma x_i^2}\right)}\right] = 0.95$$

Hence our 95 per cent confidence interval for $\alpha + \beta X_0$ is:

$$\left\{37.57 - (2.228)(1.8375)\sqrt{\left(\frac{1}{12} + \frac{(50 - 40.236)^2}{9111.5204}\right)},\right.$$

$$\left.37.57 + (2.228)(1.8375)\sqrt{\left(\frac{1}{12} + \frac{(50 - 40.236)^2}{9111.5204}\right)}\right\}$$

that is

$$\{37.57 - 4.094\sqrt{0.093796},\ 37.57 + 4.094\sqrt{0.093796}\}$$

that is

(36.32, 38.82).

Thus we can be 95 per cent confident that the average consumption expenditure of all households with an income of £50 per week is contained within the interval from £36.32 per week to £38.82 per week. Similarly, from (7.91) a 95 per cent confidence interval for the consumption of a *particular* household with an income of £50 per week is given by:

$$\{37.57 - 4.094\sqrt{(1 + 0.093796)},\ 37.57 + 4.094\sqrt{(1 + 0.093796)}\}$$

that is

(33.29, 41.85).

Thus we can be 95 per cent confident that the weekly consumption expenditure of a particular household with an income of £50 per week is contained within the interval £33.29 per week to £41.85 per week.

7.13 Summary

This chapter has been concerned with the most important and most widely used application of statistics in economics, that of investigating relationships between variables. We considered the simplest of such hypothesised relationships, where one variable is hypothesised to be related linearly to one other variable. We began by considering the type of observations we would expect to find if the theoretical proposition was in fact correct. Making certain simplifying assumptions we then went on to consider how we might estimate and test the relationship given only a sample of all possible observations. We found that the least squares method of estimation provided us with best linear unbiased estimators of the population parameters (the coefficients of the true line). Given a further assumption about the distribution of the (true) deviations, we were then able to form confidence intervals for, and test hypotheses about, the population parameters, given sample evidence. At this stage we were assuming knowledge of the variance of the true deviations. If this was unknown, we found that a modification of the variance of the estimated deviations provided an unbiased estimator of the variance of the true deviations. Consideration of the distribution of this estimator enabled us to form confidence intervals for, and test hypotheses about, the population parameters, even when the variance of the true deviations was unknown.

We then applied these results to the investigation of the consumption function using two sets of data: cross-section household data obtained from the U.K. *Family Expenditure Survey*, and time-series aggregate data from the U.S. national accounts. Finally, we considered how we may use the estimated relationship to predict the value of the dependent variable for a given value of the independent variable.

The analysis of the whole of this chapter was based on certain simplifying assumptions about the form of the relationship and the distributions of the true deviations. It is crucial that you fully appreciate the meaning and significance of these assumptions. The next chapter considers how we may investigate their validity, and how the analysis of this chapter may need to be modified if it appears that these assumptions do not, in fact, hold.

7.14 Exercises

7.1. Refer to section 7.2. Do you think it likely that the variance of household consumption expenditure will be the same at all levels of income?

7.2. Prove result (7.19).

7.3. By consulting section 7.3, find which assumptions were used to prove the unbiasedness of the least squares estimators.

7.4. Which assumptions were used in section 7.3 in order to derive the variances of the least squares estimators?

7.5. Verify, for the illustrative example of section 7.4, the following results:

(a) $E(\hat{\beta} - \beta)\bar{u} = 0$

(b) $E\hat{\beta}\bar{u} \quad = 0$

(c) $Es^2\bar{u} \quad = 0$

(d) $Es^2(\hat{\beta} - \beta) = 0$

(e) $\text{cov}(\hat{\alpha}, \hat{\beta}) = -\sigma^2 \bar{X} / \sum_1^n x_i^2$.

7.6. Refer to section 7.6. Show that fitting methods (2) (3) and (4) always give unbiased estimators of α and β if assumptions (A1) and (A4) hold.

7.7. Again referring to section 7.6, investigate the properties of a fitting method of your choice.

7.8. Rationalise intuitively the ranking (by efficiency) of the four fitting methods of section 7.6.

7.9. Investigate whether the least squares estimators of α and β are consistent.

7.10. Is s^2 a consistent estimator of σ^2? Is d^2?

7.11. A regression of Y on X gives a coefficient of determination of zero. What then is the value of the regression coefficient $(\hat{\beta})$?

7.12. Find the least squares estimator of β for the equation

$$Y_i = \beta X_i + u_i \qquad\qquad (i = 1, 2, \ldots, n).$$

Find the variance of this estimator.

7.13. An investigator intends to derive the least squares estimator $\hat{\beta}$ of β in the equation:

$$Y_i = \beta X_i + u_i \qquad\qquad (i = 1, 2, \ldots, n).$$

However, unknown to him, the true relationship between Y and X is:

$$Y_i = \alpha + \beta X_i + v_i \qquad\qquad (i = 1, 2, \ldots, n).$$

Will his $\hat{\beta}$ be an unbiased estimator of β? Calculate the variance of $\hat{\beta}$ and discuss whether its expected squared error is smaller than that of the least squares estimator of β using the correct formulation.

7.14. An economist wishes to test his theory that Y and X are related by:

$$Y_i = \alpha + \beta X_i + u_i \qquad\qquad (i = 1, 2, \ldots, n)$$

where *a priori* reasoning leads him to believe that α is twice as large as β. How would he test this believed relationship between α and β?

   If his belief about the relationship between α and β is actually correct would a straightforward least squares fit of the equation be the best procedure?

7.15. Let b be the regression coefficient of X when Y is regressed against X, and let b' be the regression coefficient of Y when X is regressed against Y (using the same set of data). Show that

$$bb' = R^2.$$

What are the implications of this result? Do they disturb you?

7.16. The least squares estimators of α and β in the relation

$$Y_i = \alpha + \beta X_i + u_i \qquad\qquad (i = 1, 2, \ldots, n)$$

are the best linear unbiased estimators if assumptions (A1) to (A4) hold. For each of these four assumptions give an example of an economic relationship for which the assumption is unlikely to hold.

7.17. For the (hypothetical) consumption–income data of chapter 2, derive the least squares estimators of the coefficients of a (linear) consumption function. Are the coefficients significantly different from zero? Derive confidence intervals for the true coefficients.

7.18. Two economists are arguing about the income elasticity of demand for durables. One thinks that the income elasticity is unity, the other that it is less than unity. They agree to use the data below and the fact that the income elasticity of demand can be estimated as the coefficient of X in the regression of Y on X where X, Y are defined by:

X: the logarithm of family income in 1969
Y: the logarithm of family expenditure on durables in 1969.

They decide to test the null hypothesis that the elasticity is 1 against the alternative that it is less than 1 using a 10 per cent significance level. What is the outcome of the test?

Family i	Y_i	X_i
1	2.447	3.176
2	2.519	3.602
3	2.602	3.796
4	2.748	3.942
5	2.771	4.097
6	2.892	4.301

Source: Adapted from *1970 Survey of Consumer Finances*, University of Michigan, Table 5–4.

7.19. It is often suggested that low employment and a low rate of wage inflation cannot co-exist. Examine the U.K. evidence below (1957–66) and discuss whether it supports the above contention.

Year	Unemployment (%)	Change in wages (%)
1957	1.6	5.0
1958	2.2	3.2
1959	2.3	2.7
1960	1.7	2.1
1961	1.6	4.1
1962	2.1	2.7
1963	2.6	2.9
1964	1.7	4.6
1965	1.5	3.5
1966	1.6	4.4

Does the recent U.K. experience (1967–70) agree with your discussion above?

Year	Unemployment (%)	Change in wages (%)
1967	2.5	4.0
1968	2.5	7.7
1969	2.5	5.7
1970	2.7	9.5

7.20. The least squares estimators of α and β in the relation

$$Y_i = \alpha + \beta X_i + u_i \qquad\qquad (i = 1, 2, \ldots, n)$$

are the best linear unbiased estimators if assumptions (A1) to (A4) hold. Discuss, for each of the three following examples, whether these four assumptions are likely to be valid:

(a) Y is aggregate consumption (in real terms);
    X is aggregate income (in real terms);
    i is across time for one country;

(b) Y is household consumption;
    X is household income;
    i is across households at a point in time;

(c) Y is quantity purchased of peanuts;
    X is price of peanuts;
    i is across time for one market.

7.21. Suppose two variables X and Y are related by

$$Y_i = \alpha + \beta X_i + u_i$$

where assumptions (A1) to (A4) hold. Nine observations produced the following (X, Y) values:

$$(1, 2), (2, 3), (3, 4), (4, 5), (5, 4), (6, 6), (7, 6), (8, 7), (9, 8).$$

(a) Estimate the least squares regression of Y on X.
(b) Assuming σ^2 is known and equal to unity, test the hypotheses
    (i)   the coefficient of X is zero;
    (ii) the intercept is zero;
    (iii) the coefficient of X is 0.5;
    (iv) the intercept is 2.0.
(c) Still assuming σ^2 known and equal to unity, find
    (i)   a 95 per cent confidence interval for β;
    (ii) a 99 per cent confidence interval for α.
(d) Assuming now that σ^2 is *unknown* rework (b) and (c) above.
(e) Do the data support the hypothesis that $\sigma^2 = 1$?

7.22. Using *quarterly* data on aggregate consumption and aggregate income estimate the aggregate marginal propensity to consume, and find a 95 per cent confidence interval for it.

Comment *at length* on whether the assumptions made are likely to be valid.

7.23. Refer to section 7.11. Discuss *at length* the likely validity or otherwise of assumptions (A1) to (A5) in these two examples.

7.24. By finding suitable data from published government statistics test the following hypotheses:
(a) aggregate investment is not affected by the rate of interest;
(b) the elasticity of demand for labour is less than one;
(c) the demand for money is affected by the rate of interest;
(d) the share of profits in national income is not declining;
(e) the level of unemployment does not affect the rate of change of real wages.
Criticise *at length* the validity of the assumptions you have employed in carrying out this test. What part(s) of your derivation would have been affected by your criticisms?

8

Further Regression Analysis

8.1 Introduction

The last chapter was concerned exclusively with estimating the parameters of a simple linear relationship between two variables, Y and X, namely:

$$Y_i = \alpha + \beta X_i + u_i$$

where certain assumptions were made about the (true) deviation term u_i, and the values of the independent variable X_i. We were able to show that, of all linear unbiased estimators of α and β, the last squares estimators were the best in the sense of minimum variance. We derived the variances of these estimators, and found also an unbiased estimator of the (unknown) variance of the true deviations, thus enabling us to form confidence intervals for, and test hypotheses about, the parameters α and β. These results depended on some or all of our assumptions (A1) $-$ (A5) about the nature of the u_i and the X_i being true. Now, in practice, we cannot be sure *a priori* whether these assumptions are true, so we need to be able to *test* for their validity. This we investigate in sections 8.4 and 8.5. If the results of our tests show that one or more of our assumptions is not, in fact, valid, the results of the previous chapter may need to be modified. The necessary modifications will be discussed in section 8.6.

First, however, we will discuss how we can generalise the results of the previous chapter to the case where we wish to investigate the effect of more than one variable on the dependent variable Y. If economic theory leads us to believe that other variables besides X influence Y, and we can observe these variables, then we need to know how to proceed to test and estimate this theoretical effect. We will for the moment maintain the same assumptions about the nature of the (true) deviation term.

8.2 Multiple Regression.

Let us consider first the case where we wish to investigate the effect of two variables on a third. Suppose economic theory leads us to expect that the variable Y depends on variables X and Z. If we further assume that this relationship is linear, we can write it as:

$$Y_i = \alpha + \beta X_i + \gamma Z_i + u_i \quad i = 1, 2, \ldots, n. \tag{8.1}$$

Let us make the same assumptions as before about the (true) deviation term u_i:

$$Eu_i = 0 \qquad i = 1, 2, \ldots, n \qquad (A1)$$

$$Eu_i^2 = \sigma^2 \qquad i = 1, 2, \ldots, n \qquad (A2)$$

$$Eu_i u_j = 0 \qquad i \neq j; i, j = 1, 2, \ldots, n \qquad (A3)$$

We need to generalise assumption (A4) as follows:

$$\left. \begin{array}{ll} X_i \text{ is fixed} & i = 1, 2, \ldots, n \\ Z_i \text{ is fixed} & i = 1, 2, \ldots, n \end{array} \right\} \quad (A4)$$

Let us interpret these assumptions and the specified relationship (8.1) using our previous illustration. Suppose Y is household consumption expenditure, X is household income and Z is household 'size'.

Before we continue, let us look more closely at how we might define our variable Z. Obviously a household with two adults and two children is larger than a household with only two adults. Similarly, a household with two adults is larger than a household with only one adult. In order to measure quantitatively the sizes of households of different compositions we need to decide how much weight to give to persons of different ages. As an initial naive suggestion, incorporating the notion that less is spent on children than on adults, we could count each child as half an adult. Obviously more sophisticated measures could be used. Thus, on our definition of size, a household with two adults and two children has a size of three ('equivalent adults'), while a household with only two adults has a size of two ('equivalent adults'). Notice that on this definition, two households, one consisting of one adult and two children, and the other consisting of two adults, both have the same size. Of course, this definition, which makes all adults, irrespective of age, equivalent, and each child, irrespective of age, equivalent to half an adult, should itself be subject to test; this definition implies something about expenditure on people of differing ages. (We might also want to give different weights to different sexes).

Let us now return to our interpretation of the model (8.1). Suppose that we could find all households in the U.S. with an income $X_1 = 30$ (dollars per day). Suppose we then look at the sub-population of these households, all of which have the same size, say $Z_1 = 2$ ('equivalent adults'). For illustrative purposes, suppose there are 40 such households, that is all with the same income ($X_1 = 30$), and all with the same size ($Z_1 = 2$). Obviously not all these 40 households will have the same consumption expenditure. However we could calculate the mean expenditure for these 40 households; in keeping with our previous notation, call Y_1 the consumption expenditure of a household with income X_1 and size Z_1; so this mean expenditure can be denoted by:

$$E(Y_1 \mid X_1, Z_1).$$

We could also calculate the variance of expenditure for these 40 households; denote this by $\text{var}(Y_1 \mid X_1, Z_1)$. Some of these households will spend more than the average for their income and size group, some less. Let us denote by u_1 the difference between a particular household's expenditure and the mean expenditure for all households with the same income (X_1) and the same size (Z_1). That is, u_1 is defined by:

$$u_1 = Y_1 - E(Y_1 \mid X_1, Z_1).$$

Thus the average value of the (true) deviation u_1 as denoted by $E(u_1 \mid X_1, Z_1)$, or simply Eu_1 as no ambiguity results, is:

$$Eu_1 = E(u_1 \mid X_1, Z_1) = E(Y_1 \mid X_1, Z_1) - E(Y_1 \mid X_1, Z_1) = 0.$$

Thus the mean value of u_1 is zero (virtually by definition). Also the variance of u_1 is obviously $\text{var}\, u_1 = \text{var}(Y_1 \mid X_1, Z_1)$.

Suppose we now look at a second sub-population of those households with an income of 30, namely those with size $Z_2 = 3$ ('equivalent adults'). Suppose there are 30 such households. To distinguish this second sub-population from the first, we use the subscript 2 on all variables, and call the income of these households X_2. X_2 and X_1 are in fact the same (that is $X_1 = X_2 = 30$) but it will create less confusion if we rename the income for this second sub-population. Thus our second sub-population consists of all households in the U.S. with income $X_2 = 30$ (dollars per day) and size $Z_2 = 3$ ('equivalent adults'). Once again, not all of these 30 households will have the same expenditure; some will spend more than the average for their income and size group, some less. Denote the mean expenditure of all these 30 households by, of course,

$$E(Y_2 \mid X_2, Z_2).$$

Denote the deviation of a particular household from the mean by:

$$u_2 = Y_2 - E(Y_2 \mid X_2, Z_2).$$

Then, as before,

$$Eu_2 = 0$$

and

$$\text{var}\, u_2 = \text{var}(Y_2 \mid X_2, Z_2).$$

Looking at the mean expenditure in each of these two sub-populations, we would expect that:

$$E(Y_2 \mid X_2, Z_2) > E(Y_1 \mid X_1, Z_1)$$

since the only difference between these two groups is that the second group has a larger household size; that is, we would expect that, on average, expenditure would increase with size. Thus $E(Y \mid X, Z)$ is likely to increase with Z.

Finally, let us see what would happen if we took a third sub-population, namely those with income $X_3 = 20$ (dollars per day) and with size $Z_3 = 3$ ('equivalent adults'). We would expect that:

$$E(Y_3 \mid X_3, Z_3) < E(Y_2 \mid X_2, Z_2)$$

since the only difference between the second and the third sub-populations is that income is lower in the latter, and so mean expenditure is likely to be lower in the third group than the second. In general, we would expect $E(Y \mid X, Z)$ to increase with X.

We see that $E(Y \mid X, Z)$ is expected to increase as X and Z increase. If we assume that the relationship is linear, we can write it as:

$$E(Y_i \mid X_i, Z_i) = \alpha + \beta X_i + \gamma Z_i \quad (\beta > 0, \gamma > 0). \tag{8.2}$$

Notice that we are not only postulating that the effect of X and Z on $E(Y \mid X, Z)$ is linear, but also that X and Z influence $E(Y \mid X, Z)$ independently; that is, an increase of Z by one (equivalent adult) leads to an increase in $E(Y \mid X, Z)$ of γ irrespective of the size of X. Similarly it implies that an increase in income (X) by one (dollar) leads to an increase in $E(Y \mid X, Z)$ of β irrespective of the household size (Z). If we feel that this formation is unrealistic, we must search for a relationship between $E(Y \mid X, Z)$ and X and Z that embodies our *a priori* ideas. This is a problem we will turn to in later sections. For the moment we will assume that the relationship is linear.

Let us now combine (8.2) with our general definition of u_i:

$$u_i = Y_i - E(Y_i \mid X_i, Z_i). \tag{8.3}$$

We get:

$$Y_i = \alpha + \beta X_i + \gamma Z_i + u_i. \tag{8.4}$$

This is, of course, just our original formulation (8.1).

We can now see what is meant by our assumptions (A1) to (A4). Assumption (A1), that each u_i has zero mean, follows immediately from the definition of u_i given in (8.3). Thus Eu_i will be non-zero only if $E(Y_i \mid X_i, Z_i) \neq \alpha + \beta X_i + \gamma Z_i$, that is if our linearity assumption is untrue. Assumption (A2), that the variance of each u_i is the same, is equivalent to the assumption that $\mathrm{var}(Y_i \mid X_i, Z_i)$ is the same for all i; that is, the variance of expenditure is the same for each of our sub-populations. Thus, if we look, for example, at the variance of household consumption expenditure for the sub-population consisting of all households with an income $30 a day and size 2 equivalent adults, this variance should be the same as the variance of household consumption expenditure for the sub-population consisting of all households with income $20 a day and size 3 equivalent adults. Economic theory sheds little light as to whether this assumption is likely to be correct; and, as before, we will maintain this assumption for the time being, and investigate how to test for its validity in later sections.

Assumptions (A3) and (A4) are more concerned with the nature of our sampling procedure than with the nature of the population. Let us therefore turn from the above discussion of the population distribution to discuss the sampling procedure. In practice, as argued several times before, we may find it too costly, impossible, or unnecessary to investigate the whole population, and we thus only investigate a sample of households. Suppose we were to obtain a sample of n households. Denote the actual consumption expenditure, the income and the size of the ith household in the sample by Y_i, X_i and Z_i respectively. Suppose $X_i = 30$ and $Z_i = 3$. Then this household can be considered as a sample of one from all households with income 30 and size 3. Denote as usual the difference between this household's expenditure (Y_i) and the mean for all households with income 30 and size 3 by u_i. In practice, as we have only one such household we do not know whether its particular u_i is positive or negative, nor do we know anything about the magnitude of its u_i. All we do know is that it is one observation on the random variable u_i which has mean zero and variance σ^2.

Suppose now we consider the following conceptual experiment. Suppose we could pick a second example of n households, choosing them in such a way that the first household in the second sample had the same income and the same size as the first household in the first sample, the second household in the second sample had the same income and the same size as the second household in the first sample, ..., the nth household in the second sample had the same income and the same size as the nth household in the first sample. Writing this more succinctly, we can say that the second sample is such that, for each $i = 1, 2, \ldots, n$, the ith household in the second sample has the same income and the same size as the ith household in the first sample.

Thus the u_i of the ith household in the first sample and the u_i of the ith household in the second sample would constitute two realisations on the random variable u_i which has zero mean and variance σ^2.

Continuing this conceptual experiment, we suppose that we collect many samples of size n, each sample being such that the ith household has the same income and the same size as the ith household in the first sample. Each sample would constitute one realisation on u_i, for each i, and the mean value of u_i across all possible samples would be zero, and the variance of u_i across all samples would be σ^2.

This conceptual experiment is what we mean by assumption (A4); that is, we could conceptually take many samples with the same (X_1, Z_1), $(X_2, Z_2), \ldots, (X_n, Z_n)$. The only things that vary over these conceptual samples are $Y_1, Y_2, \ldots, Y_n$ and, of course, $u_1, u_2, \ldots, u_n$.

Assumption (A3) should now be clear. Our sampling procedure should be such that the value u_i takes in our sample should be unaffected by the value that u_j takes in our sample, and vice versa, for all $i, j = 1, 2, \ldots, n$ $(i \neq j)$. We gave an example in our simple regression analysis of where this assumption may break down. This illustration can be extended to cover this multiple regression case, and you should refer back to see how it might be done.

We now return to our problems of estimation. Our formulation and assumptions are as follows:

$$Y_i = \alpha + \beta X_i + \gamma Z_i + u_i \qquad\qquad i = 1, 2, \ldots, n \qquad\qquad (8.4)$$

$$Eu_i = 0 \qquad\qquad\qquad i = 1, 2, \ldots, n \quad (A1)$$

$$Eu_i{}^2 = \sigma^2 \qquad\qquad\qquad i = 1, 2, \ldots, n \quad (A2)$$

$$E\, u_i u_j = 0 \qquad\qquad i \neq j; i, j = 1, 2, \ldots, n \quad (A3)$$

$$X_i \text{ is fixed} \qquad\qquad i = 1, 2, \ldots, n \;\Big\}$$
$$Z_i \text{ is fixed} \qquad\qquad i = 1, 2, \ldots, n \;\Big\} \;(A4)$$

We have a sample of n observations $(X_1, Y_1, Z_1), (X_2, Y_2, Z_2), \ldots,$ (X_n, Y_n, Z_n) and wish to estimate the parameters α, β, γ and σ^2 of the true relationship (8.4). Let us denote the estimated (or fitted) relationship between Y and X and Z by:

$$Y_i = \hat{\alpha} + \hat{\beta} X_i + \hat{\gamma} Z_i + e_i \qquad\qquad\qquad (8.5)$$

where $\hat{\alpha}, \hat{\beta}$ and $\hat{\gamma}$ are estimators of α, β and γ respectively, and $e_1, e_2, \ldots,$ e_n are the estimated deviations.* The fitted relationship:

$$\hat{Y}_i = \hat{\alpha} + \hat{\beta} X_i + \hat{\gamma} Z_i \qquad\qquad\qquad (8.6)$$

is, of course, a plane (in three dimensions) and the e_i, given by

$$e_i = Y_i - \hat{Y}_i$$

are the distances of the observations from the plane measured in the Y direction.

*Remember that 'estimated deviation' is an abbreviation of 'deviation from the estimated relationship'.

We now have the same problem as in the simple case considered in the last chapter; that is, how do we use the observations to estimate, in some optimal sense, the parameters of our relationship? How do we use the observations (X_1, Y_1, Z_1), (X_2, Y_2, Z_2), . . . , (X_n, Y_n, Z_n) to provide estimators $\hat{\alpha}$, $\hat{\beta}$ and $\hat{\gamma}$, which are in some sense optimal, of α, β and γ respectively? In geometrical terms we want to fit a plane to our observations in a three-dimensional diagram so that the fitted plane is an optimal estimator of the true plane.

We do not intend to describe the procedure in detail, as it merely involves an extension of the methods of the last chapter. We will therefore give an outline of the procedure, and point out the analogies with our previous discussion.

Our method for estimating the parameters of our model is the same as before, namely to choose $\hat{\alpha}$, $\hat{\beta}$ and $\hat{\gamma}$ to minimise the sum of squared deviations of the observations from the fitted relationship. It can be shown, by an exactly analagous procedure to that used in chapter 7, that such a method provides us with unbiased estimators of α, β and γ respectively, and that of all (linear) unbiased estimators these estimators are the best, in the sense of minimum variance.

The sum of squared (estimated) deviations is, as before, $\Sigma_1^n e_i^2$, and we choose $\hat{\alpha}$, $\hat{\beta}$ and $\hat{\gamma}$ to minimise this. Now

$$e_i = Y_i - \hat{\alpha} - \hat{\beta} X_i - \hat{\gamma} Z_i \quad \text{from (8.5)}$$

and so

$$\Sigma_1^n e_i^2 = \Sigma_1^n (Y_i - \hat{\alpha} - \hat{\beta} X_i - \hat{\gamma} Z_i)^2 . \tag{8.7}$$

Denote $\Sigma_1^n e_i^2$ by S; then to minimise S with respect to $\hat{\alpha}$, $\hat{\beta}$ and $\hat{\gamma}$ we set the partial derivatives of S with respect to $\hat{\alpha}$, $\hat{\beta}$ and $\hat{\gamma}$ equal to zero, and solve.

Thus, from (8.7)

$$\frac{\partial S}{\partial \hat{\alpha}} = -2 \sum_1^n (Y_i - \hat{\alpha} - \hat{\beta} X_i - \hat{\gamma} Z_i)$$

which gives (setting $\partial S / \partial \hat{\alpha} = 0$):

$$\sum_1^n Y_i = n\hat{\alpha} + \hat{\beta} \sum_1^n X_i + \hat{\gamma} \sum_1^n Z_i$$

or, dividing through by n, and defining $\bar{X}$, $\bar{Y}$ and $\bar{Z}$ as usual by:

$$\bar{Y} = \frac{1}{n} \sum_1^n Y_i \qquad \bar{X} = \frac{1}{n} \sum_1^n X_i \qquad \bar{Z} = \frac{1}{n} \sum_1^n Z_i \quad \text{gives}$$

$$\bar{Y} = \hat{\alpha} + \hat{\beta} \bar{X} + \hat{\gamma} \bar{Z} \tag{8.8}$$

(compare this with (7.11)).

Further, from (8.7):

$$\frac{\partial S}{\partial \hat{\beta}} = -2\sum_{1}^{n}(Y_i - \hat{\alpha} - \hat{\beta}X_i - \hat{\gamma}Z_i)X_i$$

and

$$\frac{\partial S}{\partial \hat{\gamma}} = -2\sum_{1}^{n}(Y_i - \hat{\alpha} - \hat{\beta}X_i - \hat{\gamma}Z_i)Z_i.$$

Thus, setting $\partial S/\partial \hat{\beta}$ and $\partial S/\partial \hat{\gamma}$ equal to zero and rearranging gives:

$$\sum_{1}^{n}Y_iX_i = \hat{\alpha}\sum_{1}^{n}X_i + \hat{\beta}\sum_{1}^{n}X_i^2 + \hat{\gamma}\sum_{1}^{n}X_iZ_i \tag{8.9}$$

$$\sum_{1}^{n}Y_iZ_i = \hat{\alpha}\sum_{1}^{n}Z_i + \hat{\beta}\sum_{1}^{n}X_iZ_i + \hat{\gamma}\sum_{1}^{n}Z_i^2 \tag{8.10}$$

Equations (8.8), (8.9) and (8.10) give us three linear equations to solve for the three unknowns $\hat{\alpha}$, $\hat{\beta}$ and $\hat{\gamma}$ which minimise S. To solve these we proceed as follows. Multiplying (8.8) by $n\bar{X}$ ($=\sum_{1}^{n}X_i$) and subtracting from (8.9) gives:

$$\sum_{1}^{n}Y_iX_i - n\bar{Y}\bar{X} = \hat{\beta}\left(\sum_{1}^{n}X_i^2 - n\bar{X}^2\right) + \hat{\gamma}\left(\sum_{1}^{n}X_iZ_i - n\bar{X}\bar{Z}\right)$$

and defining x_i, y_i and z_i as usual by

$$x_i = X_i - \bar{X} \qquad y_i = Y_i - \bar{Y} \qquad z_i = Z_i - \bar{Z}$$

we get

$$\sum_{1}^{n}y_ix_i = \hat{\beta}\sum_{1}^{n}x_i^2 + \hat{\gamma}\sum_{1}^{n}x_iz_i. \tag{8.11}$$

Similarly, multiplying (8.8) by $n\bar{Z}$ ($= \sum_{1}^{n}Z_i$) and subtracting from (8.10) gives

$$\sum_{1}^{n}y_iz_i = \hat{\beta}\sum_{1}^{n}x_iz_i + \hat{\gamma}\sum_{1}^{n}z_i^2. \tag{8.12}$$

To solve these for $\hat{\beta}$, we eliminate $\hat{\gamma}$ by multiplying (8.12) by $\sum_{1}^{n}x_iz_i$ and subtracting it from (8.11) multiplied by $\sum_{1}^{n}z_i^2$. Thus:

$$\left(\sum_{1}^{n}y_ix_i\right)\left(\sum_{1}^{n}z_i^2\right) - \left(\sum_{1}^{n}y_iz_i\right)\left(\sum_{1}^{n}x_iz_i\right) = \hat{\beta}\left\{\left(\sum_{1}^{n}x_i^2\right)\left(\sum_{1}^{n}z_i^2\right) - \left(\sum_{1}^{n}x_iz_i\right)^2\right\} \tag{8.13}$$

that is,

$$\hat{\beta} = \frac{\left(\sum_{1}^{n}y_ix_i\right)\left(\sum_{1}^{n}z_i^2\right) - \left(\sum_{1}^{n}y_iz_i\right)\left(\sum_{1}^{n}x_iz_i\right)}{\left(\sum_{1}^{n}x_i^2\right)\left(\sum_{1}^{n}z_i^2\right) - \left(\sum_{1}^{n}x_iz_i\right)^2} \tag{8.14}$$

(assuming that $(\Sigma_1^n x_i^2)(\Sigma_1^n z_i^2) \neq (\Sigma_1^n x_i z_i)^2$).

Similarly,

$$\hat{\gamma} = \frac{\left(\sum_1^n y_i z_i\right)\left(\sum_1^n x_i^2\right) - \left(\sum_1^n y_i x_i\right)\left(\sum_1^n x_i z_i\right)}{\left(\sum_1^n x_i^2\right)\left(\sum_1^n z_i^2\right) - \left(\sum_1^n x_i z_i\right)^2} \tag{8.15}$$

(again assuming that $(\Sigma_1^n x_i^2)(\Sigma_1^n z_i^2) \neq (\Sigma_1^n x_i z_i)^2$).

Finally, from (8.8) $\hat{\alpha}$ is given by:

$$\hat{\alpha} = \bar{Y} - \hat{\beta}\bar{X} - \hat{\gamma}\bar{Z}. \tag{8.16}$$

The first point to notice is that $\hat{\alpha}$, $\hat{\beta}$ and $\hat{\gamma}$ are all linear functions of Y_1, $Y_2, \ldots, Y_n$; they are obviously not linear in the X_i and the Z_i.

The following results can be proved, by methods analagous to those used in chapter 7. The actual proofs are omitted as they are algebraically tedious (unless matrix algebra is used), and as they add nothing to our statistical understanding.

$$E\hat{\alpha} = \alpha \quad E\hat{\beta} = \beta \quad E\hat{\gamma} = \gamma \tag{8.17}$$

that is, these least squares estimators are unbiased.

$$\text{var } \hat{\alpha} = \frac{\sigma^2\left\{\left(\sum_1^n X_i^2\right)\left(\sum_1^n Z_i^2\right) - \left(\sum_1^n X_i Z_i\right)^2\right\}}{n\left\{\left(\sum_1^n x_i^2\right)\left(\sum_1^n z_i^2\right) - \left(\sum_1^n x_i z_i\right)^2\right\}}$$

$$\text{var } \hat{\beta} = \frac{\sigma^2 \sum_1^n z_i^2}{\left\{\left(\sum_1^n x_i^2\right)\left(\sum_1^n z_i^2\right) - \left(\sum_1^n x_i z_i\right)^2\right\}} \left.\tag{8.18}\right\}$$

$$\text{var } \hat{\gamma} = \frac{\sigma^2 \sum_1^n x_i^2}{\left\{\left(\sum_1^n x_i^2\right)\left(\sum_1^n z_i^2\right) - \left(\sum_1^n x_i z_i\right)^2\right\}}$$

$$\text{cov } (\hat{\alpha}, \hat{\beta}) = \frac{\sigma^2\left\{\left(\sum_1^n Z_i\right)\left(\sum_1^n X_i Z_i\right) - \left(\sum_1^n X_i\right)\left(\sum_1^n Z_i^2\right)\right\}}{n\left\{\left(\sum_1^n x_i^2\right)\left(\sum_1^n z_i^2\right) - \left(\sum_1^n x_i z_i\right)^2\right\}}$$

$$\text{cov } (\hat{\alpha}, \hat{\gamma}) = \frac{\sigma^2\left\{\left(\sum_1^n X_i\right)\left(\sum_1^n X_i Z_i\right) - \left(\sum_1^n Z_i\right)\left(\sum_1^n X_i^2\right)\right\}}{n\left\{\left(\sum_1^n x_i^2\right)\left(\sum_1^n z_i^2\right) - \left(\sum_1^n x_i z_i\right)^2\right\}} \left.\tag{8.19}\right\}$$

$$\text{cov } (\hat{\beta}, \hat{\gamma}) = \frac{-\sigma^2 \sum\limits_{1}^{n} x_i z_i}{\left\{ \left(\sum\limits_{1}^{n} x_i{}^2 \right) \left(\sum\limits_{1}^{n} z_i{}^2 \right) - \left(\sum\limits_{1}^{n} x_i z_i \right)^2 \right\}} .$$

Results (8.17) depend on assumptions (A1) and (A4) holding, while (8.18) and (8.19) require, in addition, that assumptions (A2) and (A3) hold. If all assumptions hold, we can again prove that, of all linear unbiased estimators of α, β and γ, the least squares estimators $\hat{\alpha}$, $\hat{\beta}$ and $\hat{\gamma}$, given by (8.14), (8.15) and (8.16), are the best in the sense of minimum variance.

We can show further that an unbiased estimator of σ^2, the variance of the true deviations, is given by:

$$s^2 = \frac{1}{n-3} \sum\limits_{1}^{n} e_i{}^2 \tag{8.20}$$

where the e_i are the estimated deviations.

If we now make the further assumption:

$$u_i \text{ is normal} \quad i = 1, 2, \dots, n \tag{A5}$$

then we can show that $\hat{\alpha}$, $\hat{\beta}$ and $\hat{\gamma}$ are also normal.

Combining this with our previous results (8.17) and (8.18) gives:

$$\frac{\hat{\alpha} - \alpha}{\sigma \sqrt{\left[\dfrac{\left\{ \left(\sum\limits_{1}^{n} X_i{}^2 \right) \left(\sum\limits_{1}^{n} Z_i{}^2 \right) - \left(\sum\limits_{1}^{n} X_i Z_i \right)^2 \right\}}{n \left\{ \left(\sum\limits_{1}^{n} x_i{}^2 \right) \left(\sum\limits_{1}^{n} z_i{}^2 \right) - \left(\sum\limits_{1}^{n} x_i z_i \right)^2 \right\}} \right]}} \quad \text{is } N(0, 1)$$

$$\frac{\hat{\beta} - \beta}{\sigma \sqrt{\left[\dfrac{\sum\limits_{1}^{n} z_i{}^2}{\left\{ \left(\sum\limits_{1}^{n} x_i{}^2 \right) \left(\sum\limits_{1}^{n} z_i{}^2 \right) - \left(\sum\limits_{1}^{n} x_i z_i \right)^2 \right\}} \right]}} \quad \text{is } N(0, 1) \tag{8.21}$$

$$\frac{\hat{\gamma} - \gamma}{\sigma \sqrt{\left[\dfrac{\sum\limits_{1}^{n} x_i{}^2}{\left\{ \left(\sum\limits_{1}^{n} x_i{}^2 \right) \left(\sum\limits_{1}^{n} z_i{}^2 \right) - \left(\sum\limits_{1}^{n} x_i z_i \right)^2 \right\}} \right]}} \quad \text{is } N(0, 1)$$

Further, we can show that, if (A1) to (A5) are true, then $(n-3)s^2/\sigma^2$ has a chi-square distribution with $(n-3)$ degrees of freedom. (8.22)

At this point, it may be useful to point out the nature of the unbiased estimator of σ^2, and the degrees of freedom associated with the distribution of this estimator.

When, in chapter 5, we derived an unbiased estimator of the population variance σ^2, we found it to be $s^2 = \Sigma_1^n x_i^2 /(n-1)$, and we found that $(n-1)s^2 /\sigma^2$ had a chi-square distribution with $(n-1)$ degrees of freedom.

When, in chapter 7, we derived an unbiased estimator of the variance of the true deviations σ^2, we found it to be $s^2 = \Sigma_1^n e_i^2 /(n-2)$, and we found that $(n-2)s^2 /\sigma^2$ had a chi-square distribution with $(n-2)$ degrees of freedom. In this chapter, when we introduced a further explanatory variable, we found an unbiased estimator of the variance of the true deviations σ^2 to be $s^2 = \Sigma_1^n e_i^2 /(n-3)$, and we found that $(n-3)s^2 /\sigma^2$ had a chi-square distribution with $(n-3)$ degrees of freedom. Intuitively we can see what is happening: in chapter 5, we based our estimator of σ^2 on the sum of squares of x_i from 1 to n. However the n values of x_i are not independent since $\Sigma_1^n x_i = 0$, and thus given $(n-1)$ of the x_i we can calculate the nth. There is, therefore, one constraint imposed on the sum of squares, and so it has only $(n-1)$ degrees of freedom. In chapter 7 we based our estimator of σ^2 on the sum of squares of e_i from 1 to n. However the n values of the e_i are not independent, since $\Sigma_1^n e_i = 0$ (from (7.54)) and $\Sigma_1^n e_i X_i = 0$ (from (7.66)); thus given $(n-2)$ of the e_i we can calculate the remaining 2. This imposes two constraints on the sum of squares, and so it has only $(n-2)$ degrees of freedom.* In this chapter, the e_i were constrained to satisfy three conditions, $\Sigma_1^n e_i = 0$, $\Sigma_1^n e_i X_i = 0$, and $\Sigma_1^n e_i Z_i = 0$ (see equations (8.8), (8.9) and (8.10)), which imposes three constraints on the sum of squares, and so it has only $(n-3)$ degrees of freedom. We represent this schematically in table 8.1.

Thus, if we generalised further, and considered a model with $(k-1)$ variables explaining Y, there would be k constraints on the sum of squares (one for each of the k parameters to be estimated), and our estimate of the residual variance would be

$$s^2 = \frac{1}{n-k} \sum_1^n e_i^2$$

and we would find that $(n-k)s^2 /\sigma^2$ had a chi-square distribution with $(n-k)$ degrees of freedom.

---

*Note that the two conditions $\Sigma e_i = 0$ and $\Sigma e_i X_i = 0$ define the normal equations (7.11) and (7.12) which give the least squares estimators.

Table 8.1   Distributions of estimators of variances

Model	Number of parameters to be estimated	Names of parameters to be estimated (other than variance σ^2)	Estimators of parameters	Unbiased estimator of variance	Variable with chi-square distribution	Number of degrees of freedom of chi-square
$(i = 1, 2, \ldots, n)$						
X_i is $N(\mu, \sigma^2)$ (chapter 5)	1	μ	$\bar{X}$	$s^2 = \dfrac{\Sigma x_i^2}{n-1}$	$(n-1)\dfrac{s^2}{\sigma^2}$	$(n-1)$
$E(Y_i \mid X_i)$ is $N(\alpha + \beta X_i, \sigma^2)$ (chapter 7)	2	α β	$\hat{\alpha}$ $\hat{\beta}$	$s^2 = \dfrac{\Sigma e_i^2}{n-2}$	$(n-2)\dfrac{s^2}{\sigma^2}$	$(n-2)$
$E(Y_i \mid X_i, Z_i)$ is $N(\alpha + \beta X_i + \gamma Z_i, \sigma^2)$ (chapter 8)	3	α β γ	$\hat{\alpha}$ $\hat{\beta}$ $\hat{\gamma}$	$s^2 = \dfrac{\Sigma e_i^2}{n-3}$	$(n-3)\dfrac{s^2}{\sigma^2}$	$(n-3)$

Let us return from this brief digression to combine results (8.21) and (8.22) in the usual way to give:

$$\frac{\hat{\alpha} - \alpha}{s \sqrt{\dfrac{\left[\left(\sum_1^n X_i^2\right)\left(\sum_1^n Z_i^2\right) - \left(\sum_1^n X_i Z_i\right)^2\right]}{n\left[\left(\sum_1^n x_i^2\right)\left(\sum_1^n z_i^2\right) - \left(\sum_1^n x_i z_i\right)^2\right]}}}$$

has a t-distribution with $(n-3)$ degrees of freedom

$$\frac{\hat{\beta} - \beta}{s \sqrt{\dfrac{\sum_1^n z_i^2}{\left[\left(\sum_1^n x_i^2\right)\left(\sum_1^n z_i^2\right) - \left(\sum_1^n x_i z_i\right)^2\right]}}}$$

has a t-distribution with $(n-3)$ degrees of freedom   (8.23)

$$\frac{\hat{\gamma} - \gamma}{s \sqrt{\dfrac{\sum_1^n x_i^2}{\left[\left(\sum_1^n x_i^2\right)\left(\sum_1^n z_i^2\right) - \left(\sum_1^n x_i z_i\right)^2\right]}}}$$

has a t-distribution with $(n-3)$ degrees of freedom.

With the information contained in (8.23), we are now in a position to construct confidence intervals for, and test hypotheses about, the parameters α, β and γ.

Note that the methods used in this multiple regression are a straightforward generalisation of the methods used in the simple regression of the previous chapter. The algebra has become more complicated, but the statistical ideas and concepts are unchanged. Obviously we could generalise these methods to cover cases where three or more variables are used to explain the behaviour of Y; the algebra used would become even more complicated, but we would gain little in statistical understanding. A considerable simplification of the algebra is achieved by using matrix algebra, but this book does not propose to give a full treatment of multiple regression. For a complete treatment the reader is referred to any econometrics text, for example Johnston (13) chapter 5. In practice, however, there is little need for the user of multiple regression to be able to follow the algebra involved in the derivation of the necessary formulae; and in any case, most institutions interested in regression techniques will have a computer program to calculate the necessary results. What is much more important is that the user should clearly understand the assumptions underlying the results, be able to appreciate the importance of testing for the validity of the assumptions, and be able to know how his results need to be modified if the assumptions do not appear to hold. These important problems will be investigated in sections 8.4 and 8.5, and are common to simple regression as well as multiple regression. Two points peculiar to multiple regression should, however, be mentioned at this stage.

The first concerns the relationship between the two explanatory variables X and Z. Suppose that, in our observations, X and Z are perfectly related, that is:

$$X_i = a + bZ_i \quad i = 1, 2, \ldots, n \tag{8.24}$$

where a and b are known constants.
If (8.24) holds then

$$\bar{X} = a + b\bar{Z} \tag{8.25}$$

and so, subtracting (8.25) from (8.24):

$$x_i = bz_i \quad i = 1, 2, \ldots, n \tag{8.26}$$

hence

$$\sum_1^n x_i^2 = b^2 \sum_1^n z_i^2$$

$$\sum_1^n y_i x_i = b \sum_1^n y_i z_i$$

$$\sum_1^n x_i z_i = b \sum_1^n z_i^2$$

hence

$$\left(\sum_1^n x_i^2\right)\left(\sum_1^n z_i^2\right) - \left(\sum_1^n x_i z_i\right)^2 = b^2\left(\sum_1^n z_i^2\right)^2 - b^2\left(\sum_1^n z_i^2\right)^2 = 0$$

and

$$\left(\sum_1^n y_i x_i\right)\left(\sum_1^n z_i^2\right) - \left(\sum_1^n y_i z_i\right)\left(\sum_1^n x_i z_i\right)$$

$$= b\left(\sum_1^n y_i z_i\right)\left(\sum_1^n z_i^2\right) - b\left(\sum_1^n y_i z_i\right)\left(\sum_1^n z_i^2\right) = 0$$

and thus we are unable to solve (8.13) for $\hat\beta$.

Nor can we solve (8.11) and (8.12) for either $\hat\beta$ or $\hat\gamma$. (Notice that if (8.26) holds then (8.11) and (8.12) are the same.)

Hence, if X and Z are perfectly related we cannot find $\hat\beta$ and $\hat\gamma$. What we could do is to substitute (8.24) into (8.4) to give:

$$Y_i = \alpha + \beta(a + bZ_i) + \gamma Z_i + u_i \quad i = 1, 2, \ldots, n$$

that is:

$$Y_i = (\alpha + \beta a) + (\beta b + \gamma)Z_i + u_i \quad i = 1, 2, \ldots, n \tag{8.27}$$

and regress Y on Z to give us estimates of $(\alpha + \beta a)$ and $(\beta b + \gamma)$. However, even though we know a and b, we cannot estimate α, β and γ from estimates of $(\alpha + \beta a)$ and $(\beta b + \gamma)$.

This problem, when X and Z are perfectly related, is an extreme case of *multicollinearity*. We are unable to estimate β and γ; that is, we are unable to distinguish the separate effects of X and Z on Y. This should be intuitively obvious.

A less extreme case of multicollinearity, but one which may still cause problems, is when X and Z are highly, though not perfectly, related. We can investigate the consequences of this by expressing our results in terms of r^2, the square of the correlation coefficient between X and Z, as given by:

$$r^2 = \frac{\left(\sum_1^n x_i z_i\right)^2}{\left(\sum_1^n x_i^2\right)\left(\sum_1^n z_i^2\right)}.$$

We know that if X and Z are perfectly related, then $r^2 = 1$. If there is no relationship then $r^2 = 0$. Otherwise $0 < r^2 < 1$. Thus $(\Sigma_1^n x_i^2)(\Sigma_1^n z_i^2) \geq (\Sigma_1^n x_i z_i)^2$ with the equality holding only if $r^2 = 1$, that is if X and Z are exactly related. Hence if $r^2 < 1$ we can solve (8.11) and (8.12) for $\hat\beta$ and $\hat\gamma$. Only in the extreme case of $r^2 = 1$ are we unable to solve for $\hat\beta$ and $\hat\gamma$.

However from (8.18) we see that:

$$\text{var } \hat{\beta} = \frac{\sigma^2}{\sum\limits_{1}^{n} x_i^2 (1 - r^2)} \qquad \text{var } \hat{\gamma} = \frac{\sigma^2}{\sum\limits_{1}^{n} z_i^2 (1 - r^2)}. \qquad (8.28)$$

Thus the nearer r^2 is to unity the larger are the variances of our estimators $\hat{\beta}$ and $\hat{\gamma}$. Hence, if r^2 is near to unity, the precision of our estimators will be low, resulting in wide confidence intervals, and a strong tendency to accept the null hypothesis that the true coefficients of X and Z are zero. If r^2 is zero the variance of $\hat{\beta}$ is the same as under a simple regression. Thus the greater the multicollinearity the less precise our results. The problem of multicollinearity often arises in the analysis of economic data, particularly with time series data as many economic variables move together over time. Consider, for example, the problem of using time-series data to estimate an aggregate demand curve for some commodity, where we wish to use aggregate income and price as explanatory variables. In many cases it will be found that price and income are correlated, as both have steadily increased over the observation period. Similarly, capital and labour, as explanatory variables in a production function, may well be correlated as they both increase through time. However, the problem is by no means necessarily confined to time-series analysis; consider the possible correlation of household income and the age of the head of the household as explanatory variables in a cross-section household consumption function.

A second problem associated particularly with multiple regression concerns our measure of goodness of fit of the fitted relationship to the observations. In our simple regression case we defined a goodness of fit indicator R^2 by:

$$R^2 = 1 - \frac{\sum\limits_{1}^{n} e_i^2}{\sum\limits_{1}^{n} y_i^2}$$

that is, R^2 is the proportion of the total variance of Y 'explained' by the fitted relationship. In our multiple regression analysis we can define an exactly similar measure and interpret it in the same way. In the previous chapter we discussed in some detail the use of R^2 as a measure of goodness of fit. A further problem arises when we introduce a second variable to explain Y. It should be obvious that we cannot 'explain' *less* of Y when we introduce a further variable. Suppose we have already carried out a simple regression of Y on X; and then we carry out a multiple regression of Y on X and Z, where Z is a second variable we think may explain the variation in Y. Suppose that the true value of the coefficient, γ, of Z is zero; that is, Z does not in fact influence Y. Because we have only a sample of observations, it is possible that our estimated coefficient, $\hat{\gamma}$, is not exactly zero. We can, of course, test to see if it is significantly different from zero using the usual methods. If we found that $\hat{\gamma}$ was not significantly different from zero, we would conclude (unless X and Z were highly correlated) that Z did not in fact influence Y, and hence would exclude Z from our relationship. However, if $\hat{\gamma}$ is not exactly zero, then the estimated deviations of the multiple regression of Y on X and Z will have a smaller variance than the estimated deviations of the simple regression of Y on X. In other words, the inclusion of Z will 'explain' more of Y (if $\hat{\gamma} \neq 0$), even though, if $\hat{\gamma}$ is not *significantly* different from zero, the inclusion of Z does not 'explain' *significantly* more. Thus, if $\hat{\gamma} \neq 0$, the variance of the estimated deviations will be smaller under the multiple regression than under the simple regression. Hence, if $\hat{\gamma} \neq 0$, R^2 will be larger. Thus the question as to whether $\hat{\gamma}$ is significantly different from zero is the same as the question as to whether R^2 is *significantly* larger. As mentioned in section 7.10 there are available tests as to the significance of R^2, even though this book does not investigate this topic.

Fortunately, there is a way round this problem of R^2 necessarily increasing (or staying the same if $\hat{\gamma}$ is exactly zero). We can 'correct' R^2 for the number of variables used to explain the dependent variable. Let us look again at the definition of R^2. We defined R^2 by:

$$R^2 = 1 - \frac{\sum\limits_{1}^{n} e_i^2}{\sum\limits_{1}^{n} y_i^2}$$

that is,

$$R^2 = 1 - \frac{\dfrac{1}{n}\sum\limits_{1}^{n} e_i^2}{\dfrac{1}{n}\sum\limits_{1}^{n} y_i^2}. \tag{8.29}$$

Hence R^2 is one minus the ratio of the variance of the estimated deviations (the 'unexplained' variance) to the variance of Y. However we know that $\Sigma_1^n e_i^2 /n$ is a biased estimator of the variance of the true deviations and $\Sigma_1^n y_i^2 /n$ is a biased estimator of the variance of Y. If we have $(k-1)$ variables explaining Y, we know that $\Sigma_1^n e_i^2 /(n-k)$ is an unbiased estimator of the variance of the true deviations, and that $\Sigma_1^n y_i^2 /(n-1)$ is an unbiased estimator of the variance of Y. Let us therefore define a new ('corrected') measure of goodness of fit, $\bar{R}^2$, by:

$$\bar{R}^2 = 1 - \frac{\sum_1^n e_i^2/(n-k)}{\sum_1^n y_i^2/(n-1)}. \tag{8.30}$$

Thus $\bar{R}^2$ is one minus the ratio of an unbiased estimator of the variance of the true deviations (the 'unexplained' variance) to an unbiased estimator of the variance of Y.

We see therefore that, although the inclusion of a further variable may decrease $\sum_1^n e_i^2$, $\bar{R}^2$ may fall since $(n-k)$ has decreased by one. It can be shown that inclusion of a further variable will increase $\bar{R}^2$ only if the estimated coefficient is greater than its estimated standard deviation (that is, if its t-ratio is greater than unity). Hence $\bar{R}^2$ can still increase even though the estimated coefficient is not significantly different from zero, but $\bar{R}^2$ increases by less than R^2.

The last sentence should warn the reader not to attach too much weight to the value of $\bar{R}^2$ (or R^2) in a particular regression. To test whether the inclusion of a further variable improves the explanation of the dependent variable it would appear more sensible to test directly the coefficient of the included variable for significance.

We can see the relation between R^2 and $\bar{R}^2$ as follows:

From (8.30)

$$1 - \bar{R}^2 = \left(\frac{n-1}{n-k}\right) \frac{\sum_1^n e_i^2}{\sum_1^n y_i^2} = \left(\frac{n-1}{n-k}\right)(1 - R^2) \quad \text{(from (8.29))}.$$

Thus

$$\bar{R}^2 = 1 - \left(\frac{n-1}{n-k}\right)(1 - R^2). \tag{8.31}$$

These two points, multicollinearity and goodness of fit measures, cover the major conceptual problems that arise in the generalisation of our simple regression model to the multiple regression model. We now give a numerical example of the methods discussed in this section.

8.3 Example of Multiple Regression

We continue our example explaining household consumption. Suppose we wish to investigate the effect of household 'size' on household consumption. Using the definition of size made in the previous section, the necessary data can be obtained from the *Family Expenditure Survey*. From table 1 of the *Survey*, we see that of the 622 households with an

Table 8.2   Cross-section data on household consumption, income and size

'Household number' i	Weekly expenditure Y_i	Weekly income X_i	Size Z_i
1	9.73	7.89	1.089
2	14.14	12.40	1.697
3	18.78	17.55	2.023
4	22.34	22.52	2.263
5	25.58	27.49	2.409
6	28.87	32.50	2.654
7	30.77	37.45	2.631
8	33.61	42.52	2.811
9	38.19	47.42	2.937
10	42.12	54.69	3.051
11	48.31	68.14	3.163
12	70.92	112.26	3.345

income less than £10 a week the average number of adults per household was 1.073, and the average number of children per household was 0.033; thus, counting each child as equivalent to half an adult, the average number of equivalent adults per household was $1.073 + \frac{1}{2}(0.033) = 1.089$. Remember that we are interpreting these 622 households with average income £7.89 and average expenditure £9.73 per week as being one household. Thus we are interpreting this one household as having size 1.089 equivalent adults.* Similarly our second 'household', with income £12.40 and expenditure £14.14, has size $1.586 + \frac{1}{2}(0.221) = 1.697$ equivalent adults. Continuing in this way, we get table 8.2. Our postulated relationship is:

$$Y_i = \alpha + \beta X_i + \gamma Z_i + u_i \quad i = 1, 2, \ldots, n$$

where we are assuming that assumptions (A1) to (A5) hold. We calculate our estimated relationship:

$$Y_i = \hat{\alpha} + \hat{\beta} X_i + \hat{\gamma} Z_i + e_i \quad i = 1, 2, \ldots, n$$

using least squares estimation. The necessary calculations (see (8.14), (8.15), (8.16)) are given in table 8.3. From table 8.3, we have:

$$\bar{Y} = 31.947 \quad \bar{X} = 40.236 \quad \bar{Z} = 2.5061$$

and, using (8.14):

*We could, of course, interpret this size as being the average size of one household over 622 fortnightly periods.

Table 8.3   Calculation of least squares coefficients

i	Y_i	x_i	z_i	$y_i x_i$	$y_i z_i$	$x_i z_i$	x_i^2	z_i^2
1	−22.217	−32.346	−1.4171	718.631	31.4837	45.8375	1,046.264	2.00817
2	−17.807	−27.836	−0.8091	495.676	14.4076	22.5221	774.843	0.65464
3	−13.167	−22.686	−0.4831	298.707	6.3610	10.9596	514.655	0.23339
4	−9.607	−17.716	−0.2431	170.198	2.3355	4.3068	313.857	0.05910
5	−6.367	−12.746	−0.0971	81.154	0.6182	1.2376	162.461	0.00943
6	−3.077	−7.736	0.1479	23.804	−0.4551	−1.1442	59.846	0.02187
7	−1.177	−2.786	0.1249	3.279	−0.1470	−0.3480	7.762	0.01560
8	1.163	2.284	0.3049	3.798	0.5070	0.6964	5.217	0.09296
9	6.243	7.184	0.4309	44.850	2.6901	3.0956	51.610	0.18567
10	10.173	14.454	0.5449	147.041	5.5433	7.8760	208.918	0.29692
11	16.363	27.904	0.6569	456.593	10.7489	18.3301	778.633	0.43152
12	38.973	72.024	0.8389	2,806.991	32.6944	60.4209	5,187.457	0.70375
Σ	−0.004*	−0.002*	−0.0002*	5,250.772	106.7876	173.7904	9,111.523	4.71302

*Slight rounding error

$$\hat{\beta} = \frac{(\Sigma y_i x_i)(\Sigma z_i^2) - (\Sigma y_i z_i)(\Sigma x_i z_i)}{(\Sigma x_i^2)(\Sigma z_i^2) - (\Sigma x_i z_i)^2}$$

$$= \frac{(5250.722)(4.71302) - (106.7876)(173.7904)}{(9111.523)(4.71302) - (173.7904)^2}$$

$$= 0.48573.$$

Using (8.15)

$$\hat{\gamma} = \frac{(\Sigma y_i z_i)(\Sigma x_i^2) - (\Sigma y_i x_i)(\Sigma x_i z_i)}{(\Sigma x_i^2)(\Sigma z_i^2) - (\Sigma x_i z_i)^2}$$

$$= \frac{(106.7876)(9111.523) - (5250.722)(173.7904)}{(9111.523)(4.71302) - (173.7904)^2}$$

$$= 4.7468$$

and, using (8.16)

$$\hat{\alpha} = \bar{Y} - \hat{\beta}\bar{X} - \hat{\gamma}\bar{Z}$$

$$= 31.947 - (0.48573)(40.236) - (4.7468)(2.5061)$$

$$= 0.5072.$$

Our estimated relationship is thus:

$$Y_i = 0.51 + 0.4857 X_i + 4.7468 Z_i + e_i.$$

Table 8.4   Calculation of estimated deviations

i	Y_i	$\hat{Y}_i = 0.51 + 0.4857X_i + 4.7468Z_i$	$e_i = Y_i - \hat{Y}_i$	e_i^2	y_i^2
1	9.73	9.51	0.22	0.0484	493.595
2	14.14	14.59	−0.45	0.2025	317.089
3	18.78	18.64	0.14	0.0196	173.370
4	22.34	22.19	0.15	0.0225	92.294
5	25.58	25.30	0.28	0.0784	40.539
6	28.87	28.89	−0.02	0.0004	9.468
7	30.77	31.19	−0.42	0.1764	1.385
8	33.61	34.51	−0.90	0.8100	2.766
9	38.19	37.48	0.71	0.5041	38.975
10	42.12	41.56	0.56	0.3136	103.490
11	48.31	48.62	−0.31	0.0961	267.748
12	70.92	70.91	0.01	0.0001	1,518.895
Σ	383.36	383.39	−0.03*	2.2721	3,059.614

*Slight rounding error

Table 8.4 contains the calculations needed to derive the estimated deviations. Hence,

$$R^2 = 1 - \frac{\Sigma e_i^2}{\Sigma y_i^2} = 1 - \frac{2.2721}{3059.614} = 0.99926$$

and

$$\bar{R}^2 = 1 - \left(\frac{n-1}{n-k}\right)(1 - R^2) \quad \text{(from (8.29))}$$

$$= 1 - \frac{11}{9}(0.00074) = 0.99910.$$

The estimated variance of the true deviations s^2 is given by (from (8.20))

$$s^2 = \frac{1}{n-3}\Sigma e_i^2 = \frac{1}{9}(2.2721) = 0.25246.$$

Thus

$$s = 0.50245.$$

Hence the estimated standard deviations of the coefficients are, using (8.18):

estimated s.d. $(\hat{\alpha}) = 0.78049$

estimated s.d. $(\hat{\beta}) = 0.00966$

estimated s.d. $(\hat{\gamma}) = 0.42493.$

Thus our estimated relationship can be written:

$$Y_i = 0.51 + 0.4857X_i + 4.7468Z_i + e_i \quad R^2 = 0.9993 \quad (8.32)$$
$$(0.7) \quad (50.6) \quad (11.2) \quad \bar{R}^2 = 0.9991.$$

(The numbers in brackets under the estimated coefficients are their t-ratios.)

You should check that both $\hat{\beta}$ and $\hat{\gamma}$ are significantly different from zero at the 1 per cent level. Comparing (8.32) with our simple regression (7.77) we see that the proportion of the variance of Y explained has increased from 0.9890 to 0.9993. The inclusion of Z, as shown by the t-ratio of its coefficient, has significantly improved the explanation. Notice, however, that with the inclusion of Z, the estimated marginal propensity to consume has fallen from 0.5763 to 0.4857. This is a consequence of the correlation of X and Z (r^2 between X and Z is 0.7034), so that X in our simple regression was being imputed with the effect of Z on Y. We note that the effect of household size on expenditure is such as to imply an increase of £4.75 per week for each equivalent adult at all levels of household income.

At this stage we might consider whether the age of the head of the household has any significant effect on household consumption expenditure. Theoretical justification for the age of the head of the household to be an important factor can be found in the lifecycle theories of consumers' behaviour, one of the best-known accounts being that by Modigliani and Brumberg (20). We do not propose to present the detailed algebra involved in the estimation, but merely to present the results. The required data are again found in the *Family Expenditure Survey*, table 1. The inclusion of age yields the following estimated relationship:

$$Y_i = -0.78 + 0.4826X_i + 5.0231Z_i + 0.0145W_i + e_i \quad (8.33)$$
$$(27.9) \quad (3.8) \quad (0.2)$$

$$R^2 = 0.9993$$
$$\bar{R}^2 = 0.9990.$$

(The numbers in brackets under the estimated coefficients are their t-ratios.) W_i is the age of the head of the ith household; all other variables are as defined before.

We notice that the coefficient on W, as shown by its t-ratio, is not significantly different from zero (even at the 10 per cent level). We also notice that R^2 has remained unchanged at 0.9993, and, as a consequence, $\bar{R}^2$ has fallen. The evidence suggests that W does not significantly improve the explanation, and that we should prefer (8.32) to (8.33) as an explanation of household consumption expenditure.

8.4 The Assumptions and Economic Theory

Chapter 7 was concerned with the estimation and testing of the relationship:

$$Y_i = \alpha + \beta X_i + u_i \qquad i = 1, 2, \ldots, n \qquad (8.34)$$

where the following assumptions were made:

$$
\begin{array}{lll}
Eu_i = 0 & i = 1, 2, \ldots, n & \text{(A1)} \\
Eu_i^2 = \sigma^2 & i = 1, 2, \ldots, n & \text{(A2)} \\
Eu_i u_j = 0 & i \neq j; i, j = 1, 2, \ldots, n & \text{(A3)} \\
X_i \text{ is fixed} & i = 1, 2, \ldots, n & \text{(A4)} \\
u_i \text{ is normal} & i = 1, 2, \ldots, n & \text{(A5)}
\end{array}
$$

Notice that the multiple regression analysis of the previous two sections made exactly the same assumptions, except for the generalisation of (A4) necessary to allow for the inclusion of the extra variables.

Given these assumptions, we were able to show that the least squares estimators of α and β were unbiased, we were able to find their variances, and we were able to show that of all linear unbiased estimators of α and β, the least squares estimators were the best, in the sense of minimum variance. The validity of these results depended crucially on one or more of the above assumptions being true.

In particular, the proof of unbiasedness required that assumptions (A1) and (A4) were true; the derivation of the variances of the least squares estimators required in addition that assumptions (A2) and (A3) were true; and the proof that these least squares estimators are the best linear unbiased estimators required all four assumptions (A1) to (A4) to be true. Notice also that assumptions (A1) to (A4) were needed to show that s^2 is an unbiased estimator of σ^2. Finally, in order to show that the least squares estimators were normally distributed, and that $(n-2)s^2/\sigma^2$ had a chi-square distribution, we needed, in addition, assumption (A5).

Thus the validity of our results, and the validity of the application of least squares estimation to economic relationships, depends critically on the validity of our assumptions. We need, therefore, to be able to check whether our assumptions do, in fact, hold in any particular application. Also, if one or more of our assumptions do not appear to hold, we need to know how to modify our analysis.

Unfortunately, we are unlikely to be able to verify directly the validity of our assumptions in practice. In order to test directly assumption (A2), for example, we would need to have information on Y and on X for every member of the population; by the very nature of our problem, we are unable, or unwilling, to sample the whole population (and if we did, the validity of our assumptions would no longer be of any interest, since if we have sampled the whole population then we have no need for statistical inference). In many economic problems, we are prohibited by the fact that we cannot obtain experimental data as physical scientists can, but have to rely on data gathered by the workings of the economic system. To give an example of this problem, suppose we wish to estimate the parameters of the U.S. aggregate consumption function; we cannot ask the government to set National Income at $981 billion for the next twenty years so that we can get some information about the distribution of aggregate consumption at a fixed income level. If income last year happened to be $981 billion, then the aggregate U.S. consumption figure for last year is one observation from the distribution of aggregate consumption corresponding to an income level of $981 billion. If we *could* hold income constant, we would get a whole series of aggregate consumption figures, from which we could find the distribution of the (true) deviation term u. However, we are unable to hold income constant, so we have to rely on this one observation. Note carefully the distinction between the *practical* impossibility of holding a variable constant, and the *conceptual* impossibility. We are arguing above that, even if the government *could* hold income constant, they would be unwilling to do so for obvious reasons. A more important point, to which we will return later, is whether, even if they wanted to hold income constant, they could in fact do so.

Thus in practice we are unlikely to have a whole series of observations on each u_i, and are thus unable to test *directly* the validity of the assumptions (A1) to (A5). We rely, therefore, on indirect tests of our assumptions.

One potential source of indications (rather than proofs) as to whether our assumptions are valid is economic theory itself. We notice, however, that economic theories are very rarely formulated in the form of (8.34), containing an explicit deviation term. We introduced the deviation term to account for those factors which influence Y other than X. Economic theory often accounts for these other factors by assuming that they are held constant, that is by introducing a *ceteris paribus* assumption. This may be acceptable on a theoretical level, but it raises problems when we come to estimate and test the theory empirically. As mentioned before, for empirical work we generally need to rely on data that have been generated by the workings of the economic system. We cannot, in general, hold these other factors constant. For example, suppose we wish to investigate the theoretical proposition that the quantity of (say) potatoes demanded depends upon the price of potatoes. This proposition is usually qualified by a *ceteris paribus* statement, which means that it is meant to hold given that all other factors influencing the demand for potatoes are held constant. These other factors would possibly include the income of consumers, the prices of substitutes for potatoes and the tastes of consumers. We, as economists, can hardly take over a potato market and insist that all producers of substitutes keep their prices constant, that all consumers have their income held constant, and that they do not change their tastes, while we vary the price of potatoes and study the response! Even if we could, other problems would arise: if we had not taken over all sources of potato supply, consumers would just go to another supplier or grow their own. Even if these last two eventualities did not happen, how could we be sure that consumers were not reacting to our experiment in that when they saw us put up the price, they deliberately refrained from buying as they anticipated a price reduction later?

Going back to our original example of chapter 7, that of studying the influence of household consumption expenditure, similar problems arise. There we introduced the deviation term to account for factors other than income influencing consumption expenditure; for example, the household's assets, the availability of credit to the household, the size (and composition) of the household, the age of the head of the household, its tastes and so on. It is just conceptually possible, given enough time and money, for us to collect a large enough sample of households where some of these factors were the same for each household in the sample. However, we could never get a sample of households all completely alike except for their incomes; for example, how would you decide whether two households had the same tastes? However carefully we sample, and however great our resources, there will virtually inevitably be some factor which influences consumption that we have been unable to account for, and to hold constant. In other words, there will always be a deviation term. The mere fact that we are trying to account for human behaviour is justification enough for a deviation term in our empirical analysis.

These remarks above are intended to show that, however carefully we specify our relationship, and however carefully we select our sample data, there will always be some deviation factor that we are unable to explain. We cannot theorise away the deviation term. Notice that we can reduce the variance of the deviation term by introducing more variables, in addition to X, into our estimated equation for Y. If we thought that factors Z and W also influence Y (and we could measure Z and W) we could reformulate our relationship:

$$Y_i = \alpha + \beta X_i + u_i \quad i = 1, 2, \ldots, n \tag{8.35}$$

as:

$$Y_i = \alpha + \beta X_i + \gamma Z_i + \delta W_i + v_i \quad i = 1, 2, \ldots, n. \tag{8.36}$$

However, as we have argued above, (8.36) still contains a deviation term v_i. The variance of each v_i may well be smaller than the variance of each u_i

(if Z and W do in fact influence Y). However, we have already noted that the validity of the multiple regression analysis also depends crucially on the assumptions made about the deviation term v. Thus, whether we do a simple regression or a multiple regression, we require methods for checking the validity of our assumptions about the deviation term.

Let us, therefore, return from this attempt to use economic theory to explain away the deviation term, and see if economic theory can shed any light on the validity of our assumptions about the deviation term.

We look first at assumption (A1), namely $Eu_i = 0$ $(i = 1, 2, \ldots, n)$. In a sense this is not an assumption *per se*, but follows immediately from the definition of u_i; namely:

$$u_i = Y_i - E(Y_i \mid X_i).$$

Taking expected values it follows immediately that:

$$Eu_i = E(u_i \mid X_i) = E(Y_i \mid X_i) - E(Y_i \mid X_i) = 0.$$

Thus, by writing

$$Y_i = \alpha + \beta X_i + u_i$$

then $Eu_i = 0$ follows, given that:

$$E(Y_i \mid X_i) = \alpha + \beta X_i.$$

Hence the assumption that $Eu_i = 0$ for all i follows from

$$E(Y_i \mid X_i) = \alpha + \beta X_i \quad \text{for all } i.$$

Thus assumption (A1) depends upon the linearity of the relationship between the mean value of Y_i (given X_i) and X_i. Does economic theory tell us whether a relationship is linear or not? If we look at any standard economics textbook, we notice that very rarely is the actual form of the relationship specified. If the theory does not specify the form of the relationship, we will have to rely on statistical tests of assumption (A1); these we will investigate shortly. For cases when a particular non-linear form is specified by economic theory, we will consider ways of transforming them to a linear form, which we can estimate and test using the methods described in chapter 7 and section 8.2.

Let us now look at assumption (A2), namely that $Eu_i^2 = \sigma^2$ $(i = 1, 2, \ldots, n)$. This assumption, of *homoscedasticity* of the deviations, is that the variance of the population about the relationship is the same for all values of X. This would break down if the population were distributed as in figure 8.1.

We may get some clues as to the validity of this assumption from

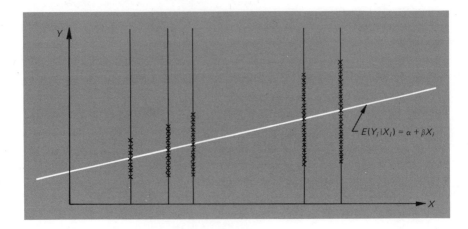

Fig. 8.1   Illustration of heteroscedastic deviations

economic theory. For example, in our income-consumption relationship, we might feel it theoretically reasonable that low-income households might show less variability in their expenditures than high-income households. In the demand-for-potatoes example, we might expect that when the price became very high, the variability of demand would decrease (if only to avoid a negative quantity being demanded). We illustrate this possibility in figure 8.2.

Notice though, that here again economic theory provides us only with indications rather than proofs as to the validity of the assumption. We rely in the last resort on statistical tests; these we will describe shortly.

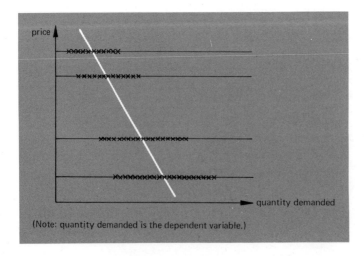

Fig. 8.2   A further illustration of heteroscedastic deviations

However it is important that we do look at economic theory in order to avoid making mistakes and logical errors.

Assumption (A3) is rather more difficult. In chapter 7 we gave an illustration of where this assumption may break down. Assumption (A3), namely $Eu_i u_j = 0$ $(i \neq j; i, j = 1, 2, \ldots, n)$, is that the observation we get on u_i does not influence the observation we get on u_j, and vice versa. Our example of chapter 7 illustrating where this assumption may break down was if we sampled a whole street that indulged in 'keeping up with the Joneses'; there, if one household spent above average for its income group (that is, if its u was positive), then it was likely that its neighbour also spent above average for *its* income group (that is, *its* u was positive). This kind of dependence might be avoided by careful sampling, but in other cases we might be unable to avoid it. Consider, for example, the following situation: suppose we wish to investigate the relationship between U.S. aggregate consumption and U.S. aggregate income. We have available a series of quarterly observations for some twenty years, and propose to estimate and test the relationship:

$$C_t = \alpha + \beta Y_t + u_t \quad t = 1, 2, \ldots, n \tag{8.37}$$

where C_t is aggregate consumption in the U.S. in period t, and Y_t is aggregate income in the U.S. in period t. Suppose however that an important determinant of consumption this period is consumption last period; that is, there is some kind of habit effect in consumption. Then u_t contains C_{t-1} amongst other things. Now, lagging (8.37) one period, we get:

$$C_{t-1} = \alpha + \beta Y_{t-1} + u_{t-1}$$

and thus C_{t-1} depends on u_{t-1}, and so u_t depends on u_{t-1}. Therefore there is dependence between successive values of the deviation term. Notice that we could get round this by reformulating (8.37) by taking into account this omitted factor, and writing the relationship as:

$$C_t = \alpha + \beta Y_t + \gamma C_{t-1} + v_t \quad t = 1, 2, \ldots, n. \tag{8.38}$$

This *may* remove the dependence between successive deviation terms (and we say 'may' since there may be another variable Z_t contained in v_t which is dependent on Z_{t-1}), but it raises a problem that violates one of the other assumptions (you should try and work out which).

Thus, here again economic theory may provide us with indications as to whether assumption (A3) is likely to be valid; but we need to rely on statistical tests for verification.

Assumption (A4), that each X_i is fixed, means that, at least conceptually, we could collect many samples all with the same values for each X_i. We can replace this by a weaker assumption for cases when we have no control over what each X_i is. In these cases we have to rely on X_i values given to us by the workings of the economic system, and thus the X_i are random variables, not fixed (or chosen) by us. This weaker assumption is that:

X_i and u_j are independent for all i and j; $i, j = 1, 2, \ldots, n$ (A4)$'$

We will show later how the results of chapter 7 still hold with this weaker assumption. For the moment, let us anticipate this modification, and investigate under what conditions even this weaker assumption may be violated.

Consider the problem of estimating the aggregate consumption function:

$$C_t = \alpha + \beta Y_t + u_t \quad t = 1, 2, \ldots, n. \tag{8.39}$$

We know that the Y_t values we observe were not fixed by us; further, we know that *in practice* we would be unpopular if we tried to fix income at Y_1 to get a second observation on C_1, at Y_2 to get a second observation on C_2, and so on. Let us see if *conceptually* we could get a second sample with the same $Y_1, Y_2, \ldots, Y_n$. We know that (8.39) is but one equation purporting to explain the workings of the economic system. We could form a 'mini-model' by adding to equation (8.39) the income identity (8.40):

$$Y_t = C_t + I_t \tag{8.40}$$

where I_t is all non-consumption expenditure. Let us suppose that I_t is exogenous, that is, its value is determined outside the economic system.

Then (8.39) says that income Y_t determines consumption C_t, while (8.40) says that consumption C_t is a component of income Y_t. Hence Y_t depends on C_t (from (8.40)), and C_t depends on u_t (from (8.39)). Thus Y_t depends on u_t. Thus the value that Y_t takes is not only not fixed, but it also depends on the value that u_t takes. Thus, in (8.39) even our weaker version of assumption (A4) is violated. Even conceptually we are unable to hold Y_t constant, independent of u_t.

We can now see why equation (8.38) violates the weaker version of assumption (A4). We note that (8.38) is a multiple regression, but here we still require that Y_t and C_{t-1} (the variables explaining the dependent variable) can be either fixed or, if they are random variables, independent of the deviation v_s for all s and t. We have already seen that Y_t depends on v_t, but also C_{t-1} depends on $v_{t-1}, v_{t-2}, \ldots$. Thus the so-called independent variables (Y_t and C_{t-1} are *not* independent of the deviation (v_s) for all values of s and t.

Thus in the case of assumption (A4) and its weaker version (A4)′, economic theory *can* shed light onto its validity. This is very useful since, as we shall see, statistical testing of assumption (A4) or (A4)′ is impossible.

Finally, we look at assumption (A5), that each u_i is normal. Economic theory usually tells us very little about the validity of this assumption, and once again we need to rely on statistical testing.

In the next section we investigate how we might test statistically the assumptions, and indicate how our analysis may need to be modified if the assumptions do not hold.

8.5 Testing the Assumptions

Four of the five assumptions ((A1) to (A3) and (A5)) are directly concerned with the nature of the deviation term. The weaker version of the fifth assumption ((A4)′) is concerned with the independence of the deviation term and the 'independent' variable(s). As discussed previously, we do not have observations on the true deviations (the u_i); what we do have is observations on the estimated deviations (the e_i). We have already shown how we can use the estimated deviations to provide us with an estimator of the variance of the true deviations; let us now see whether we can use the estimated deviations to test the assumptions made about the true deviations.

We will take the assumptions one at a time, and consider how to test them singly. Throughout the testing of each assumption, we will assume that the other assumptions do hold. Considerable problems arise if we try and test several assumptions simultaneously.

We look first at assumption (A1), that $Eu_i = 0$, for all i. We have already argued that this assumption arises as a result of the assumed linearity between $E(Y_i \mid X_i)$ and X_i. Suppose the population is not scattered around a linear relationship between $E(Y_i \mid X_i)$ and X_i but is, in fact, scattered around a non-linear relationship between $E(Y_i \mid X_i)$ and X_i. Then the population would be distributed as in figure 8.3 (assuming that the variance of the (true) deviations was the same for each X_i).

Suppose we mistakenly assumed that the true relationship was linear when in fact the true relationship was non-linear. Figure 8.4 illustrates how the population would be distributed about the assumed-true (linear) relationship.

Thus, in our example, only Eu_2 and Eu_{n-2} would be zero, while Eu_1, Eu_{n-1} and Eu_n would be negative, and $Eu_3, Eu_4, \ldots, Eu_{n-3}$ would be positive; where the u's measure the deviations of the population

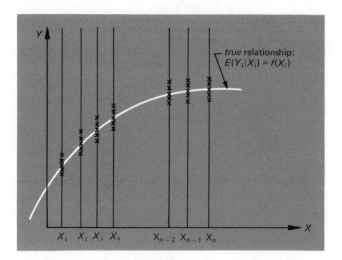

Fig. 8.3   Non-linear true relationship

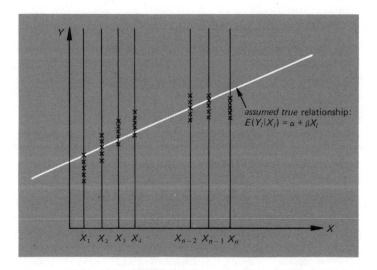

Fig. 8.4   Distribution of population around assumed-true relationship

members from the assumed-true relationship $E(Y_i \mid X_i) = \alpha + \beta X_i$. Thus, if the true relationship is non-linear, but we assume that it is linear, then Eu_i will not be zero for all i. Hence the least squares estimators of α and β will be biased – this is hardly surprising as α and β are not the parameters of the true relationship (strictly speaking, therefore, 'bias' has little meaning in this context).

Now, in practice, we have only one observation on Y_1 corresponding to

X_1, one observation on Y_2 corresponding to X_2, . . . , one observation on Y_n corresponding to X_n; so we are unable to test directly whether $Eu_i = 0$ or, equivalently, whether the true relationship is linear. Nor do we know the actual values of $u_1, u_2, . . . , u_n$ corresponding to the $Y_1, Y_2, . . . , Y_n$ in our sample. What we do have is an estimate, e_1, of u_1; an estimate, e_2, of u_2; . . . ; an estimate, e_n, of u_n. Can we use these estimates to test $Eu_i = 0$ $(i = 1, 2, . . . , n)$? Before answering this question, let us see whether we could test $Eu_i = 0$ $(i = 1, 2, . . . , n)$ if we did know the actual values of $u_1, u_2, . . . , u_n$ in our sample. Let us consider first the problem of using our one observation on u_1 to test $Eu_1 = 0$.
Our null hypothesis is:

$H_0 : Eu_1 = 0.$

Our alternative hypothesis is:

$H_1 : Eu_1 \neq 0.$

Assuming, for both the null and the alternative hypotheses, that (A2) and (A5) are correct, then, if H_0 is correct u_1 is $N(0, \sigma^2)$, while, if H_1 is correct, then u_1 is normally distributed with non-zero mean and variance σ^2. (Note that, if (A1) does not hold, then (A2) must be modified if it is still meant to imply that the variance of each u_i is the same.) Thus, if we knew σ^2, we could test H_0 using the one observation on u_1 in the usual way (rejecting H_0 at the 5 per cent level if $| u_1 | > 1.96\sigma$). However, in practice, we do not know σ^2, and can hardly estimate it using one observation on u_1. We could use all n observations on the u_i to estimate σ^2, but then to be consistent we ought then to use all n observations u_1, $u_2, . . . , u_n$ to test the joint hypothesis $Eu_i = 0$ for all i. This joint hypothesis states that $u_1, u_2, . . . , u_n$ are n (independent) observations on a random variable with mean 0 (and unknown variance σ^2). Thus, we could form $\bar{u}$ and $s^2 = \Sigma(u_i - \bar{u})^2/(n - 1)$ and test the null hypothesis $H_0: Eu_i = 0$ (for all i) in the usual way (that is, rejecting H_0 if $| \bar{u} |/s > t_{n-1,\alpha}$, where $t_{n-1,\alpha}$ is the appropriate critical value from the tables of the t-distribution).

Now, in practice, we do not know the values of the true deviations u_1, $u_2, . . . , u_n$. We have estimates of them: $e_1, e_2, . . . , e_n$ respectively. Thus it would appear reasonable that a test as to whether $\bar{e}$ is significantly different from zero might give us some indication as to whether $\bar{u}$ is significantly different from zero. However, this avenue of exploration is fruitless since, as we have already seen, one consequence of the least squares method of estimation is that $\bar{e}$ is always zero (see (7.54)).

Let us pause for a moment, and think how we might proceed if we could devise a test of the joint hypothesis $Eu_i = 0$ (for all i). Suppose we found the test significant, and thus rejected the null hypothesis; how might we modify our analysis, or, equivalently, what alternative hypothesis do we have in mind when we test the above null hypothesis? If there is a relationship between $E(Y_i \mid X_i)$ and X_i, but we mis-specify the *form* of the relationship, there should be some relationship between Eu_i and X_i. Consider the position if the true relationship between $E(Y_i \mid X_i)$ and X_i is quadratic, that is:

$$E(Y_i \mid X_i) = a + bX_i + cX_i^2 \qquad\qquad i = 1, 2, \ldots, n$$

Thus

$$Y_i = a + bX_i + cX_i^2 + v_i \qquad\qquad i = 1, 2, \ldots, n \qquad (8.41)$$

where

$$Ev_i = 0 \qquad\qquad i = 1, 2, \ldots, n$$

Suppose we mis-specify the form of the relationship, and think that it is linear, that is:

$$E(Y_i \mid X_i) = \alpha + \beta X_i \qquad\qquad i = 1, 2, \ldots, n.$$

Thus our (incorrect) specification is:

$$Y_i = \alpha + \beta X_i + u_i \qquad\qquad i = 1, 2, \ldots, n. \qquad (8.42)$$

Then

$$u_i = Y_i - \alpha - \beta X_i.$$

But

$$Y_i = a + bX_i + cX_i^2 + v_i \quad \text{(from (8.41))}$$

and so

$$u_i = a + bX_i + cX_i^2 + v_i - \alpha - \beta X_i$$

that is

$$u_i = (a - \alpha) + (b - \beta)X_i + cX_i^2 + v_i$$

and so

$$Eu_i = E(u_i \mid X_i) = (a - \alpha) + (b - \beta)X_i + cX_i^2 \quad \text{(for all } i) \text{ since } Ev_i = 0.$$

Hence there is a relationship between Eu_i and X_i.

If we look at figure 8.4, which illustrated the population scatter around a mis-specified relationship, we see that for small X, Eu is negative; Eu increases as X increases, through 0 at $X = X_2$; Eu reaches a maximum at around $X = X_4$; from where Eu decreases, through 0 at $X = X_{n-2}$, and becomes negative again for $X > X_{n-2}$.

Thus we have the important result that if there is a relationship between $E(Y_i \mid X_i)$ and X_i but we have mis-specified the form of the relationship, then there will be a relationship between Eu_i and X_i, where the u_i are the deviations of the observations from the mis-specified relationship. Therefore, in order to check whether we have correctly specified the form of the relationship, it would seem reasonable to investigate whether there is any *pattern* in the estimated deviations. Thus, if our mis-specification is of the form previously illustrated, we would expect our observations and our fitted line to look as in figure 8.5.

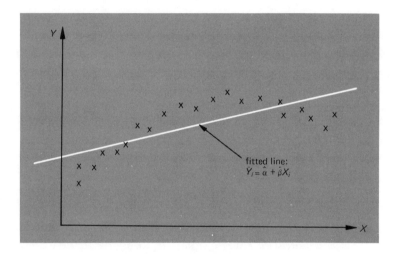

Fig. 8.5   Hypothetical scatter for mis-specified relationship

Thus, if we have mis-specified the form of the relationship, we would expect to see some kind of *pattern* in the estimated deviations (reflecting the pattern, or relationship, of Eu_i with X_i). Therefore, a useful first check as to whether we have specified the form of the relationship correctly is to examine visually the estimated deviations to see if any pattern exists. We could also check statistically for such a pattern, by seeing if there is any relationship between e_i and X_i. Note, however, that we will never find a *linear* relationship between e_i and X_i, since, if we estimate a and b in the relation between

$$e_i = a + bX_i + w_i$$

we would find our least squares estimators $\hat{a}$ and $\hat{b}$ both zero. This is since:

$$\hat{b} = \frac{\Sigma(e_i - \bar{e})x_i}{\Sigma x_i^2} = \frac{\Sigma e_i x_i}{\Sigma x_i^2} \quad (\text{since } \bar{e} = 0, \text{ from (7.54)})$$

$$= 0 \qquad\qquad \text{(from (7.66))}$$

and so

$$\hat{a} = \bar{e} - \hat{b}\bar{X} = 0.$$

Alternatively, if we feel that the relationship between $E(Y_i \mid X_i)$ and X_i is non-linear, we could fit the (X, Y) observations directly to a non-linear function. For example, if we felt that the relationship was quadratic, we could estimate the parameters of:

$$Y_i = \alpha + \beta X_i + \gamma X_i^2 + v_i$$

by the usual multiple regression analysis. We could then test whether the relationship was quadratic or linear by testing whether the estimated coefficient of X^2 was significantly different from zero or not. The problem with this procedure is that it tells us which of the two forms (linear and quadratic) fits better; but there may, of course, be other functional forms that fit the data better than either of these.

It should be noted at this stage that, as there are an infinite number of different functional forms that are mathematically possible, we ought to exercise caution over the types that we try, and as far as possible be guided by economic theory in our choice. This word of caution is necessary since, with enough effort, we could always find a functional form that fits the observations exactly (if we have n points, the functional form expressing Y as a polynomial of degree $(n - 1)$ in X can always be made to fit the data exactly: a straight line can be made to fit exactly through two points; a quadratic can be made to fit exactly through three points, and so on), but the functional form so obtained may be unsatisfactory as far as economic theory is concerned, and it may well perform badly when we come to use it for prediction purposes.

An alternative procedure that can be employed to investigate the linearity assumption for cases when we have no particular alternative form in mind is the following: arrange the observations in ascending order of X, renumbering them if necessary, so that:

$$X_1 \leqslant X_2 \leqslant X_3 \leqslant \ldots \leqslant X_n.$$

Then divide the observations up into two parts, with, say, the first m observations in the first part, and the remaining $(n - m)$ in the second part. Then fit a linear relationship to the first m observations: (X_1, Y_1), $(X_2, Y_2), \ldots, (X_m, Y_m)$, denoting this fitted line by:

$$Y_i = \hat{\alpha}_1 + \hat{\beta}_1 X_i + e_i \quad i = 1, 2, \ldots, m. \tag{8.43}$$

Similarly, fit a line to the remaining $(n - m)$ observations: $(X_{m+1}, Y_{m+1}), (X_{m+2}, Y_{m+2}), \ldots, (X_n, Y_n)$; denote this by:

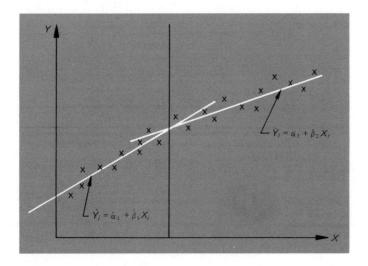

Fig. 8.6 Testing assumption (A1)

$$Y_i = \hat{\alpha}_2 + \hat{\beta}_2 X_i + e_i \quad (i = m + 1, m + 2, \ldots, n). \tag{8.44}$$

We illustrate this procedure in figure 8.6.

We then test whether these two lines are significantly different, by testing whether $\hat{\alpha}_1$ and $\hat{\alpha}_2$ are significantly different and whether $\hat{\beta}_1$ and $\hat{\beta}_2$ are significantly different, in the usual way. For example, consider the null hypothesis that the true relationship is linear, that is that the two lines are both estimates of the true line

$$Y_i = \alpha + \beta X_i + u_i \quad (i = 1, 2, \ldots, n).$$

Then if this null hypothesis is true

$$\hat{\beta}_1 \text{ is } N\left(\beta, \frac{\sigma^2}{\sum\limits_{1}^{m}(X_i - \bar{X}_1)^2}\right) \quad \text{where} \quad \bar{X}_1 = \frac{1}{m}\sum\limits_{1}^{m} X_i$$

and

$$\hat{\beta}_2 \text{ is } N\left(\beta, \frac{\sigma^2}{\sum\limits_{m+1}^{n}(X_i - \bar{X}_2)^2}\right) \quad \text{where} \quad \bar{X}_2 = \frac{1}{n-m}\sum\limits_{m+1}^{n} X_i$$

and so, since $\hat{\beta}_1$ and $\hat{\beta}_2$ are independent, then

$$\hat{\beta}_1 - \hat{\beta}_2 \text{ is } N\left\{0, \sigma^2\left(\frac{1}{\sum\limits_{1}^{m}(X_i - \bar{X}_1)^2} + \frac{1}{\sum\limits_{m+1}^{n}(X_i - \bar{X}_2)^2}\right)\right\}.$$

σ^2 can be estimated by s^2 where

$$s^2 = \frac{\sum\limits_{1}^{m} e_i^2 + \sum\limits_{m+1}^{n} e_i^2}{(m-2)+(n-m-2)} = \frac{\sum\limits_{1}^{m} e_i^2 + \sum\limits_{m+1}^{n} e_i^2}{(n-4)}$$

(where e_i, $i = 1, 2, \ldots, m$ are the deviations from the first line and e_i, $i = m+1, m+2, \ldots, n$ are the deviations from the second line) and it can be shown that $[(n-4)s^2]/\sigma^2$ has a chi-square distribution with $(n-4)$ degrees of freedom.

Hence, if the null hypothesis is correct, then

$$\frac{\hat{\beta}_1 - \hat{\beta}_2}{s\bigg/\left\{\left(\sum\limits_{1}^{m}(X_i - \bar{X}_1)\right)^{-2} + \left(\sum\limits_{m+1}^{n}(X_i - \bar{X}_2)\right)^{-2}\right\}} \tag{8.45}$$

has a t-distribution with $(n-4)$ degrees of freedom, and we can test the observed value of this expression against the appropriate critical values of the t-distribution with $(n-4)$ degrees of freedom. (Note: we are assuming that (A2)-(A5) are correct throughout this discussion of the validity of assumption (A1).)

We could of course divide up the observations into more than two parts, and fit lines separately to each part, testing for significant differences in the usual way.

Let us now illustrate some of these ideas for testing the validity of our assumptions in the context of our simple regression of section 7.11 of household consumption against household income. The (estimated) deviations, e_i, are plotted against X_i in figure 8.7 (see table 7.28).

There appears to be a very clear pattern of the type demonstrated earlier as expected to occur if the correct relationship was non-linear. Let us test this by splitting the data into two halves, the first half containing the first six observations as ranked in order of the magnitude of X, and the second half containing the last six observations. Fitting a line by least squares to the first six observations gives us the following relationship:

$$Y_i = 4.46 + 0.77021X_i \quad R^2 = 0.9922 \tag{8.46}$$
$$(6.0) \qquad (22.6) \qquad i = 1, 2, \ldots, 6$$

(t-ratios in brackets).

Fitting a line to the last six observations gives us:

$$Y_i = 12.08 + 0.52806X_i \quad R^2 = 0.9955 \tag{8.47}$$
$$i = 7, 8, \ldots, 12$$

(t-ratios in brackets).

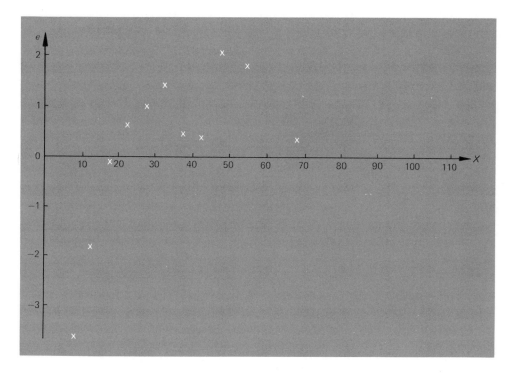

Fig. 8.7   Deviations of (7.72)

These are shown graphically in figure 8.8.

We see clearly a marked difference (.24215) between the coefficient of
X in (8.46) and the coefficient of X in (8.47). Calculating the expression
in (8.45) we find it to be 5.170, while the appropriate critical values from
the t-tables (with $n - 4 = 8$ degrees of freedom) for a two-tailed test at the
1 per cent significance level are ±3.355. We thus conclude that there is a
significant difference between the estimated marginal propensities to
consume from the two halves of the data. It would appear reasonable then
to reject the linear relationship specified between household consumption
expenditure and household income. The problem now is what alternative
form may be preferable. The quadratic form appears to be a likely
candidate on purely statistical grounds, but would probably be rejected on
economic grounds for the following reasons.

Consider the quadratic form

$$E(Y \mid X) = \alpha + \beta Y + \gamma X^2 .$$

Differentiating

$$\frac{d}{dX} E(Y \mid X) = \beta + 2\gamma X$$

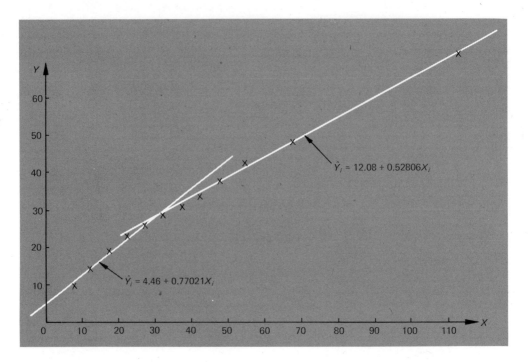

Fig. 8.8   Testing assumption (A1)

thus

$$\frac{d}{dX}E(Y\mid X) \gtreqqless 0 \quad \text{according as } \beta + 2\gamma X \gtreqqless 0.$$

Further $(d^2/dX^2)E(Y\mid X) = 2\gamma$, and so the marginal propensity to consume $((d/dX)E(Y\mid X))$ increases with X if γ is positive and decreases with X if γ is negative. The latter is the more theoretically reasonable case (otherwise the M.P.C. may become greater than unity). If $\gamma < 0$, then $(d/dX)E(Y\mid X) \gtreqqless 0$ according as $X \lesseqqgtr -\beta/2\gamma$. Hence for X sufficiently large (greater than $-\beta/2\gamma$) the M.P.C. becomes negative. Theory would reject this possibility.

A more theoretically adequate form for the relationship is log-linear, that is log $Y_i = \alpha + \beta$ log $X_i + u_i$ $i = 1, 2, \ldots, n$. Now β measures the *elasticity* of consumption with respect to income; and if β lies between 0 and 1, the M.P.C. declines as income increases (though always remaining strictly non-negative).

Let us investigate this possibility. The data, derived from table 7.26, are given in table 8.5.

Carrying out our least squares regression of log Y on log X in the usual

Table 8.5   The logarithms
of household income and
consumption

i	$\log Y_i$	$\log X_i$
1	2.2752	2.0656
2	2.6490	2.5177
3	2.9328	2.8651
4	3.1064	3.1144
5	3.2418	3.3138
6	3.3628	3.4812
7	3.4265	3.6230
8	3.5148	3.7500
9	3.6426	3.8590
10	3.7405	4.0017
11	3.8776	4.2216
12	4.2616	4.7208

way we get the following estimated relationship:

$$\log Y_i = 0.7959 + 0.7339 \log X_i + e_i \qquad R^2 = 0.9982 \qquad (8.48)$$
$$ (22.8) \quad\;\; (74.1)$$

(t-ratios in brackets).
We note that both the intercept and the coefficient on $\log X$ are significantly different from zero. Note also* that the R^2 is higher in (8.48) than in the linear specification (7.77). The (estimated) deviations e_i, of (8.48), are graphed against $\log X_i$ in figure 8.9.

There are now fewer signs of a pattern. Fitting the two halves of the data separately we get:

$$\log Y_i = 0.7102 + 0.7666 \log X_i + e_i \qquad R^2 = 0.9982$$
$$ (14.7) \quad\;\; (46.5) \qquad\qquad i = 1, 2, \ldots, 6$$

$$\left.\begin{array}{l}\\ \\ \\ \\\end{array}\right\} \quad (8.49)$$

$$\log Y_i = 0.7069 + 0.7537 \log X_i + e_i \qquad R^2 = 0.9964$$
$$ (7.7) \quad\;\; (33.1) \qquad\qquad i = 7, 8, \ldots, 12.$$

The difference between the two estimated coefficients of $\log X$ is now only 0.0129, and the statistic in (8.45) is only 0.46, well inside the critical values of ± 3.355. It appears therefore that the log-linear formulation is more adequate.

---

*Although (8.48) explains more of the variance of $\log Y$ than (7.77) explains of the variance of Y, this does not imply that (8.48) explains more of Y than does (7.77), nor does it imply that (7.77) explains less of $\log Y$ than does (8.48). Strictly speaking, therefore, a comparison of R^2 for different dependent variables is meaningless.

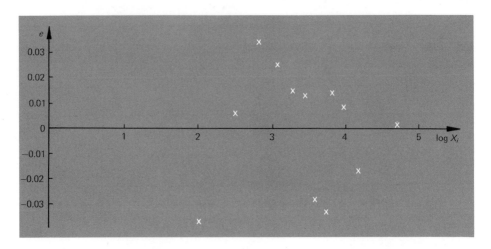

Fig. 8.9   Deviations of log-linear consumption function

Converting our estimated equation (8.48):

$$\log \hat{Y}_i = 0.7959 + 0.7339 \log X_i$$

into an expression in Y_i and X_i, we get (taking antilogs):

$$\hat{Y}_i = 2.216 X_i^{0.7339}$$

which is shown graphically in figure 8.10.

It should be noted that the coefficient of $\log X$ is not the marginal propensity to consume, but the elasticity of consumption with respect to income. The marginal propensity to consume declines with income in this formulation. Evaluated at the sample mean $(\bar{X}, \bar{Y})$ the M.P.C. is given by:

$$\frac{dY}{dX} = \frac{dY/\bar{Y}}{dX/\bar{X}} \times \frac{\bar{Y}}{\bar{X}} = \frac{d(\log Y)}{d(\log X)} \frac{\bar{Y}}{\bar{X}} = 0.7339 \times \frac{31.947}{40.236} = 0.5827.$$

Finally, let us now re-estimate our multiple regression (8.32) using a log-linear relationship. We get:

$$\log Y_i = 0.8598 + 0.6844 \log X_i + 0.1223 \log Z_i \quad R^2 = 0.9987$$
$$(18.7) \quad (24.8) \qquad\qquad (1.9) \qquad\qquad \bar{R}^2 = 0.9984$$

$$(8.50)$$

or, taking antilogs,

$$\hat{Y}_i = 2.363 \, X_i^{0.6844} \, Z_i^{0.1223}.$$

The coefficient of $\log Z$ is only just significantly different from zero at the 5 per cent level on a one-tailed test. Comparing the linear formulation (8.32) with the log-linear formulation (8.49) we see that the former has a

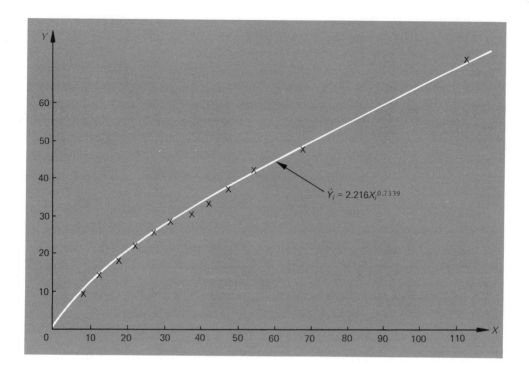

The equation shown in the figure is:

$$\hat{Y}_i = 2.216 X_i^{0.7339}$$

Fig. 8.10   Log-linear consumption function

slightly higher* $\bar{R}^2$. However the log-linear formulation implies, for example, that the effect of an additional equivalent adult on household consumption is greater the higher is income. This seems more appropriate on theoretical grounds.

Let us now turn our attention to testing the validity of assumption (A2). While we are testing (A2) we will assume that the other assumptions do in fact hold. Assumption (A2) is that the variance of each (true) deviation term, u_i, is the same for all i. This assumption is referred to as *homoscedasticity* of the deviations. If it does not hold the deviations are said to be *heteroscedastic*. Suppose that this assumption was not true, but that the variance of the u's increased with X (see figure 8.1). We would then expect our observations to look like those in figure 8.11. That is, we would expect to see the estimated deviations increasing in magnitude as X increases. Thus the first obvious test of assumption (A2) is to graph

*Once again (see previous footnote) this comparison is, strictly speaking, meaningless. Such comparisons should, therefore, only be used as very rough guides; no substantial conclusions can be inferred from them.

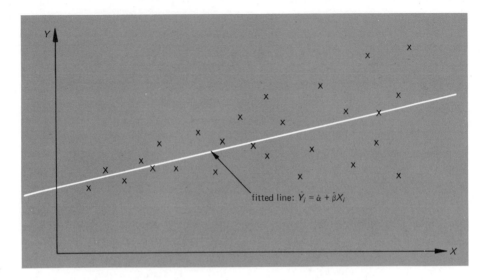

Fig. 8.11   Hypothetical scatter of sample from relationship with heteroscedastic deviations

the *magnitude* of the (estimated) deviations against X and to check visually for any obvious association. We could also test statistically for association, by regressing the magnitude of the (estimated) deviations $(|e_i|)$ against X_i or some function of X_i, the particular function depending on the type of heteroscedasticity believed possible. An alternative procedure would be to regress e_i^2 against X_i or some function of X_i, again the particular function depending on the type of heteroscedasticity believed possible.

We could also follow a similar procedure to that outlined when testing assumption (A1). We could split the data in two parts, ordered in terms of increasing X, estimate a fitted line for each part separately, and find for each part the estimated variance of the true deviations. Suppose s_1^2 was the estimated variance of the true deviations for the first part of the data, and s_2^2 the estimated variance of the true deviations for the second part; we could then test to see if s_1^2 and s_2^2 were significantly different. If we found they were, it would indicate that the variances of the true deviations were not all equal. We have not covered in this book the necessary statistical material for testing two variances for significant difference, but the procedure can be found in any econometrics book (see for example Johnston (14) p. 218).

We can of course split the data into more than two parts. However the problem with this procedure is the same as that encountered using a similar procedure to test assumption (A1): it can only test the null hypothesis that (A2) is correct against the alternative that it is not. We will see in section (8.6) that if (A2) is not true, we can proceed in estimating our relationship only if we replace (A2) by another *specific* assumption. So if (A2) is not valid, we need to be able to replace it by a specific alternative assumption concerning the variances of the true deviations before we can proceed with our estimation. Thus if we feel that the variance of the true deviations may increase with X (i.e. var $u_i = \sigma^2 X_i$) we will get *direct* evidence as to the validity of this by regressing e_i^2 on X_i and testing for a significant relationship.

Following this procedure for our time-series regression of aggregate U.S. Personal Consumption Expenditures on U.S. Gross National Product, and using table 7.31, we estimate the following relationship between e_i^2 and X_i:

$$e_i^2 = 131.3 - 0.0064 X_i + v_i \qquad R^2 = 0.0002 \qquad (8.51)$$
$$(-0.1)$$

We note* that the coefficient on X is not significantly different from zero at the 5 per cent level, nor at the 10 per cent level. Morever we would be unhappy at accepting (8.51) since it implies that e^2 will be zero when $X = 20,515$, and that e^2 will become negative for larger X. The statistical evidence and economic theory suggest that we accept assumption (A2) as holding in this application.

Let us now consider assumption (A3). While we are considering ways of testing (A3) we will assume that the other assumptions do in fact hold. Assumption (A3) says that the (true) deviations are independent, or less restrictively that the deviation u_i has zero covariance with the deviation u_j for all $i, j = 1, 2, \ldots, n, i \neq j$; that is

$$Eu_i u_j = 0 \quad i \neq j; i, j = 1, 2, \ldots, n$$

How might we test this assumption? Suppose we actually knew the values of $u_1, u_2, \ldots, u_n$ for the observations in our sample. u_1 is, of course, just one observation from the distribution of all possible u_1's. Similarly u_2 is just one observation from the distribution of all possible u_2's. Obviously we would need several observations on u_1 and several observations on u_2 before we could investigate whether there was any relationship between u_1 and u_2. As it is we only have an estimate, e_1, of u_1 and an estimate, e_2, of u_2. Thus we are obviously unable to use our observed e_1 and e_2 to test for a relationship between u_1 and u_2. Generalising, we see that we cannot use our observed e_i and e_j to test whether $Eu_i u_j = 0$ for all $i, j = 1, 2, \ldots, n, i \neq j$.

*These significance tests are carried out ignoring the non-independence of the e's (which we show shortly). Strictly speaking, this dependence ought to be taken into account.

Let us therefore continue our discussion using the previous framework and consider what we might replace (A3) by if we found it did not hold. We need therefore to consider what kind of relationship may obtain between u_i and u_j if they are in fact not independent. In our discussion of our cross-section household income-consumption regression, we noted that dependence would exist between u_i and u_j if there was some dependence between the ith household sampled and the jth household sampled. Thus if we can choose our sample so that households are chosen independently, we should be able to ensure that assumption (A3) does in fact hold. In this context we should expect to be able to sample in an independent manner. However in our time-series aggregate income-consumption regression, we did not choose the sample — it was generated by the workings of the economic system. In order to avoid confusion, let us reiterate our interpretation of the data as being a sample of possible observations. In 1966 (for example) U.S. Gross National Product was $981 billion and Personal Consumption Expenditures was $586.1 billion. We are asserting that if G.N.P. was $981 billion in several years, each year we would observe a different value for P.C.E. Our observed value of $586.1 billion for P.C.E. in 1966 is just one observation from the distribution of all possible values for P.C.E. corresponding to a G.N.P. of $981 billion. The difference (true deviation) between this particular year's P.C.E. and the average of all possible values of P.C.E. corresponding to a G.N.P. of $981 billion is, as usual, denoted by u_{21} ($i = 21$ for 1966). We do not know the value of u_{21} in 1966, but we have an estimate of it, i.e. e_{21}. The question now as to the validity of assumption (A3) in this context is whether the value of u_{20} influences the value of u_{21} — that is, if P.C.E. is above average (for the G.N.P. prevailing) in 1965, will this imply that P.C.E. will be above average (for the G.N.P. prevailing) in 1966? We have already discussed in the previous section how such a dependence between successive deviations may occur. This could fruitfully be reread at this stage. This discussion implies that the deviation in period $(t - 1)$ may well influence the deviation in period t; that is u_t may depend on u_{t-1}. Notice that this also implies that u_t depends on u_{t-2}, u_{t-3}, etc., indirectly. Thus, particularly in time-series analysis, we may expect that, if deviations are not independent, they may be related by

$$u_t = \rho u_{t-1} + \epsilon_t \qquad t = 2, 3, \ldots, n \qquad (8.52)$$

where

$$E\epsilon_t = 0 \quad E\epsilon_t^2 = \sigma_\epsilon^2 \quad E\epsilon_t\epsilon_s = 0 \quad t \neq s; s, t = 1, 2, \ldots, n.$$

(and $|\rho| < 1$ so that the variance of u_t remains finite).

(8.52) says that u_t is dependent on u_{t-1}, but not completely dependent unless the deviation term, ϵ_t, is zero for all t. Notice that we say 'particularly in time-series analysis', as, with time, there is a natural ordering of observations. Such a natural ordering is unlikely to be found in cross-section data (i.e., the first household's behaviour is unlikely to influence the second household's behaviour, the second unlikely to influence the third, etc., unless, of course, the households live in order along the street, and they all 'keep up with the Joneses' the 'Joneses' being their lowest-numbered neighbours).

Thus, in the analysis of time-series data, a theoretically plausible alternative hypothesis to the null hypothesis of independence of the (true) deviations is that specified by (8.52). How can we test between these two alternatives? We do not know the u_i, but we do have estimates of them — the e_i. Therefore to test for a relationship between successive u's, it would appear reasonable to look for a relationship between successive e's, that is to look for a *pattern* in the (estimated) deviations. (Notice the similarity with the procedure involved in testing assumption (A1) — but notice also the difference; there we were looking for a pattern in the (estimated) deviations when they were ordered in terms of increasing X; here we are looking for a pattern in the (estimated) deviations when they are ordered in terms of increasing *time*. Obviously a problem arises when, as it often will, X increases with time; in such a case we may have difficulty differentiating a test of assumption (A1) from a test of assumption (A3).) So to test whether the true deviations satisfy (8.52) we could regress e_t on e_{t-1}, to obtain an estimated relationship:

$$e_t = \hat{\rho} e_{t-1} + v_t \quad t = 2, 3, \ldots, n \tag{8.53}$$

and test whether $\hat{\rho}$ is significantly different from zero. The problem with this procedure is that $\hat{\rho}$ in (8.53) is not an unbiased estimator of ρ in (8.52). Even if we knew the u's and regressed u_t on u_{t-1}, this would not give us an unbiased estimator of ρ (since u_{t-1} in (8.52) is not independent of ϵ_s for all s; u_{t-1} depends on $\epsilon_{t-1}, \epsilon_{t-2}, \ldots$,; that is, assumption (A4) and its weaker alternative (A4)' is violated). In fact it can be shown that $\hat{\rho}$ is a downward biased (and inconsistent) estimator of ρ. Our least squares estimator of ρ in (8.53) is given by:

$$\hat{\rho} = \frac{\sum_{2}^{n} e_t e_{t-1}}{\sum_{2}^{n} e_{t-1}^2} \quad \text{(see exercise 8.9)}.$$

To remedy these deficiences of $\hat{\rho}$ as an estimator of ρ, and to provide a more satisfactory test of the null hypothesis $\rho = 0$ (i.e. u_t and u_{t-1} are independent; i.e. assumption (A3) is true), the following test statistic, named the Durbin—Watson statistic after its originators (see reference (7)) is often used:

$$d = \frac{\sum_{2}^{n}(e_t - e_{t-1})^2}{\sum_{1}^{n}e_t^2}. \tag{8.54}$$

By expanding the numerator of this expression, d can be expressed as

$$d = \frac{\sum_{2}^{n}e_t^2}{\sum_{1}^{n}e_t^2} + \frac{\sum_{2}^{n}e_{t-1}^2}{\sum_{1}^{n}e_t^2} - \frac{2\sum_{2}^{n}e_te_{t-1}}{\sum_{1}^{n}e_t^2}. \tag{8.55}$$

We see that the first two terms in (8.55) are each just a little under unity, since $\sum_{2}^{n}e_t^2$ differs from $\sum_{1}^{n}e_t^2$ by only e_1^2, and $\sum_{2}^{n}e_{t-1}^2$ differs from $\sum_{1}^{n}e_t^2$ by only e_n^2. The numerator of the third term in (8.55) is approximately the sample covariance between e_t and e_{t-1}. Thus if e_t and e_{t-1} are unrelated, their sample covariance will be near zero, and so d will be near 2 (that is $1 + 1 - 0$). If e_t and e_{t-1} are highly positively related their correlation will be near to unity and $\sum_{2}^{n}e_te_{t-1}$ will be almost

$$\sqrt{\left(\sum_{2}^{n}e_t^2 \sum_{2}^{n}e_{t-1}^2\right)}.$$

Hence the third term in (8.55) will be near -2, and so d will be almost zero. Similarly if e_t and e_{t-1} are highly negatively related, then the third term in (8.55) will be near $+2$, and d will be almost 4. d is of course a sample statistic, and is thus a random variable.

Durbin and Watson calculated the distribution of d under the null hypothesis of no relationship between successive (true) deviations, that is under the null hypothesis that ρ is zero in (8.52). They found the expected value of d to be 2; this we have intuitively rationalised above. Unfortunately they found that the distribution of d around its mean depended on the particular values of the X_i $(i = 1, 2, \ldots, n)$ in the original relationship. Thus there are no *unique* critical values (for a particular significance level) against which one can compare the observed value of d to test for acceptance or rejection of the null hypothesis of independence. Let us illustrate this by reference to the tables of the Durbin—Watson statistic. We see from the tables that for a regression involving one explanatory variable (our simple regression) and 29 observations used for estimation, there are two entries in the table for a 5 per cent significance level: $d_L = 1.34$, $d_U = 1.48$.

We use these critical values as follows: if our observed value of d is less than $d_L = 1.34$, d is significantly lower (at the 5 per cent level) than would be expected if there were no dependence between successive deviations: there is evidence of significant positive dependence between successive deviations, usually termed positive *autocorrelation* (i.e., ρ in (8.52) is positive). If our observed value of d lies between $d_L = 1.34$ and $d_U = 1.48$, the test is inconclusive: it indicates neither positive auto-correlation nor no autocorrelation. If our observed value of d lies between $d_U = 1.48$ and 2.52 (i.e. a symmetrical interval about the expected value 2) then we can accept (at the 5 per cent significance level) the null hypothesis of no autocorrelation. Similarly, if d lies between 2.52 and 2.66 the test is inconclusive; if d is greater than 2.66 there is significant evidence (at the 5 per cent level) of negative autocorrelation (i.e., ρ in (8.52) is negative).

Let us illustrate the use of the Durbin-Watson test to investigate the

Table 8.6 Calculation of the Durbin—Watson statistic

1	8.8		77.44		
2	19.3	8.8	372.49	10.5	110.25
3	14.0	19.3	196.00	−5.3	28.09
4	19.3	14.0	372.49	5.3	28.09
5	11.2	19.3	125.44	−8.1	65.61
6	−11.0	11.2	120.00	−22.2	492.84
7	−15.9	−11.0	252.81	−4.9	24.01
8	−16.9	−15.9	285.61	−1.0	1.00
9	−5.2	−16.9	27.04	11.7	136.89
10	−6.2	−5.2	38.44	−1.0	1.00
11	−3.6	−6.2	12.96	2.6	6.76
12	−2.6	−3.6	6.76	1.0	1.00
13	2.6	−2.6	6.76	5.2	27.04
14	0.0	2.6	0.00	−2.6	6.76
15	1.4	0.0	1.96	1.4	1.96
16	−0.7	1.4	0.49	−2.1	4.41
17	−6.9	−0.7	47.61	−6.2	38.44
18	−7.8	−6.9	60.84	−0.9	0.81
19	−7.3	−7.8	53.29	0.5	0.25
20	−9.5	−7.3	90.25	−2.2	4.84
21	−15.2	−9.5	231.04	−5.7	32.49
22	−14.5	−15.2	210.25	−0.7	0.49
23	−11.4	−14.5	129.96	3.1	9.61
24	−5.9	−11.4	34.81	5.5	30.25
25	9.7	−5.9	94.09	15.6	243.36
26	13.0	9.7	169.00	3.3	10.89
27	15.1	13.0	228.01	2.1	4.41
28	10.2	15.1	104.04	−4.9	24.01
29	17.6	10.2	309.76	7.4	54.76
Σ			3,660.64		1,390.32

validity of assumption (A3) with respect to our aggregate income-consumption relationship estimated in section 7.11. From table 7.31 we find the estimated deviations. We can thus complete table 8.6 and thus calculate the value of d (the Durbin–Watson statistic). (Note that it is often useful to plot the estimated deviations, and check visually for any pattern.) Hence:

$$d = \frac{\sum\limits_{2}^{n}(e_t - e_{t-1})^2}{\sum\limits_{2}^{n}e_t^2} = \frac{1,390.32}{3,660.64} = 0.38$$

The critical values for the Durbin–Watson statistic chosen earlier for illustrative purposes are in fact the relevant ones for this analysis (29 observations, one explanatory variable). We see that our observed value of $d = 0.38$ is considerably below the lower critical value at the 5 per cent significance level. Thus this statistical test confirms as significant the pattern in the (estimated) deviations that is immediately apparent in a visual check. So it would appear that we should reject the null hypothesis of no independence between successive deviations, and thus reject the validity of assumption (A3) in this application. We will discuss in the next section how our results are modified when assumption (A3) does not hold.

Let us now turn our attention briefly to investigating the validity of assumption (A4) or its weaker alternative (A4)'. (A4) says that the X_i can be kept fixed in repeated sampling, while (A4)' says that in situations where we are unable to hold the X_i constant in repeated sampling, then the X_i observed are independent of the u_j for all i and j. The obvious test of this if we knew the value of the u's would be to calculate the correlation between the u's and the X's to see if any significant association exists. Equally obvious is that, as we do not know the u's, we could use their estimates, the e's, and see if any correlation exists between the e's and the X's. However, we already know that this avenue of exploration would be fruitless since the covariance, and hence the correlation, between the e's and the X's is zero:

$$\sum\limits_{1}^{n}(e_i - \bar{e})(X_i - \bar{X}) = \sum\limits_{1}^{n}e_i(X_i - \bar{X}) \quad \text{since } \bar{e} = 0 \text{ from (7.54)}$$

$$= \sum\limits_{1}^{n}e_i x_i$$

$$= 0 \quad \text{from (7.67)}.$$

Thus we are unable to use the estimated deviations to test assumption (A4) or (A4)'. We need therefore to rely on economic theory to tell us whether assumption (A4) (or (A4)') is likely to be valid in practice. We will look in the next section and in chapter 9 at the consequences for our estimation procedures if these assumptions do not hold.

Finally, let us turn to assumption (A5). Again the obvious test as to whether the u_i are normal is to see whether the e_i are normal. As a preliminary visual check we could present the e's in a frequency distribution, and see if it approximates to normality. There are available statistical tests to see if a frequency distribution is significantly non-normal. We have not described such tests in this book, but they are based on the use of the chi-square distribution, and the interested reader should be able to follow the necessary procedure, as described, for example, in Hoel (13). As might be envisaged, problems of testing (A5) arise if some of the other assumptions are violated; such considerations lie outside the scope of this book.

8.6 Consequences of Violated Assumptions

We now turn to the problem of identifying where modifications need to be made in our analysis of chapter 7 if the assumptions that were made there do not hold. We will take one assumption at a time, and consider the effects on our estimation procedure when the assumption is no longer valid; as we consider the violation of each assumption in turn, we will suppose that the remaining assumptions are in fact valid. Considerable complexities of analysis occur if we try to consider the effects of the violation of two or more assumptions at the same time. We also intend to give only an insight into the consequences of violated assumptions — a full treatment would constitute a book in itself. The subject matter of econometrics is concerned with this full analysis.

In order to investigate the effects of the violation of the assumptions, it would be useful first to trace through the analysis of chapter 7 to identify the main stages in the analysis, and the assumptions used in each stage.

Stage 1   derivation of the least squares estimators, $\hat{\alpha}$ and $\hat{\beta}$
(equations (7.21) and (7.20); section 7.3)
no assumptions used;

Stage 2   proof of unbiasedness of the least squares estimators
(equations (7.33) and (7.31); section 7.3)
assumptions (A1) and (A4) used;

Stage 3   derivation of the variances of the least squares estimators
(equations (7.39) and (7.35); section 7.3)
assumptions (A1), (A2), (A3), and (A4) used;

Stage 4   proof that of all linear unbiased estimators the least squares estimators are the best in the sense of minimum variance
(section 7.5)
assumptions (A1), (A2), (A3) and (A4) used;

Stage 5   demonstration that the least squares estimators are normally distributed
(equations (7.49) and (7.48); section 7.7)
assumption (A5) used;

Stage 6   proof that s^2 is an unbiased estimator of σ^2
(equation (7.59); section 7.8)
assumptions (A1), (A2), (A3) and (A4) used;

Stage 7   demonstration that $(n-2)s^2/\sigma^2$ has a chi-square distribution with $(n-2)$ degrees of freedom
(equation (7.62); section 7.8)
assumptions (A1), (A2), (A3), (A4) and (A5) used;

Stage 8   derivation of necessary information for forming confidence intervals for, and testing hypotheses about, the true parameters α and β
(equations (7.64) and (7.63); section 7.9)
assumptions (A1), (A2), (A3), (A4) and (A5) used.

You should check back to remind yourself where each necessary assumption was used in each stage.

The first point to notice is that no assumptions at all were used in stage 1, the derivation of the least square estimators. This should be immediately apparent, since the least squares estimators are the parameters of the line fitted to the data by the method of least squares. Given a scatter of observations, we can always fit a line to them by this method. (This statement requires qualification in the case of a multiple regression; see section 8.2.)

The second point to notice is that only assumptions (A1) and (A4) are needed in stage 2. In other words, as long as (A1) and (A4) are valid, then the least squares estimators are unbiased, even if (A2), (A3) and (A5) are invalid.

A final point to notice at this stage is that assumption (A5) is used only at the stages involving the *distribution* of our estimators. If (A5) (which, of course, is an assumption about the *distribution* of the true deviations) does not hold, but all the other assumptions do hold, then only stages 5, 7 and 8 will be in need of modification.

Let us now examine our assumptions one by one, starting with assumption (A1). We see that the validity of (A1) is crucial to all but stages 1 and 5. Things begin to go wrong at stage 2 — we are unable to prove unbiasedness if (A1) does not hold. Let us run through the discussion starting at equation (7.29):

$$\hat{\beta} = \beta + \sum_1^n w_i u_i \quad \text{where} \quad w_i = \frac{x_i}{x_1^2 + x_2^2 + \ldots + x_n^2} \quad \text{for all } i.$$

If assumption (A4) holds each w_i is constant in repeated sampling, and so $E(w_i u_i) = w_i E u_i$ for all i. Hence:

$$E\hat{\beta} = \beta + (w_1 E u_1 + w_2 E u_2 + \ldots + w_n E u_n). \tag{8.56}$$

Thus if (A1) does not hold, i.e. $E u_i \neq 0$ for all i, then the term in brackets in (8.56) will not in general be zero, and so $E\hat{\beta} \neq \beta$. Similarly we can show that $\hat{\alpha}$ is biased. (However, as we have already noted, 'bias' has little meaning in this context, as α and β are not the parameters of the true relationship).

We have discussed at length in sections 8.4 and 8.5 that the validity of assumption (A1) relies crucially on the correct specification of the functional form of the relationship. The only remedy therefore for the violation of assumption (A1) is to search for the correct specification, this search, of course, being guided by economic theory and by the statistical appearance of the departure from linearity.

Assumption (A2) is not used in the proof of unbiasedness. Thus, as long as (A1) and (A4) hold, assumption (A2) can be violated without destroying the unbiasedness property of the least squares estimators. However, assumption (A2) is used in stage 3, that of finding the variances of the least squares estimators. Let us take up the derivation starting at equation (7.34):

$$\text{var } \hat{\beta} = E(w_1^2 u_1^2 + w_2^2 u_2^2 + \ldots + w_n^2 u_n^2 + 2w_1 w_2 u_1 u_2 +$$
$$2w_1 w_3 u_1 u_3 + \ldots + 2w_{n-1} w_n u_{n-1} u_n).$$

If assumptions (A3) and (A4) hold, the expected value of all the terms $2w_i w_j u_i u_j$ $(i \neq j)$ will be zero, as explained in the discussion following equation (7.34). Also if (A4) holds, we can still write $E(w_i^2 u_i^2) = w_i^2 E u_i^2$ for all i, and so

$$\text{var } \hat{\beta} = w_1^2 E u_1^2 + w_2^2 E u_2^2 + \ldots + w_n^2 E u_n^2. \tag{8.57}$$

However if (A2) does not hold, and $E u_i^2 \neq \sigma^2$ for all i, we cannot complete the remaining steps to derive the variance formula for $\hat{\beta}$ (equation (7.35)).

If we replace assumption (A2) by some specific alternative assumption about the variances of the u_i we could derive the variance of $\hat{\beta}$ under this alternative. For illustrative purposes, suppose the variance of the u_i increases with X_i^2; that is,

$$\text{var } u_i = Eu_i^2 = \sigma^2 X_i^2. \tag{8.58}$$

Substituting this in (8.57) we obtain:

$$\text{var } \hat{\beta} = \sigma^2 (w_1^2 X_1^2 + w_2^2 X_2^2 + \ldots + w_n^2 X_n^2).$$

After some algebraic simplification this becomes

$$\text{var } \hat{\beta} = \sigma^2 \frac{\sum_{1}^{n} x_i^2 X_i^2}{\left(\sum_{1}^{n} x_i^2 \right)^2}. \tag{8.59}$$

Summarising our discussion so far, we have shown that if (A1), (A3) and (A4) hold, but (A2) is replaced by (8.58) (i.e., if the variance of the true deviations is not the same for all X, but increases as the square of X), then the least squares estimators are still unbiased, but their variances are no longer given by (7.39) and (7.35). In fact the correct variance of $\hat{\beta}$ is now given by (8.59). (A correct variance for $\hat{\alpha}$ could also be calculated.) We have therefore reworked stage 3 for the case when assumption (A2) is not valid, but is replaced by (8.58).

However we encounter a further problem when we come to stage 4. If assumption (A2) does not hold, the least squares estimators, although unbiased, may not now be the best, in the sense of minimum variance, of all linear unbiased estimators. In other words, if assumption (A2) is not valid, there may be some alternative unbiased estimator which has a smaller variance than the least squares estimator, and which therefore is preferred to it. How might we find this alternative estimator?

To answer this question, we reiterate that if we have a relationship that obeys assumptions (A1) to (A4), then least squares estimators are the best linear unbiased estimators. Can we derive a second relationship from our original one so that the second relationship obeys all the usual assumptions?

Consider our situation:

$$Y_i = \alpha + \beta X_i + u_i \quad i = 1, 2, \ldots, n. \tag{8.60}$$

This satisfies (A1), (A3), (A4), (A5) but (A2) is replaced by

$$Eu_i^2 = \sigma^2 X_i^2. \tag{8.61}$$

Consider the variable v_i, defined in terms of u_i by

$$v_i = \frac{u_i}{X_i} \quad i = 1, 2, \ldots, n. \tag{8.62}$$

Hence

$$Ev_i = E\left(\frac{u_i}{X_i}\right) = \frac{1}{X_i} Eu_i \quad \text{by assumption (A4)}$$

$$= 0 \quad \text{for all } i, \text{ by assumption (A1)}$$

Further,

$$Ev_i^2 = E\left(\frac{u_i}{X_i}\right)^2 = \frac{1}{X_i^2} Eu_i^2 \quad \text{by (A4)}$$

$$= \frac{\sigma^2}{X_i^2} X_i^2 \quad \text{by (8.61)}$$

$$= \sigma^2, \text{ for all } i.$$

Also,

$$Ev_i v_j = E\left(\frac{u_i}{X_i}\frac{u_i}{X_j}\right) = \frac{1}{X_i X_j} Eu_i u_j \quad \text{by (A4)}$$

$$= 0 \quad \text{by (A3) for all } i, j, i \neq j.$$

(Also, if u_i is normal, then so is v_i.)
Hence this v_i obeys all the usual assumptions required for the true deviation of a relationship. Thus the obvious procedure is to divide (8.60) throughout by X_i. This gives

$$\frac{Y_i}{X_i} = \alpha \frac{1}{X_i} + \beta + \frac{u_i}{X_i} \quad i = 1, 2, \dots, n$$

that is,

$$Y_i' = \beta + \alpha X_i' + v_i \quad i = 1, 2, \dots, n \quad (8.63)$$

where

$$Y_i' = Y_i/X_i; \quad X_i' = 1/X_i; \quad v_i = u_i/X_i.$$

Thus the relationship in (8.63) satisfies all the assumptions necessary to show that least squares estimators are the best linear unbiased estimators. Hence we can derive best linear unbiased estimators of α and β by estimating (8.63) by least squares regression of Y' on X'. Notice that, in our derived (or transformed) relationship (8.63), β is now the intercept and α the coefficient of the variable.

In general then, if we have a relationship where all the assumptions except (A2) are valid, and we know the specific alternative assumption about the variances of the true deviations, we can derive a transformed relationship from the original in such a way that the transformed relationship obeys *all* the assumptions; and thus least squares estimation of the transformed equation provides us with best linear unbiased estimators of the parameters.

Let us illustrate this discussion in the context of our simple model of section 7.4. There we postulated that the true relationship was given by

$$Y_i = \alpha + \beta X_i + u_i \quad i = 1, 2, 3. \tag{8.64}$$

where

$$\alpha = 2 \quad \beta = 0.5.$$

Let us modify this illustrative example in such a way that the variances of the u's are not the same for each i, but increase with X^2 as given by (8.61). Table 8.7 gives distributions of the u_i which satisfy (8.61). You should verify from table 8.7 that $Eu_1 = Eu_2 = Eu_3 = 0$, and that

Table 8.7   Heteroscedastic deviations

u_1	$P[u_1]$	u_2	$P[u_2]$	u_3	$P[u_3]$
−1	¼	−2	¼	−3	¼
0	½	0	½	0	½
1	¼	2	¼	3	¼

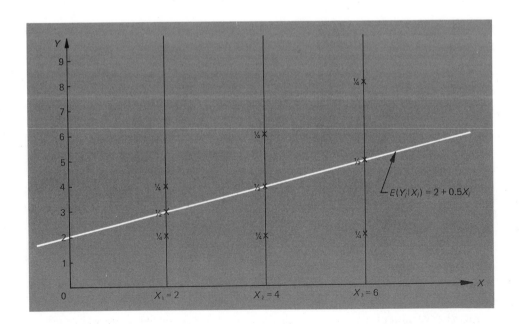

Fig. 8.12   Heteroscedastic deviations

var $u_1 = 1/2$, var $u_2 = 2$, var $u_3 = 9/2$; i.e. var $u_i = \sigma^2 X_i^2$ where $\sigma^2 = 1/8$. Figure 8.12 illustrates the population distribution. Our transformed relationship is derived by dividing (8.64) throughout by X_i to give

$$Y_i' = \beta + \alpha X_i' + v_i \quad i = 1, 2, 3.$$

$$\beta = 0.5 \quad \alpha = 2$$

where

$$Y_i' = Y_i/X_i \quad X_i' = 1/X_i \quad v_i = u_i/X_i \quad \text{all } i.$$

Thus (for example)

when $\quad Y_1 = 2 \quad Y_1' = Y_1/X_1 = 2/2$

when $\quad Y_1 = 3 \quad Y_1' = Y_1/X_1 = 3/2$ $\Bigg\}$ $X_1' = 1/X_1 = 1/2$

when $\quad Y_2 = 6 \quad Y_2' = Y_2/X_2 = 6/4 \quad X_2' = 1/X_2 = 1/4$

and so on.

Similarly (for example)

when $\quad u_1 = -1 \quad v_1 = u_1/X_1 = -\frac{1}{2}$

when $\quad u_1 = \;\; 0 \quad v_1 = u_1/X_1 = 0$

when $\quad u_2 = \;\; 2 \quad v_2 = u_2/X_2 = \frac{1}{2}$

Table 8.8   Deviations of transformed equation

v_1	$P[v_1]$	v_2	$P[v_2]$	v_3	$P[v_3]$
$-\frac{1}{2}$	$\frac{1}{4}$	$-\frac{1}{2}$	$\frac{1}{4}$	$-\frac{1}{2}$	$\frac{1}{4}$
0	$\frac{1}{2}$	0	$\frac{1}{2}$	0	$\frac{1}{2}$
$\frac{1}{2}$	$\frac{1}{4}$	$\frac{1}{2}$	$\frac{1}{4}$	$\frac{1}{2}$	$\frac{1}{4}$

Thus we can construct the probability distributions of the v_i; these are given in table 8.8.

You should check that $Ev_i = 0$ all i, $Ev_i^2 = 1/8$ all i. Our transformed relationship has the population as given in figure 8.13.

Now, suppose, as before, we take samples of size 3, one corresponding to $X_1 = 2$, one to $X_2 = 4$, one to $X_3 = 6$. One possible sample we may get is

$$Y_1 = 2 \quad Y_2 = 6 \quad Y_3 = 5$$

$$X_1 = 2 \quad X_2 = 4 \quad X_3 = 6.$$

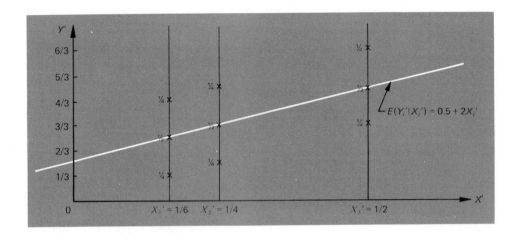

Fig. 8.13   Transformed relationship

As there is a one-to-one correspondence between the (X_i, Y_i) and the (X_i', Y_i'), this sample above is equivalent to:

$$Y_1' = 1 \qquad Y_2' = 3/2 \qquad Y_3' = 5/6$$
$$X_1' = 1/2 \qquad X_2' = 1/4 \qquad X_3' = 1/6.$$

Fitting a line by least squares to the scatter in the (X, Y) plane, we get the fitted relationship:

$$Y_i = \frac{4}{3} + \frac{3}{4} X_i + e_i \qquad i = 1, 2, 3.$$

i.e. an estimate of α of 4/3, and estimate of β of 3/4.

Fitting a line by least squares to the scatter in the (X', Y') plane, we get the fitted relationship,

$$Y_i' = \frac{59}{52} - \frac{1}{13} X_i' + e_i' \qquad i = 1, 2, 3$$

that is

$$\left(\frac{Y_i}{X_i}\right) = \frac{59}{52} - \frac{1}{13} \frac{1}{X_i} + e_i'$$

that is

$$Y_i = -\frac{1}{13} + \frac{59}{52} X_i + e_i$$

that is, an estimate of α of $-1/13$, an estimate of β of 59/52. The scatter of observations and the two fitted lines are shown graphically in figure 8.14.

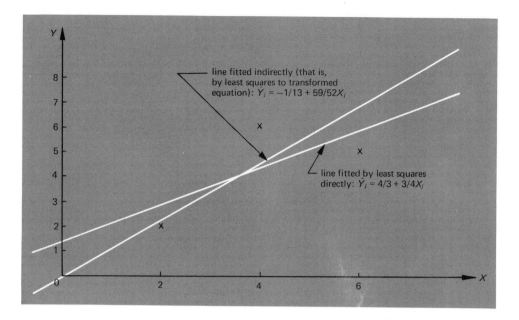

Fig. 8.14   Lines fitted by two methods

You can see what has happened — the line fitted indirectly goes nearer to the (X_1, Y_1) observation, and further from the (X_3, Y_3) observation. That is, the line is fitted to give relatively less weight to the (X_3, Y_3) observation than the (X_1, Y_1) observation, since the former is more likely to be further from the true line than the latter.

Repeating this procedure for all 27 possible samples, and denoting by α^* and β^* the estimates obtained indirectly from the transformed relationship, we get table 8.9.

(Note that for each sample,

$$X_1 = 2 \qquad X_2 = 4 \qquad X_3 = 6$$
$$X_1' = 1/2 \qquad X_2' = 1/4 \qquad X_3' = 1/6.)$$

The distributions of $\hat{\alpha}$ and $\hat{\beta}$ are thus given as in table 8.10. You should verify that $E\hat{\alpha} = 2$ ($= \alpha$) and $E\hat{\beta} = 0.5$ ($= \beta$). You should also check that

$$\text{var } \hat{\alpha} = 28/9$$

$$\text{var } \hat{\beta} = 5/16 \left(= \sigma^2 \frac{\Sigma x_i^2 X_i^2}{(\Sigma x_i^2)^2} \quad \text{from (8.59)} \right).$$

The distributions of α^* and β^* (the estimates obtained from the transformed equation) are given in table 8.11. You should verify that $E\alpha^* = 2$

Table 8.9   Lines fitted by two methods to all 27 samples

Sample number	Prob-ability (× 64)	u_1	u_2	u_3	Y_1	Y_2	Y_3	$\hat{\alpha}$ (× 3)	$\hat{\beta}$ (× 4)	Y'_1 (× 2)	Y'_2 (× 4)	Y'_3 (× 6)	α^* (× 13)	β^* (× 52)
1	1	−1	−2	−3	2	2	2	6	0	2	2	2	26	0
2	2	−1	−2	0	2	2	5	0	3	2	2	5	11	27
3	1	−1	−2	3	2	2	8	−6	6	2	2	8	−4	54
4	2	−1	0	−3	2	4	2	8	0	2	4	2	20	16
5	4	−1	0	0	2	4	5	2	3	2	4	5	5	43
6	2	−1	0	3	2	4	8	−4	6	2	4	8	−10	70
7	1	−1	2	−3	2	6	2	10	0	2	6	2	14	32
8	2	−1	2	0	2	6	5	4	3	2	6	5	−1	59
9	1	−1	2	3	2	6	8	−2	6	2	6	8	−16	86
10	2	0	−2	−3	3	2	2	10	−1	3	2	2	47	−17
11	4	0	−2	0	3	2	5	4	2	3	2	5	32	10
12	2	0	−2	3	3	2	8	−2	5	3	2	8	17	37
13	4	0	0	−3	3	4	2	12	−1	3	4	2	41	−1
14	8	0	0	0	3	4	5	6	2	3	4	5	26	26
15	4	0	0	3	3	4	8	0	5	3	4	8	11	53
16	2	0	2	−3	3	6	2	14	−1	3	6	2	35	15
17	4	0	2	0	3	6	5	8	2	3	6	5	20	42
18	2	0	2	3	3	6	8	2	5	3	6	8	5	69
19	1	1	−2	−3	4	2	2	14	−2	4	2	2	68	−34
20	2	1	−2	0	4	2	5	8	1	4	2	5	53	−7
21	1	1	−2	3	4	2	8	2	4	4	2	8	38	20
22	2	1	0	−3	4	4	2	16	−2	4	4	2	62	−18
23	4	1	0	0	4	4	5	10	1	4	4	5	47	9
24	2	1	0	3	4	4	8	4	4	4	4	8	32	36
25	1	1	2	−3	4	6	2	18	−2	4	6	2	56	−2
26	2	1	2	0	4	6	5	12	1	4	6	5	41	25
27	1	1	2	3	4	6	8	6	4	4	6	8	26	52

(The particular sample we illustrated in figure 8.14 was number 8.)

Table 8.10   Distribution of $\hat{\alpha}$ and $\hat{\beta}$

Sample numbers	$\hat{\alpha}$ (× 3)	$P[\hat{\alpha}]$ (× 64)	Sample numbers	$\hat{\beta}$ (× 4)	$P[\hat{\beta}]$ (× 64)
3	−6	1	19, 22, 25	−2	4
6	−4	2	10, 13, 16	−1	8
9, 12	−2	3	1, 4, 7	0	4
2, 15	0	6	20, 23, 26	1	8
5, 18, 21	2	7	11, 14, 17	2	16
8, 11, 24	4	8	2, 5, 8	3	8
1, 14, 27	6	10	21, 24, 27	4	4
4, 17, 20	8	8	12, 15, 18	5	8
7, 10, 23	10	7	3, 6, 9	6	4
13, 26	12	6			
16, 19	14	3	Σ		64
22	16	2			
25	18	1			
Σ		64			

Table 8.11 Distributions of α^* and β^*

Sample numbers	α^* (x 13)	$P[\alpha^*]$ (x 64)	Sample numbers	β^* (x 52)	$P[\beta^*]$ (x 64)
9	−16	1	19	−34	1
6	−10	2	22	−18	2
3	−4	1	10	−17	2
8	−1	2	20	−7	2
5, 18	5	6	25	−2	1
2, 15	11	6	13	−1	4
7	14	1	1	0	1
12	17	2	23	9	4
4, 17	20	6	11	10	4
1, 14, 27	26	10	16	15	2
11, 24	32	6	4	16	2
16	35	2	21	20	1
21	38	1	26	25	2
13, 26	41	6	14	26	8
10, 23	47	6	2	27	2
20	53	2	7	32	1
25	56	1	24	36	2
22	62	2	12	37	2
19	68	1	17	42	4
			5	43	4
Σ		64	27	52	1
			15	53	4
			3	54	1
			8	59	2
			18	69	2
			6	70	2
			9	86	1
			Σ		64

$(=\alpha)$ and $E\beta^* = 0.5$ $(=\beta)$. You should also check that

$$\text{var } \alpha^* = 27/13 \ (= \sigma^2/\Sigma x_i'^2)$$

$$\text{var } \beta^* = 49/208 \left(= \frac{\sigma^2 \Sigma X_i'^2}{n\Sigma x_i'^2} \right).$$

(You should note very carefully that the intercept in the transformed equation is the estimate of β, and the coefficient of X' in the transformed equation is the estimate of α.)

Comparing our two estimators of α, we see that both are unbiased, and that α^* has the smaller variance.

$$\text{var } \alpha^* = \frac{27}{13} < \frac{28}{9} = \text{var } \hat{\alpha}.$$

Comparing our two estimators of β, we see that both are unbiased, and that $\beta*$ has the smaller variance.

$$\text{var } \beta* = \frac{49}{208} < \frac{65}{208} = \frac{5}{16} = \text{var } \hat{\beta}.$$

Thus the estimators obtained by least squares from the transformed relationship are more efficient than the ones obtained by least squares from the original relationship. We knew this result would occur as the transformed relationship obeys all the conditions necessary for least squares estimators to be best linear unbiased, while the original relationship does not.

It must be noted, however, that in order to obtain the transformed equation we needed to know what specific alternative assumption replaces (A2); that is, we need to know exactly how the variances of the u_i vary. In practice we may not know this. However, we will be able to get some information as to whether the variances are the same for each i or, if this is not true, in what way they vary, by the methods described in the last section.

Let us now move on to consideration of the consequences of the violation of assumption (A3). If assumption (A3) does not hold, but (A1), (A2) and (A4) do, we can still complete stage 2 of our analysis and show that least squares estimators are unbiased. However, (as in the violation of assumption (A2)), problems arise when we come to stage 3; viz. the derivation of the variances of the least squares estimators. Let us once again take up the derivation, starting at equation (7.34):

$$\text{var } \beta = E(w_1{}^2 u_1{}^2 + w_2{}^2 u_2{}^2 + \ldots + w_n{}^2 u_n{}^2 + 2w_1 w_2 u_1 u_2$$

$$+ 2w_1 w_3 u_1 u_3 + \ldots + 2w_{n-1} w_n u_{n-1} u_n).$$

If assumptions (A2) and (A4) hold, each term $w_i{}^2 u_i{}^2$ in the above expression has expected value $w_i{}^2 \sigma^2$. However, if assumption (A3) does not hold, the terms in $w_i w_j u_i u_j$ no longer have zero expected value for all $i, j, (i \neq j)$. Unless we know the specific alternative assumption replacing (A3), we will be unable to evaluate the variance of $\hat{\beta}$. On the other hand, if u_i and u_j are not independent (i.e. (A3) is violated), and we know the specific way that u_i and u_j are dependent, then we can evaluate $Eu_i u_j$ for all i, j and thus evaluate the above expression for the variance of our least squares estimator $\hat{\beta}$.

Let us consider the specific alternative to (A3) that we discussed before as likely to occur in time-series analysis:

$$u_t = \rho u_{t-1} + \epsilon_t \qquad t = 1, 2, \ldots, n \tag{8.65}$$

where

$$E\epsilon_t = 0 \quad E\epsilon_t^2 = \sigma_\epsilon^2 \quad E\epsilon_t\epsilon_s = 0 \quad t \neq s; t, s = 1, 2, \ldots, n.$$

If (8.65) holds, rather than (A3), we can calculate the value of Eu_tu_s for all t and s. (We are using t as a subscript rather than i to emphasise the point that the kind of dependence between successive deviations as given by (8.65) is more likely to occur in time-series analysis than in cross-section analysis.) Thus if (8.65) holds we can evaluate the above expression for the variance of $\hat{\beta}$. This variance will, of course, be different from the variance of $\hat{\beta}$ when (A3) holds.

So, as was shown in the discussion of the violation of assumption (A2), we can modify stage 3 of our analysis for the case when (A3) no longer holds, and is replaced by a specific alternative, for example of the form of (8.65).

However, we encounter a further problem when we come to stage 4. If (A3) is violated we can no longer show that least squares estimators are the best linear unbiased estimators. In other words, if (A3) is violated, there may exist an alternative estimation method which yields linear unbiased estimators with smaller variances than the least squares estimators. You will recall that, in our discussion of the violation of assumption (A2), we got round this problem by transforming our original relationship in such a way that the transformed relationship satisfied all the conditions necessary for least squares estimators to be best linear unbiased. We will tackle the violation of assumption (A3) by a similar procedure. Before we do however, we must make an important point concerning stage 6 of our analysis.

Stage 6 involves the proof that s^2 is an unbiased estimator of σ^2, the variance of the true deviations. This was an important stage, since the variances of the least squares estimators depend on σ^2, and we needed an unbiased estimator of σ^2 before we could find unbiased estimators of the variances of our estimated coefficients. Now the proof that s^2 is an unbiased estimator of σ^2 relied, *in particular*, on the validity of assumption (A3). If (A3) does not hold, then s^2 is no longer an unbiased estimator of σ^2 (the same of course is true if any of the other assumptions (A1), (A2) and (A4) do not hold). Whether s^2 will be biased downwards or biased upwards will depend on the particular way assumption (A3) is violated. However, it can be shown that, if the deviations are related by (8.65) with a positive value for ρ (a very common case in practice), then s^2 will be biased downwards. Thus, using s^2, we will underestimate the variance of the true deviations, and thus will underestimate the variances of our estimated coefficients. This will give the impression of more precise estimation than is really the case. This is a very serious danger, and thus great care should be taken when using least squares estimation to check the validity of assumption (A3) (by methods described in the previous section); otherwise one might conclude that the estimated coefficients are more accurate than they really are.

Let us now return to the problem of transforming the original relationship in such a way that the transformed relationship satisfies all the conditions necessary for least squares estimation to give best linear unbiased estimators.

Our relationship is:

$$Y_t = \alpha + \beta X_t + u_t \quad t = 1, 2, \ldots, n \tag{8.66}$$

where (A1), (A2) and (A4) hold, but (A3) is replaced by:

$$u_t = \rho u_{t-1} + \epsilon_t \tag{8.67}$$

where the usual assumptions hold with respect to the deviation term ϵ of (8.67).

Lagging (8.66) by one period, we get

$$Y_{t-1} = \alpha + \beta X_{t-1} + u_{t-1} \quad t = 2, 3, \ldots, n. \tag{8.68}$$

Subtracting (8.68) multiplied by ρ from (8.66) we get:

$$Y_t - \rho Y_{t-1} = \alpha - \alpha\rho + \beta X_t - \rho\beta X_{t-1} + u_t - \rho u_{t-1} \quad t = 2, 3, \ldots, n.$$

Using (8.67) we get:

$$Y_t - \rho Y_{t-1} = \alpha(1 - \rho) + \beta(X_t - \rho X_{t-1}) + \epsilon_t \quad t = 2, 3, \ldots, n. \tag{8.69}$$

Defining new variables Y' and X' by:

$$Y_t' = Y_t - \rho Y_{t-1} \quad t = 2, 3, \ldots, n$$

and

$$X_t' = X_t - \rho X_{t-1} \quad t = 2, 3, \ldots, n$$

we can write (8.69) as

$$Y_t' = \alpha' + \beta X_t' + \epsilon_t \quad t = 2, 3, \ldots, n \tag{8.70}$$

where

$$\alpha' = \alpha(1 - \rho).$$

Now, equation (8.70) satisfies all the assumptions necessary for least squares estimators to be best linear unbiased. (Note that the intercept is $\alpha(1 - \rho)$, but if we know ρ we can estimate α from an estimate of α'.)

Unfortunately, as the astute reader has probably noticed, if we have n observations on Y and X, we have only $(n-1)$ observations on Y' and X' (that is, Y_1' and X_1' are not defined). The fewer observations we have, the less accurate is our estimation. This would be a particularly severe problem if n were small in the first place. However we might be able to get round this problem by defining Y_1' and X_1' in terms of our original Y and X so that (8.70) is satisfied for $t=1$ as well. This we do in the next few paragraphs. Unfortunately the required analysis is somewhat complicated for an initial reading. The reader is therefore advised to omit the next few paragraphs on a first reading, and take up the discussion at the point * on page 363.

The obvious way of defining Y_1' and X_1' so that (8.70) is satisfied for $t=1$ as well is by:

$$Y_1' = Y_1 \text{ and } X_1' = X_1.$$

Then

$$Y_1' = \alpha + \beta X_1' + u_1 \quad \text{(from (8.66))}. \tag{8.71}$$

Compare this carefully with the relationship between Y_t' and X_t' for $t = 2, 3, \ldots, n$ (that is, (8.70)).

We note that the intercept in (8.71) is different from the intercept in (8.70). We could get round this problem by rewriting our relationship as:

$$Y_t' = \alpha Z_t + \beta X_t' + \epsilon_t \quad t = 1, 2, \ldots, n \tag{8.72}$$

where

$$Z_t = \begin{cases} 1 & \text{for } t = 1 \\ (1 - \rho) & \text{for } t = 2, 3, \ldots, n \end{cases}$$

Thus (8.72) reduces to (8.70) for $t = 2, 3, \ldots, n$ and reduces to

$$Y_1' = \alpha + \beta X_1' + \epsilon_1 \quad \text{for } t = 1.$$

Note that this is not quite the same as (8.71) — the difference being that ϵ_1 replaces u_1. However, we can define ϵ_1 in (8.72) as being u_1. Note that in doing this we still maintain independence between all the ϵ_t, since u_1 does not depend on $\epsilon_2, \epsilon_3, \ldots, \epsilon_n$. We still have a problem though with (8.72) since the variance of ϵ_1 (that is u_1) will be different from σ_ϵ^2 (the variance of ϵ_t for $t = 2, 3, \ldots, n$). We can see this as follows: from (8.67)

$$u_t = \rho u_{t-1} + \epsilon_t$$

and so

$$u_{t-1} = \rho u_{t-2} + \epsilon_{t-1}$$

thus

$$u_t = \rho(\rho u_{t-2} + \epsilon_{t-1}) + \epsilon_t$$

$$= \rho^2 u_{tP2} + \epsilon_t + \rho \epsilon_{t-1}$$

Continuing in this way, we find

$$u_t = \epsilon_t + \rho \epsilon_{t-1} + \rho^2 \epsilon_{t-2} + \ldots$$

Thus

$$E u_t = E \epsilon_t + \rho E \epsilon_{t-1} + \rho^2 E \epsilon_{t-2} + \ldots$$

$$= 0 \quad \text{since each } \epsilon_t \text{ has zero expected value.}$$

And so

$$\text{var } u_t = E u_t^2 = E(\epsilon_t + \rho \epsilon_{t-1} + \rho^2 \epsilon_{t-2} + \ldots)^2$$

Thus, using the assumptions made about the ϵ_t

$$\text{var } u_t = \sigma_\epsilon^2 + \rho^2 \sigma_\epsilon^2 + \rho^4 \sigma_\epsilon^2 + \ldots$$

$$= \sigma_\epsilon^2(1 + \rho^2 + \rho^4 + \ldots)$$

that is

$$\text{var } u_t = \sigma_\epsilon^2/(1 - \rho^2) \quad (-1 < \rho < 1). \tag{8.73}$$

Thus the variance of ϵ_1 (if we define ϵ_1 as u_1) is $1/(1 - \rho^2)$ times the variance of the other ϵ_t ($t = 2, 3, \ldots, n$).

However, this now gives us the required clue as to how to define Y_1' and X_1'. Define them by:

$$Y_1' = \sqrt{(1 - \rho^2)} Y_1 \qquad X_1' = \sqrt{(1 - \rho^2)} X_1 \tag{8.74}$$

then, from (8.66)

$$Y_1' = \alpha \sqrt{(1 - \rho^2)} + \beta X_1' + \epsilon_1 \tag{8.75}$$

where

$$\epsilon_1 = \sqrt{(1 - \rho^2)} u_1 \quad \text{and so} \quad \text{var } \epsilon_1 = (1 - \rho^2) \text{var } u_1 = \sigma_\epsilon^2 \quad \text{(from (8.73)).}$$

Combining (8.70) and (8.75) our relationship can be written:

$$Y_t' = \alpha Z_t + \beta X_t' + \epsilon_t \qquad t = 1, 2, \ldots, n \tag{8.76}$$

where Z_t is defined by

$$Z_1 = \sqrt{(1 - \rho^2)}$$

$$Z_t = (1 - \rho) \qquad t = 2, 3, \ldots, n.$$

Now (8.76) is a relationship satisfying all the assumptions necessary for the least squares estimators of α and β to be the best linear unbiased estimators.

Note that (8.76) is a multiple regression. Comparing (8.70) and (8.76) we see that the only difference is that the former is a relationship on which we have $(n - 1)$ observations, while we have n observations on the latter. If n is large, the additional complications needed to derive (8.76) could be avoided by just using (8.70).

*Let us for simplicity consider the case where we have a large number of observations and thus use (8.70). In order to estimate α and β from (8.70) we need to know ρ. If we did not know ρ we would be unable to calculate the transformed variables Y' and X'.

Unfortunately, in practice we are unlikely to know the value of ρ, and thus have to use an estimate of it. ρ is defined by the coefficient of u_{t-1} in (8.67), and we have discussed the problems of estimating ρ in the previous section. Regressing our estimated deviations e_t on e_{t-1} would, in general, gives us an underestimate of ρ. The problems involved in satisfactorily estimating ρ take us into the realms of econometrics. However, even at our level it is obviously better to use the estimate of ρ obtained as above to find Y' and X', and then to estimate α and β using equation (8.70) (or (8.76)), rather than ignoring the problem altogether.

Let us illustrate the procedure using our time-series income consumption relationship. We found in the previous section that the Durbin–Watson statistic rejected the null hypothesis of an autocorrelation of the deviations in favour of the alternative hypothesis of positive autocorrelation. If we regress e_t on e_{t-1} we get an estimate of ρ given by

$$\hat{\rho} = \frac{\sum\limits_{2}^{n} e_t e_{t-1}}{\sum\limits_{2}^{n} e_{t-1}^{2}}.$$

Using the data in table 8.6, this gives $\rho = 2{,}771.88/3{,}350.88 = 0.8272$ Hence our variables Y' and X' are given by:

$$Y_2' = Y_2 - 0.8272Y_1 = 306.2 - (0.8272)(301.4) = 56.882$$
$$Y_3' = Y_3 - 0.8272Y_2 = 312.8 - (0.8272)(306.2) = 59.511$$

etc.

$$X_2' = X_2 - 0.8272X_1 = 468.3 - (0.8272)(477.6) = 73.229$$
$$X_3' = X_3 - 0.8272X_2 = 487.7 - (0.8272)(468.3) = 100.322$$

etc.

*Resumption point for readers who omitted the discussion of the definition of Y_1 and X_1.

Continuing in this way, we get the estimated relationship between Y' and X':

$$Y_t' = 1.88 + 0.6027X_t' + e_t' \qquad R^2 = 0.9498$$
$$(0.4) \quad (22.2) \qquad\qquad d = 1.29$$

(t-ratios in brackets). $\qquad\qquad t = 2, 3, \ldots, 29$

Notice that the value of the Durbin–Watson statistic for this transformed equation is 1.29. The appropriate critical values (one explanatory variable, 28 observations) are $d_L = 1.33$ and $d_U = 1.48$ at the 5 per cent significance level. Thus the calculated value of the Durbin–Watson statistic still rejects the null hypothesis of no autocorrelation of the deviations in the transformed equation at the 5 per cent significance level. However, the amount of autocorrelation has clearly been reduced considerably. These two remarks suggest that the actual value of p is higher than the 0.8272 used in the transformation (as was foreseen), and that a higher value might reduce the autocorrelation further.

Transforming back to the original form, remembering that 1.88 is an estimate of $\alpha(1 - \rho)$, that is 0.1778α, we get

$$Y_t = 10.94) + 0.6027X_t + e_t$$
$$(0.4) \quad (22.2)$$

(t-ratios in brackets).

Comparing this with our original estimate of chapter 7 (which did not take account of the autocorrelation):

$$Y_t = -0.28 + 0.6133X_t + e_t$$
$$(-0.04) \quad (67.0)$$

(t-ratios in brackets),

we see that, although the estimated coefficient of X has not changed much, its accuracy (as reflected in its t-ratio) has decreased considerably. This, as we discussed earlier, was to be expected.

Penultimately, let us consider what effects the violation of assumption (A4) has on the analysis of chapter 7. We see that the validity of (A4) is crucial to the unbiasedness property of least squares estimation. Let us take up the discussion, starting at (7.29):

$$\hat{\beta} = \beta + w_1 u_1 + w_2 u_2 + \ldots + w_n u_n$$

Our argument went as follows: if (A4) holds, then each X_i is the same in all samples, each x_i is the same in all samples, and so each w_i is the same in all samples. Thus

$$E(w_i u_i) = w_i E u_i \qquad (8.77)$$

Thus, using assumption (A1) in addition, we showed that $E\hat{\beta} = \beta$.

However, if assumption (A4) is not true, equation (8.77) is incorrect, and we are unable to show that $E\hat{\beta} = \beta$ by that method. We have said several times that we can replace (A4) by a weaker assumption (A4)$'$, namely

X_i and u_j are independent, for all $i, j = 1, 2, \ldots, n$ (A4)$'$.

Let us see, therefore, how we can prove unbiasedness using (A4)$'$ instead of (A4).

If (A4)$'$ is true, then $\overline{X}$ and u_j are independent for all j, and thus x_i and u_j are independent for all i and j. Hence each $w_i = x_i/(\Sigma_1^n x_j^2)$ is independent of u_j, for all i and j.

Now, with w_i and u_j independent for all i and j, and, in particular, for $i = j$, we can expand $E(w_i u_i)$ as

$E(w_i u_i) = (E w_i)(E u_i)$ (from (3.51))

and thus, using assumption (A1), $E(w_i u_i) = 0$ for all i. Hence $E\hat{\beta} = \beta$.

In a similar fashion, all our stages can be validated using assumption (A4)$'$ instead of (A4).

Notice that we need u_j independent of *all* the X_i, for otherwise w_j (which contains all the x_i) would depend upon u_j. Notice also that our original assumption (A4) is just a special case of our new assumption (A4)$'$.

Thus, if (A4)$'$ is violated, then our least squares estimators are not even unbiased. Obviously the rest of our analysis breaks down if assumption (A4)$'$ does not hold. The validity of assumption (A4)$'$ is thus of crucial importance. As our discussion of the testing of (A4)$'$ in the previous section showed, we are unable to test statistically for its validity; we thus rely very heavily for the usefulness of our analysis on the light that economic theory sheds as to the validity or otherwise of this assumption.

We will consider in some detail in the next chapter one of the main causes of the breakdown of this assumption, and the possible remedies. We will consider briefly in this section one case where (A4)$'$ is not true, but for which least squares estimators do have one desirable property.

Consider the following relationship:

$$Y_t = \alpha + \beta Y_{t-1} + u_t \quad t = 0, 1, 2, \ldots, n. \tag{8.78}$$

Suppose assumptions (A1), (A2) and (A3) hold, and that $|\beta| < 1$ so that the process is not explosive.

Assumption (A4) cannot hold however. The value that Y_t takes is dependent on u_t. Similarly Y_{t-1} depends on u_{t-1}; Y_{t-1} also depends on $u_{t-2}, u_{t-3}, \ldots$, via its dependence on Y_{t-2}, Y_{t-3}, $\ldots$. Thus the variable 'X_t' (in the case Y_{t-1}) is not independent of u_s for all s. 'X_t' (that is Y_{t-1}) *is* independent of $u_t, u_{t+1}, u_{t+2}, \ldots$, but is dependent on $u_{t-1}, u_{t-2}, \ldots$. Thus (A4) is violated, and the least squares estimators of α and β in (8.78) will not be unbiased. However, the least squares estimators do have one important property — they are consistent. A rigorous proof is outside the scope of this book. The following intuitive proof should, however, give some insight. Consider (7.29), as modified for the notation of this example:

$$\hat{\beta} = \beta + \frac{\sum_{1}^{n} y_{t-1} u_t}{\sum_{1}^{n} y_{t-1}^2} \qquad (8.79)$$

where

$$y_{t-1} = Y_{t-1} - \bar{Y} \qquad \bar{Y} = \frac{1}{n} \sum_{1}^{n} Y_{t-1}.$$

The second term on the right hand side of (8.79) can be written:

$$\frac{\frac{1}{n} \sum_{1}^{n} y_{t-1} u_t}{\frac{1}{n} \sum_{1}^{n} y_{t-1}^2}. \qquad (8.80)$$

The numerator of this expression is the sample covariance of Y_{t-1} and u_t; the denominator is the sample variance of Y_{t-1}. Intuitively, we can see that, as n increases, so that the sample approaches the population, the sample covariance between Y_{t-1} and u_t approaches the population covariance between Y_{t-1} and u_t. Now Y_{t-1} and u_t *are* independent (Y_{t-1} depends on u_{t-1}, u_{t-2}, ..., but not on u_t, u_{t+1}, ...), and thus their population covariance is zero. Hence the numerator of (8.80) approaches zero as the sample size (n) approaches infinity. By similar reasoning, the denominator of (8.80) approaches the population variance of Y_{t-1}, which is non-zero (and finite if $|\beta| < 1$), since Y_{t-1} is not constant. Hence the expression in (8.80) approaches zero, and thus from (8.79) $\hat{\beta}$ approaches β as the sample size approaches infinity. Thus the least squares estimator of β, although biased, is consistent. A rigorous proof of this intuitively demonstrated result can be found in Johnston (14).

Thus, in this particular example (relationship (8.78)), one particular property of least squares estimation can be salvaged from the collapse resulting from the violation of assumption (A4). That this is not always possible will be seen in the next chapter.

Finally, let us turn to assumption (A5). Notice that this was only used in stages 5, 7 and 8. If (A5) is violated, but (A1) to (A4) are valid, all the other stages would remain valid. If (A5) is not true, and the u_i are not normal, then, unless we can postulate a specific alternative distribution for the u_i, we will be unable to find the distributions of $\hat{\alpha}$, $\hat{\beta}$ or s^2. This would make it impossible for us to use our estimates for forming confidence intervals for, and testing hypotheses about, the population parameters. However, if we have a specific alternative distribution for the u_i, we will be able to derive the distribution of our estimators, and thus proceed in the usual fashion. However, very little investigation into alternative distributions of the u_i has been undertaken. Most studies use assumption (A5), and rely on the Central Limit Theorem for their justification for its use.

8.7 Summary

In many ways this chapter is the most important of the whole book, as it is concerned with the validity of the assumptions used in chapter 7. Before we discussed this important topic, we first showed how the methods of chapter 7 could be generalised to the analysis of relationships where the dependent variable was postulated to depend on several independent variables. The remainder of the chapter was then devoted to discussion of the validity of the assumptions used in regression analysis.

We began by considering whether economic theory itself could shed any light on the validity of the assumptions. We saw that, although economic theory could not prove or disprove the validity of the assumptions, it might well shed useful light on their plausibility. We then went on to consider how we may test statistically for the validity of the assumptions and saw that, for some of the assumptions, statistical tests could be derived. Finally, we considered the important problem of determining what effect violated assumptions would have on the analysis of chapter 7. Having shown that violated assumptions may alter drastically the results of chapter 7, we concluded by considering how we may modify our analysis if, in fact, these assumptions do not appear to be valid.

The important 'message' to emerge from this chapter is that conclusions based on false assumptions may be not only misleading, but also dangerously wrong. It is therefore crucial in empirical work that a constant check is kept on the validity of the assumptions, and a constant awareness is maintained of the likely consequences of false assumptions, so that the necessary modifications to the analysis may be implemented.

8.8 Exercises

8.1. It is proposed to estimate the coefficients of the following equation:

$$Y_i = \alpha + \beta X_i + \gamma Z_i + u_i \quad i = 1, 2, \ldots, n$$

Discuss the use of (ordinary) least squares in the following cases:

(a) Y is aggregate consumption (in real terms);
 X is aggregate income (in real terms);
 Z is aggregate consumption lagged one period (in real terms);
 i is over time for one country.

(b) Y is average household expenditure on beef;
 X is price of beef;
 Z is average household income;
 i is over time.

(c) Y is rate of change of money wage rates;
 X is percentage unemployment;
 Z is rate of change in prices;
 i is over time for the U.S.

8.2. By finding suitable data, estimate the relationship in exercise 8.1 for cases (a) and (c). In the light of your results investigate the validity of the assumptions you have used.

8.3. By using the data in a copy of the *Current Population Reports*, estimate a consumption function for a consumption good of your choice. As usual, investigate the validity of the assumptions, and modify your analysis if and where necessary.

8.4. Indicate how you would set about proving (8.17).

8.5. By using (8.28) show that $\hat{\beta}$ and $\hat{\gamma}$ are consistent estimators of β and γ respectively.

8.6. Suppose that X and Y are related by

$$Y_i = \alpha + \beta X_i + u_i \quad (i = 1, 2, \ldots, n)$$

and that all the usual assumptions hold except (A2) which is replaced by

$$Eu_i^2 = \sigma^2 X_i \quad (i = 1, 2, \ldots, n).$$

Discuss the consequences for the properties of the least squares estimators when (A2) is violated in this way. Find a transformed equation for which least squares estimators of its parameters retain the usual desirable properties.
   Repeat this exercise for the case when (A2) is replaced by

$$Eu_i^2 = \sigma^2 / X_i^2 \quad (i = 1, 2, \ldots, n).$$

8.7. Suppose that X and Y are related by

$$Y_t = \alpha + \beta X_t + u_t \quad (t = 1, 2, \ldots, n)$$

and that all the usual assumptions hold except (A3) which is replaced by

$$u_t = \rho_1 u_{t-1} + \rho_2 u_{t-2} + \epsilon_t$$

where the deviation (ϵ_t) of this equation obeys all the usual assumptions. Discuss the consequences for the properties of the least squares estimators when (A3) is violated in this way. Find a transformed equation for which the least squares estimators of its parameters retain the usual desirable properties.

8.8. Show that the least squares estimator $\hat{\alpha}$ of α in the relation

$$Y_i = \alpha + u_i \quad (i = 1, 2, \ldots, n)$$

is given by

$$\hat{\alpha} = \bar{Y} = \frac{1}{n} \sum_1^n Y_i.$$

Show that, if the usual assumptions hold with respect to the deviation terms, then

$$E\hat{\alpha} = \alpha \quad \text{var } \hat{\alpha} = \sigma^2 / n.$$

Relate these results to the material of chapter 5.

8.9. By referring to exercise 7.12, prove that the least squares estimator $\hat{\rho}$ in $e_t = \hat{\rho} e_{t-1} + v_t$, where we have observations e_t $(t = 1, 2, \ldots, n)$, is given by

$$\hat{\rho} = \sum_2^n e_t e_{t-1} \bigg/ \sum_2^n e_{t-1}^2.$$

8.10. Select, from the reading list for your macroeconomics course, a number of references on empirical economics. Study these references carefully, paying particular attention to the validity of the assumptions employed. Criticise the statistical methodology and suggest improvements where relevant.

8.11. Choose a specific area of economics that particularly interests you. Express the alternative theories (including your own) in testable form. Find the appropriate data and investigate which theory best explains the evidence. Pay particular attention to investigating the validity of the assumptions you have employed.

9

Estimation of Simultaneous Models

9.1 Introduction

In the last two chapters we have considered ways of testing economic theories which explain the behaviour of one variable in terms of one or more other variables. We need now to broaden our approach so that we can test economic theories that explain the behaviour of several variables. Consider for example the theory of the determination of equilibrium in the market for a particular good: the quantity supplied is postulated to be an increasing function of price and the quantity demanded to be a decreasing function of price. Equilibrium is defined where quantity supplied equals quantity demanded, and this determines equilibrium price and equilibrium quantity. We can express this theory in the form of a *model:*

$$Q_S = f(P) \tag{9.1}$$

$$Q_D = g(P) \tag{9.2}$$

$$Q_S = Q_D \tag{9.3}$$

where P is price and Q_S and Q_D are the quantity supplied and demanded respectively. (9.1) represents the supply curve, where $f(P)$ is an increasing function of P, and (9.2) represents the demand curve, where $g(P)$ is a decreasing function of P. (9.1) and (9.2) are *behavioural equations*, expressing the theories' conclusions about peoples' behaviour. (9.3) is an equilibrium condition. The equilibrium values of price and quantity are found by solving the three equations. We can see this clearly if we specify particular forms for the functions f and g. Suppose they are linear. Our model can then be written:

$$Q_S = \alpha + \beta P \tag{9.4}$$

$$Q_D = \delta - \theta P \tag{9.5}$$

$$Q_S = Q_D. \tag{9.6}$$

The parameters, α, β, δ and θ, are referred to as the *structural parameters*, or the *parameters of the structure*. The structure or *structural form* is another name for the model as given by equations (9.4) to (9.6).

Solving the model, we can find equilibrium price and quantity, as follows: substituting (9.4) and (9.5) into (9.6) gives

$$\alpha + \beta P = \delta - \theta P$$

that is,

$$P = \frac{\delta - \alpha}{\beta + \theta} \qquad (9.7)$$

Substituting this back into either (9.4) or (9.5) gives equilibrium quantity $Q = Q_S = Q_D$:

$$Q = \frac{\alpha\theta + \beta\delta}{\beta + \theta} \qquad (9.8)$$

(9.7) and (9.8) thus give the equilibrium values for price and quantity in the market represented by (9.4) to (9.6).

Suppose we now extend our simple model to allow for shifts in the supply and demand curves. One variable that shifts the supply curve is W, the wage rate. Quantity demanded will depend on consumers' income Y. Thus our model is

$$Q_S = \alpha + \beta P - \gamma W \qquad (9.9)$$

$$Q_D = \delta - \theta P + \lambda Y \qquad (9.10)$$

$$Q_S = Q_D . \qquad (9.11)$$

(All structural parameters are assumed positive.)

Now the equilibrium values of price and quantity depend upon the values of W and Y. Given the values of W and Y, the equilibrium values of Q and P are determined by the simultaneous solution of the model as given by (9.9) to (9.11). We generally assume in this simple model that the values of Q and P do not influence W and Y. Thus the values of W and Y are determined *outside* the model (9.9) to (9.11), while the values of P and Q are determined *within* the model. To distinguish these two types of variables we term W and Y as *exogenous variables*, and P and Q as *endogenous variables*.

Now, as we have remarked before, the equilibrium values of the endogenous variables depend on the values of the exogenous variables. To make this explicit, let us solve for these equilibrium values. By substituting (9.9) and (9.10) into (9.11) we get:

$$\alpha + \beta P - \gamma W = \delta - \theta P + \lambda Y$$

that is,

$$P = \frac{\delta - \alpha}{\beta + \theta} + \frac{\gamma W}{\beta + \theta} + \frac{\lambda Y}{\beta + \theta} . \qquad (9.12)$$

Substituting this back into either (9.9) or (9.10) give equilibrium quantity $Q = Q_S = Q_D$:

$$Q = \frac{\alpha\theta + \beta\delta}{\beta + \theta} - \frac{\gamma\theta W}{\beta + \theta} + \frac{\beta\lambda Y}{\beta + \theta}. \tag{9.13}$$

(9.12) and (9.13) show how the equilibrium values of the endogenous variables react to changes in the exogenous variables. We see that if W increases, equilibrium price rises and equilibrium quantity falls, and if Y increases both equilibrium price and quantity rise.

Equations (9.12) and (9.13) are termed the *reduced form equations* of the model given by (9.9) to (9.11). The reduced form equation for price (9.12) shows how price reacts to changes in the exogenous variables; similarly the reduced form equation for quantity (9.13) shows how quantity reacts to changes in the exogenous variables.

Notice that equations (9.7) and (9.8) are the reduced form equations of the model given by (9.4) to (9.6) — in that model there were no exogenous variables.

Having considered a simple theoretical model, we now consider how we might investigate it empirically. There are several things we might wish to do. First, we might wish to investigate the conclusions of the model concerning the reaction of equilibrium price and quantity to changes in the exogenous variables. To do this all we need to do is to test and estimate the reduced form equations (9.12) and (9.13). Secondly, we might wish to investigate the component parts of the theory, that is to test and estimate the individual behavioural equations (9.9) and (9.10). Thirdly, we might just be interested in one of the behavioural equations, say the demand curve (9.10). However if the data we observe are actually generated by the simultaneous solution of the model we cannot ignore the effect of the supply curve on the data; we must allow for this effect when we test and estimate the demand curve.

We now consider a second simple economic theory, the conclusions of which concern the behaviour of more than one variable. We consider the simplest model of income determination:

$$C = \alpha + \beta Y \tag{9.14}$$

$$Y = C + I. \tag{9.15}$$

Equilibrium income is determined where income equals total expenditure; consumption expenditure is an increasing function of income. I, which is non-consumption expenditure, is considered in this model to be exogenous — its value is determined outside the model as given by (9.14) and (9.15). C and Y are endogenous variables. Here again the equilibrium values of the endogenous variables depend on the value of the exogenous variable. We can see this explicitly by finding the reduced form of the model, that is expressing each endogenous variable in terms of the exogenous variable. This yields

$$C = \frac{\alpha}{1-\beta} + \frac{\beta}{1-\beta} I \qquad (9.16)$$

$$Y = \frac{\alpha}{1-\beta} + \frac{1}{1-\beta} I. \qquad (9.17)$$

The reduced form equation for Y (9.17) is the familiar multiplier equation, showing how an increase in the exogenous variable I, by ΔI leads to a (multiplied) increase in Y of $\Delta I/(1-\beta)$.

Note that the parameters of the reduced form depend upon the parameters (α and β) of the structural form (the same was true in our supply and demand model). Again let us consider how we might investigate this simple model empirically. If all we are interested in is the reaction of C and Y to changes in I we can test and estimate the reduced form equations (9.16) and (9.17). If we wish to investigate the behavioural equation of the model then we need to test and estimate the structural equation (9.14). However if the data we observe were generated by the simultaneous solution of the model, we cannot ignore the rest of the model when we investigate (9.14). We must take into account the fact that C and Y were simultaneously determined, given the value of I.

We now consider how we might investigate the model as given by (9.14) and (9.15). First however we must introduce a deviation term into our consumption function to allow for the factors other than income that influence consumption (we cannot in practice hold these constant). Our model is thus:

$$C_t = \alpha + \beta Y_t + u_t \qquad (9.18)$$
$$Y_t = C_t + I_t. \qquad (9.19)$$

Our variables are:

C_t aggregate consumption expenditure in period t
Y_t aggregate income in period t
I_t aggregate non-consumption expenditure in period t.

We assume that assumptions (A1) and (A3) hold with respect to the deviation term. Let us also clarify our assumption of I as an exogenous variable in the model. This exogeneity assumption means that the value of I in all periods is determined *outside* the model as given by (9.18) and

(9.19) – in particular this means that the value of the deviation term u_s does not influence I_t for any s or t.

We consider in the next section the properties of least squares estimators in the context of this particular simultaneous model. The same ideas carry over to other simultaneous models.

9.2   Least squares estimation in a simultaneous model

We noted above that investigating a simultaneous model could mean investigating the structural form or investigating the reduced form – which we do depending on the use to which our investigation is to be put. The structural form of our simple income determination model is

$$C_t = \alpha + \beta Y_t + u_t \tag{9.18}$$

$$Y_t = C_t + I_t. \tag{9.19}$$

The reduced form is found by solving for the equilibrium values of the endogenous variables $(C$ and $Y)$ in terms of the exogenous variable I, and the deviation term. The reduced form is

$$C_t = a + bI_t + v_t \tag{9.20}$$

$$Y_t = c + dI_t + v_t \tag{9.21}$$

where

$$\left. \begin{array}{ll} a = \dfrac{\alpha}{1-\beta} \qquad b = \dfrac{\beta}{1-\beta} \\[3mm] \qquad\qquad\qquad v_t = \dfrac{u_t}{1-\beta}. \\[3mm] c = \dfrac{\alpha}{1-\beta} \qquad d = \dfrac{1}{1-\beta} \end{array} \right\} \tag{9.22}$$

(9.20) and (9.21), the reduced form equations for C and Y respectively, are of course the same as (9.16) and (9.17) except for the inclusion of the deviation term.

Suppose first that we are interested in the reduced form equations. Consider (9.20): in this the deviation term is $v_t = u_t/(1-\beta)$. If assumptions (A1) to (A3) hold with respect to u_t they must also hold with respect to v_t. Further we have assumed that I_t is exogenous, and thus independent of u_s for all s and t. Thus I_t and v_s are independent for all values of s and t. Hence the reduced form equation for C satisfies all the assumptions necessary for least squares estimators of its parameters (a and b) to be best linear unbiased. Denote, as usual, the least squares estimators of a and b, based on a sample of n observations on (C_t, I_t) $(t = 1, 2, \ldots, n)$ by

$\hat{a}$ and $\hat{b}$. Thus, as usual $\hat{a}$ and $\hat{b}$ are given in terms of our observations by:

$$\hat{b} = \frac{\sum\limits_{1}^{n}(C_t - \bar{C})(I_t - \bar{I})}{\sum\limits_{1}^{n}(I_t - \bar{I})^2} \qquad (9.23)$$

$$\hat{a} = \bar{C} - \hat{b}\bar{I}. \qquad (9.24)$$

Thus, as (9.20) satisfies all the assumptions made in chapter 7, we have

$$E\hat{a} = a \quad E\hat{b} = b$$

$$\left.\begin{array}{cc} \text{var } \hat{a} = \dfrac{\sigma_v{}^2 \sum\limits_{1}^{n}I_t{}^2}{n\sum\limits_{1}^{n}(I_t - \bar{I})^2} & \text{var } \hat{b} = \dfrac{\sigma_v{}^2}{\sum\limits_{1}^{n}(I_t - \bar{I})^2} \end{array}\right\} \qquad (9.25)$$

where $\sigma_v{}^2$ is the variance of v_t (for all t).

Notice that $\hat{b}$ is not only an unbiased estimator of b, but it is also consistent: as it is unbiased, it is asymptotically unbiased, and also its variance tends to zero as the sample size tends to infinity (examine (9.25)).

In order to simplify the following algebra, we denote the covariance between X and Y, based on n observations $X_1, X_2, \ldots, X_n$ and $Y_1, Y_2, \ldots, Y_n$, by m_{XY} that is,

$$m_{XY} = \frac{1}{n}\sum\limits_{1}^{n}(X_t - \bar{X})(Y_t - \bar{Y})$$

in particular

$$m_{XX} = \frac{1}{n}\sum\limits_{1}^{n}(X_t - \bar{X})^2$$

Using this notation (9.23) can be written

$$\hat{b} = \frac{m_{CI}}{m_{II}}. \qquad (9.26)$$

By a similar argument the least squares estimators of the parameters of the reduced form equation for Y are best linear unbiased and consistent.

$$\hat{d} = \frac{m_{YI}}{m_{II}} \qquad \hat{c} = \bar{Y} - \hat{d}\bar{I}$$

and

$$E\hat{c} = c \qquad E\hat{d} = d. \tag{9.27}$$

Thus we see that no problems arise in estimating the reduced form equations of our model. Least squares estimation provides best linear unbiased (and consistent) estimators of the reduced form parameters.

Suppose now that we are interested in estimating the structural form of our model. Of the two equations that form the model, only one is a behavioural equation, the other being an equilibrium condition. Consider therefore the structural equation:

$$C_t = \alpha + \beta Y_t + u_t. \tag{9.28}$$

Now we are assuming that assumptions (A1) (A2) and (A3) hold with respect to the deviation term. So far so good. Now, what can we say about the validity of assumption (A4)'? In the notation of equation (9.28) the validity of assumption (A4)' requires that Y_t and u_s are independent for all s and t. However, we know that this is not true. The reduced form equation for Y (9.21) explicitly shows that u_t influences Y_t (v_t is just a multiple of u_t). In particular (9.21) shows that u_t and Y_t are positively related – when u_t increases so does Y_t. This is immediately obvious: as u_t increases, so does *ex ante* consumption expenditure and hence equilibrium between income and expenditure is attained at a higher level of income. Thus assumption (A4)' cannot be valid with respect to the consumption function (9.28). We showed in section 8.6 that if (A4)' is violated then least squares estimators are no longer unbiased. Let us work through this again. The least squares estimators of α and β in (9.28) are $\hat{\alpha}$ and $\hat{\beta}$ given by:

$$\hat{\beta} = \frac{\sum\limits_{1}^{n} c_t y_t}{\sum\limits_{1}^{n} y_t^{2}} \qquad \text{where } c_t = C_t - \bar{C} \qquad \bar{C} = \frac{1}{n}\Sigma C_t$$

$$y_t = Y_t - \bar{Y} \qquad \bar{Y} = \frac{1}{n}\Sigma Y_t. \tag{9.29}$$

$$\hat{\alpha} = \bar{C} - \hat{\beta}\bar{Y}$$

Following through the algebra from (7.22) to (7.25) we can express $\hat{\beta}$ in terms of β by (cf. (7.25)):

$$\hat{\beta} = \beta + \frac{\sum\limits_{1}^{n} y_t u_t}{\sum\limits_{1}^{n} y_t^{2}}. \tag{9.30}$$

But as Y_t and u_t are dependent it no longer follows that $E[(y_t u_t)/\Sigma y_t^2]$ is equal to $E(y_t/\Sigma y_t^2)Eu_t$, and hence in general the second term on the right-hand side of (9.30) does not have zero expected value. Hence the expected value of $\hat{\beta}$ is not equal to β, and thus the least squares estimator of β in (9.28) is biased. A similar result can be shown for the least squares estimator of α.

This bias in the least squares estimators results from the dependence of Y on u, a consequence of the simultaneous determination of C and Y. (Thus, even if we are only interested in the consumption function, we cannot ignore the rest of the model.)

In section 8.6 we considered another relationship in which assumption (A4)' was violated, and for which least squares estimation therefore provided biased estimators. However, in that case, we were able to show that the least squares estimators retained one desirable property — namely consistency. We now investigate whether this property is retained in the least squares estimators of the parameters of the consumption function.

Consider (9.30), this can be expressed in terms of our covariance notation:

$$\hat{\beta} = \beta + \frac{m_{Yu}}{m_{YY}} .$$
(9.31)

For $\hat{\beta}$ to be a consistent estimator of β we require that m_{Yu}/m_{YY} approaches zero as the sample size approaches infinity. Intuitively we can see that it does not. m_{Yu} is the sample covariance between Y_t and u_t; m_{Yu} will approach the population covariance between Y_t and u_t as the sample size approaches infinity. However this population covariance is not zero, as Y_t and u_t are dependent. Similarly, m_{YY} the sample variance of Y_t approaches the (finite) population variance of Y_t. Thus m_{Yu}/m_{YY} approaches some positive value as the sample size approaches infinity. Thus $\hat{\beta}$ does not approach β, and thus the least squares estimator of β is inconsistent. A similar result can be shown for $\hat{\alpha}$.

We can demonstrate this result in a slightly more rigorous way by expressing (9.31) in terms of I and u. Now Y is given in terms of I by the reduced form equation for Y (9.21):

$$Y_t = c + dI_t + v_t.$$

Averaging,

$$\bar{Y} = c + d\bar{I} + \bar{v}.$$

Subtracting,

$$Y_t - \bar{Y} = d(I_t - \bar{I}) + (v_t - \bar{v}).$$
(9.32)

Now

$$v_t = \frac{u_t}{1 - \beta} = du_t \text{ from (9.22) and so } v_t - \bar{v} = d(u_t - \bar{u}).$$

Thus (9.32) becomes:

$$Y_t - \bar{Y} = d(I_t - \bar{I}) + d(u_t - \bar{u}). \tag{9.33}$$

Multiplying each term in (9.33) by $(u_t - \bar{u})$ and summing gives:

$$\sum_1^n (Y_t - \bar{Y})(u_t - \bar{u}) = d\sum_1^n (I_t - \bar{I})(u_t - \bar{u}) + d\sum_1^n (u_t - \bar{u})^2.$$

Dividing through by n, and using our covariance notation, we get:

$$m_{Yu} = dm_{Iu} + dm_{uu}. \tag{9.34}$$

Similarly

$$m_{YY} = d^2 m_{II} + 2d^2 m_{Iu} + d^2 m_{uu}. \tag{9.35}$$

Hence (9.31) can be expressed in terms of I and u, using (9.34) and (9.35) (and $d = 1/(1 - \beta)$) as

$$\hat{\beta} = \beta + \frac{(1 - \beta)(m_{Iu} + m_{uu})}{m_{II} + 2m_{Iu} + m_{uu}}. \tag{9.36}$$

Now I is assumed exogenous and therefore independent of u; thus the sample covariance, m_{Iu}, between I and u will approach zero as the sample size approaches infinity. Similarly m_{II} approaches σ_I^2, the population variance of I, and m_{uu} approaches σ^2, the variance of the deviations. Thus, from (9.36), we see that as the sample size approaches infinity, $\hat{\beta}$ approaches β plus

$$\frac{(1 - \beta)\sigma^2}{\sigma_I^2 + \sigma^2}. \tag{9.37}$$

This is obviously non-zero, and thus $\hat{\beta}$ is an inconsistent estimator of β. (9.37) expresses the asymptotic bias of $\hat{\beta}$ as an estimator of β. For β lying between 0 and 1 this asymptotic bias is positive. We see that the asymptotic bias will be greater, the larger σ^2 is, the smaller σ_I^2 is and the nearer to zero the true β is.

To illustate these results consider our illustrative example of section 7.4. We will use the same parameters, but rename the variables to conform with the notation of this chapter.

Thus we suppose that the true relationship between aggregate consumption and aggregate income in period t is given by:

$$C_t = 2 + 0.5Y_t + u_t \quad \text{(for all } t)\tag{9.38}$$

where the distribution of the deviation term u_t (for all t) is as given in table 9.1.

Complete the model with the equilibrium condition:

$$Y_t = C_t + I_t.\tag{9.39}$$

The reduced form of the model given by (9.38) and (9.39) is

$$C_t = 4 + I_t + 2u_t\tag{9.40}$$

$$Y_t = 4 + 2I_t + 2u_t\tag{9.41}$$

(compare these with the general reduced forms (9.20) and (9.21)).

Consider the extreme case $I_t = 0$ for all t. In this case (9.40) and (9.41) reduce to:

$$C_t = 4 + 2u_t$$

$$Y_t = 4 + 2u_t.$$

Thus, whatever u_t is, C_t and Y_t are equal; this should be obvious by examining (9.39).

Consider the three possible values of u_t:

if $u_t = -1$ $C_t = 2$ $Y_t = 2$

if $u_t = 0$ $C_t = 4$ $Y_t = 4$

if $u_t = 1$ $C_t = 6$ $Y_t = 6$

We illustrate in figure 9.1.

The points marked with a cross in figure 9.1 are the only possible observations that we can obtain (when I_t has been zero for all t) — they are the total population of all possible observations. Thus, even if we observed this total population, a line fitted to it will have a slope of unity — not 0.5 (the marginal propensity to consume). Thus the asymptotic bias is +0.5, which agrees with (9.37) with $\beta = 0.5$, $\sigma_I^2 = 0$ (I constant).

Note carefully that *if* income could be fixed at $Y_t = 2$, and u_t equalled 1, then C_t would be 4; *but* income can *not* be fixed at 2 if consumption is 4 (for I_t zero) since then (9.39) would not be satisfied.

Note also the implication of (9.37): if I_t remains constant at any value throughout our observation period, then the line fitted to all possible values of (C_t, Y_t) will have a slope of unity, whatever the true value of β.

Let us now suppose that I_t has varied during our observation period.

Table 9.1   Distribution
of u_t (for all t)

u_t	$P[u_t]$
-1	¼
0	½
1	¼

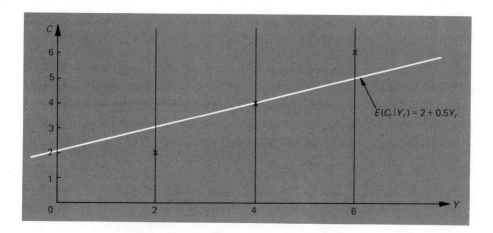

Fig. 9.1 (C, Y) observations for $I = 0$

Specifically, suppose that I_t has taken three values:

$$I_1 = 0 \quad I_2 = 1 \quad I_3 = 2.$$

In each time period, u_t may take values -1, 0 or 1 with probabilities ¼, ½, ¼ respectively. By substituting in (9.40) and (9.41) we get the following possible values for C_t and Y_t for each of the three values of I.

$I_1 = 0$ if $u_1 = -1$ $C_1 = 2$ $Y_1 = 2$

              if $u_1 = 0$ $C_1 = 4$ $Y_1 = 4$

              if $u_1 = 1$ $C_1 = 6$ $Y_1 = 6$

(Note that $C_1 + I_1 = C_1 + 0 = Y_1$ for all u_1 values.)

$I_2 = 1$ if $u_2 = -1$ $C_2 = 3$ $Y_2 = 4$

              if $u_2 = 0$ $C_2 = 5$ $Y_2 = 6$

              if $u_2 = 1$ $C_2 = 7$ $Y_2 = 8.$

(Note that $C_2 + I_2 = C_2 + 1 = Y_2$ for all u_2 values.)

$$I_3 = 2 \quad \text{if } u_3 = -1 \quad C_3 = 4 \quad Y_3 = 6$$
$$\text{if } u_3 = 0 \quad C_3 = 6 \quad Y_3 = 8$$
$$\text{if } u_3 = 1 \quad C_3 = 8 \quad Y_3 = 10.$$

(Note $C_3 + I_3 = C_3 + 2 = Y_3$ for all u_3 values.) All possible values of (C_t, Y_t) are shown graphically in figure 9.2. (The number attached to each possible observation in figure 9.2 is the probability of it being observed, given the value of I.)

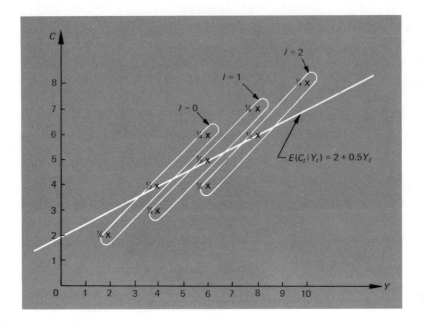

Fig. 9.2 (C, Y) observations for three values of I

If I can only take these three values, then an infinite sample of observations would consist of an infinite number of observations on these 9 possible (C, Y) pairs, such that the relative frequency of each point being observed is equal to its probability. The line fitted by least squares to this infinite sample (taking into account the differing relative frequencies of occurrence) is:

$$C_t = 5/7 + 5/7 Y_t + e_t.$$

This confirms our visual impression of an upward asymptotic bias. We can also see how a wider spread of values of I leads to a smaller asymptotic bias. Notice also the downward asymptotic bias in the least squares estimate of α.

The general result (9.37) for the asymptotic bias in the least squares estimation of β yields in this case (with $\beta = 0.5$, $\sigma_I^2 = \frac{2}{3}$, $\sigma^2 = \frac{1}{2}$) $(\frac{1}{2}\,\frac{1}{2})/(\frac{2}{3} + \frac{1}{2}) = 3/14$ which is confirmed by our results above.

Let us now illustrate how least squares estimation of the reduced form parameters yields unbiased and consistent estimators. Consider the reduced form equation for C ((9.40)):

$$C_t = 4 + I_t + 2u_t.$$

I_t takes values 0, 1 and 2, while u_t takes values -1, 0 and 1 with probabilities $\frac{1}{4}$, $\frac{1}{2}$ and $\frac{1}{4}$ respectively. The total population of all C_t values is shown graphically in figure 9.3. Thus we can see that the reduced form equation for C satisfies all the usual assumptions. For an infinite sample (that is, all 9 possible points observed with relative frequencies equal to their probabilities) the least squares fitted line is obviously:

$$C_t = 4 + I_t + e_t. \tag{9.42}$$

Thus there is no asymptotic bias in the least squares estimators of the parameters of the reduced form equation for C. The analogous result for the reduced form equation for Y could be similarly illustrated.

Summarising our discussion so far, we have shown that least squares estimation of the structural parameters of a simultaneous model yield

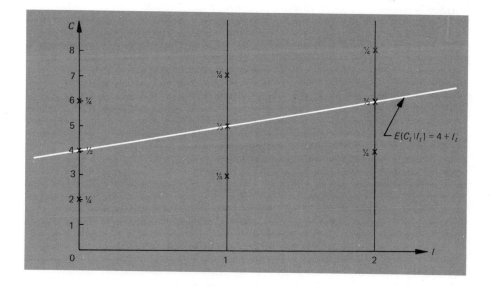

Fig. 9.3　(C, I) observations for three values of I

biased and inconsistent estimators. Further, least squares estimation of the reduced form parameters of a simultaneous model yield unbiased and consistent estimators. If all we are interested in is the reduced form, we need go no further. However, if we wish to have 'good' estimators of the structural parameters we clearly need to find an alternative estimation technique to least squares. One possible alternative technique should now be apparent. By examining (9.22) we see that the reduced form parameters (a, b, c, d) are expressed in terms of the structural parameters (α, β). Hence we should be able to express the structural parameters (α, β) in terms of the reduced form parameters (a, b, c, d). We have 'good' estimators of the reduced form parameters; let us therefore define estimators of the structural parameters in terms of the estimators of the reduced form parameters, so that the relationship between the estimators is the same as the relationship between the true parameters as given by (9.22). This method is known as *indirect least squares*. To distinguish between this method of estimating the structural parameters, and the method based on direct least squares estimation of the structural parameters, we will call this latter *ordinary least squares*. We now investigate in detail the use of indirect least squares.

9.3   Estimation by Indirect Least Squares

From (9.22) we have the following relationships between the true reduced form parameters of our model and the true structural parameters:

$$\left.\begin{array}{cc} a = \dfrac{\alpha}{1 - \beta} & b = \dfrac{\beta}{1 - \beta} \\[2em] c = \dfrac{\alpha}{1 - \beta} & d = \dfrac{1}{1 - \beta}. \end{array}\right\} \quad (9.43)$$

We can rearrange these to express the structural parameters in terms of the true reduced form parameters:

$$\left.\begin{array}{cc} \alpha = \dfrac{a}{d} & \beta = \dfrac{b}{b + 1} \\[2em] \alpha = \dfrac{c}{d} & \beta = 1 - \dfrac{1}{d}. \end{array}\right\} \quad (9.44)$$

Our least squares estimators of the reduced form parameters we have denoted by $\hat{a}$, $\hat{b}$, $\hat{c}$ and $\hat{d}$. These, as we have already shown, are unbiased and consistent estimators of a, b, c and d respectively. Let us therefore define estimators α^* and β^* of α and β by:

$$\left.\begin{array}{ll} \alpha* = \dfrac{\hat{a}}{\hat{d}} & \beta* = \dfrac{\hat{b}}{\hat{b}+1} \\[2em] \alpha* = \dfrac{\hat{c}}{\hat{d}} & \beta* = 1 - \dfrac{1}{\hat{d}}. \end{array}\right\} \quad (9.45)$$

Now it looks as though $\beta*$ defined by $b/(b+1)$ is different from $\beta*$ as defined by $1 - 1/\hat{d}$. However, we can show that these two definitions yield the same $\beta*$. Consider first $\beta*$ as defined by:

$$\beta* = 1 - \frac{1}{\hat{d}}.$$

$\hat{d}$ is given in terms of our observations by (from (9.27)):

$$\hat{d} = \frac{m_{YI}}{m_{II}}.$$

Hence

$$\beta* = 1 - \frac{m_{II}}{m_{YI}} = \frac{m_{YI} - m_{II}}{m_{YI}}. \qquad (9.46)$$

However

$$Y_t = C_t + I_t$$

and so

$$(Y_t - \bar{Y}) = (C_t - \bar{C}) + (I_t - \bar{I})$$

and thus

$$m_{YI} = m_{CI} + m_{II}. \qquad (9.47)$$

Hence from (9.46)

$$\beta* = \frac{m_{CI}}{m_{YI}}. \qquad (9.48)$$

Consider now the $\beta*$ as defined by:

$$\beta* = \frac{\hat{b}}{\hat{b}+1}.$$

From (9.26)

$$\hat{b} = \frac{m_{CI}}{m_{II}}.$$

Hence

$$\beta^* = \frac{m_{CI}/m_{II}}{m_{CI}/m_{II} + 1} = \frac{m_{CI}}{m_{CI} + m_{II}} = \frac{m_{CI}}{m_{YI}} \quad \text{(from (9.47))}. \qquad (9.49)$$

Comparing (9.48) and (9.49) we see that both definitions of β^* yield the same estimator. We can similarly show that $\hat{a} = \hat{c}$, and so the two definitions of α^* yield the same estimator, which can be expressed as

$$\alpha^* = \bar{C} - \beta^*\bar{Y}. \qquad (9.50)$$

We can now see why α^* and β^* are termed the indirect least squares estimators of α and β. They were obtained indirectly, by working back from the least squares estimators of the reduced form parameters.

The next step is to investigate the properties of these indirect least squares estimators, and to see if they are any better than ordinary least squares. First, let us investigate whether they are unbiased. We see from (9.45) that:

$$E\beta^* = E\left(1 - \frac{1}{\hat{d}}\right) = 1 - E\left(\frac{1}{\hat{d}}\right).$$

Also we know that $E\hat{d} = d$ (from (9.27)) and thus

$$1 - \frac{1}{E\hat{d}} = 1 - \frac{1}{d} = \beta \quad \text{from (9.44)}.$$

Thus the crucial factor to enable us to prove unbiasedness is whether $E(1/\hat{d})$ equals $1/E\hat{d}$. Unfortunately, these two expressions are not equal.* Thus β^* is not in general an unbiased estimator of β.

However β^* is a consistent estimator of β; we know that $\hat{d}$ is a consistent estimator of d, and so $\hat{d}$ approaches d, and thus $[1 - (1/\hat{d})]$ approaches $[1 - (1/d)]$, and thus β^* approaches β as the sample size approaches infinity. Similarly α^* can be shown to be a consistent estimator of α.

Hence the indirect least squares estimators α^* and β^* are consistent while the ordinary least squares estimators are not. Thus indirect least squares estimation appears more desirable. We must note however that both methods yield biased estimators, and in a small sample the bias *may* be less using ordinary rather than indirect least squares. In general we do not know the magnitude (or direction) of the bias for small samples. However as we do know that the bias tends to zero with indirect least

---

*Consider, for example, the random variable X which takes values 1 and 3 with equal probability. Then $EX = 2$. The variable $1/X$ takes values 1 and 1/3 with equal probability, and so $E(1/X) = 2/3$. Thus $E(1/X) = 2/3 \neq 1/2 = 1/EX$.

Table 9.2   Observations on C, Y and I

t	C_t	Y_t	I_t
1	301.4	477.6	176.2
2	306.2	468.3	162.1
3	312.8	487.7	174.9
4	320.0	490.7	170.7
5	338.1	533.5	195.4
6	342.3	576.5	234.2
7	350.9	598.5	247.6
8	364.2	621.8	257.6
9	370.9	613.7	242.8
10	395.1	654.8	259.7
11	406.3	668.8	262.5
12	414.7	680.9	266.2
13	419.0	679.5	260.5
14	441.5	720.4	278.9
15	453.0	736.8	283.8
16	462.2	755.3	293.1
17	482.9	799.1	316.2
18	501.4	830.7	329.3
19	528.7	874.4	345.7
20	558.1	925.9	367.8
21	586.1	981.0	394.9
22	603.2	1,007.7	404.5
23	633.4	1,051.8	418.4
24	655.4	1,078.8	423.4
25	668.9	1,075.3	406.4
26	691.9	1,107.5	415.6
27	733.0	1,171.1	438.1
28	766.3	1,233.4	467.1
29	759.8	1,210.7	450.9

squares, but does not with ordinary least squares, it would seem preferable to use indirect least squares.

Let us illustrate the method of indirect least squares using the data on aggregate consumption and aggregate income in section 7.11. The data are given in table 9.2.

Estimating the reduced form equation for C by least squares gives

$$C_t = 11.267 + 1.54743 I_t + e_t \qquad R^2 = 0.9607$$
$$ (0.6) \qquad (25.7) \qquad \text{Durbin-Watson statistic} = 0.37$$
$$ (19.418) \qquad (0.06024) \qquad n = 29$$

$$(9.51)$$

(the first number in brackets is the t-ratio, the second is the estimated standard deviation). Similarly, estimating the reduced form equation for Y by least squares gives:

$$Y_t = 11.267 + 2.54743 \ I_t + e_t \qquad R^2 = 0.9851$$
$$ ((0.6) \qquad (42.3) \qquad \text{Durbin-Watson statistic} = 0.37$$
$$ (19.418) \qquad (0.06024) \qquad n = 29$$

$$(9.52)$$

(the first number in brackets is the t-ratio, the second is the estimated standard deviation).

Now, from (9.45) our indirect least squares estimate $\beta*$ is given by

$$\beta* = \frac{\hat{b}}{\hat{b}+1} = \frac{1.54743}{2.54743} = 0.6074$$

or

$$\beta* = 1 - \frac{1}{\hat{d}} = 1 - \frac{1}{2.54743} = 0.6074$$

which are, as we proved earlier, the same.

Our indirect least squares estimate $\alpha*$ is given by (from (9.45)):

$$\alpha* = \frac{\hat{a}}{\hat{d}} = \frac{\hat{c}}{\hat{d}} = \frac{11.267}{2.54743} = 4.423$$

Thus, our aggregate consumption function (9.18) estimated by indirect least squares is:

$$C_t = 4.42 + 0.6074\,Y_t + e_t. \tag{9.53}$$

In order to use these estimates to provide confidence intervals for, or to test hypotheses about, α and β, we must return to the reduced form estimates. The reduced form estimate of $b(=\beta/(1-\beta))$ is $\hat{b} = 1.54743$. The estimated standard deviation of $\hat{b}$ is 0.06024 (from (9.51)). Thus, using (7.63), a 95 per cent confidence interval for b is:

$$1.54743 \quad (2.052)\,(0.06024),\ 1.54743 + (2.052)\,(0.06024));$$

that is

$$(1.42382, 1.67104). \tag{9.54}$$

Now

$$\beta = \frac{b}{b+1}.$$

Thus, (9.54) can be converted into a 95 per cent confidence interval for β given by:

$$\left(\frac{1.42382}{2.42382}, \frac{1.67104}{2.67104} \right)$$

that is,

$$(0.5874, 0.6256).$$

Comparing our estimated aggregate consumption function using indirect least squares with that using ordinary least squares in section 7.11, we see that the former gives a lower value for the marginal propensity to consume (thus correcting for the upward asymptotic bias of ordinary least squares) and a higher value for the intercept.

A test of the null hypothesis that β is zero against the alternative that it is positive, can be converted into a test concerning the parameters of the reduced form equations. Since $d = 1/(1 - \beta)$ then if $\beta = 0$ then $d = 1$, and if $\beta > 0$ then $d > 1$. Thus testing to see if $\hat{d}$ is significantly greater than 1 is equivalent to testing whether β^* is significantly greater than zero. By inspecting (9.52) we see that $\hat{d}$ is indeed significantly greater than 1 at the 1 per cent level (critical value for $(\hat{d} - 1)/(\text{estimated s.d.}(\hat{d}))$ is 2.473; actual value of $(\hat{d} - 1)/(\text{estimated s.d.}(\hat{d}))$ is 25.7).

One problem that we have ignored so far with this particular example is the significant positive autocorrelation of the residuals of both reduced form equation, as evidenced by their low values for the Durbin-Watson statistic ($= 0.37$). Thus the reduced form equations should be re-estimated using the methods described and illustrated in section 8.6. These will then give new indirect least squares estimators of the structural parameters. This is left as an exercise for the reader.

9.4 Identification

We have shown that estimation of a structural equation in a simultaneous model by ordinary least squares leads to biased and inconsistent estimators. We have also shown that estimation by indirect least squares is preferable because of its asymptotic properties. The application of indirect least squares estimation requires two stages: first, the estimation of the reduced form equations by (ordinary) least squares; secondly, the derivation of the (indirect least squares) estimators of the structural parameters from the estimators of the reduced form parameters in such a way that the relationships between the estimators of the structural parameters and the estimators of the reduced form parameters are the same as the relationships between the true structural parameters and the true reduced form parameters. We can always carry out the first of these two stages. A problem arises as to whether we can carry out the necessary second stage. This is known as the *identification problem*. We will distinguish three possible cases:

(i) when there are an infinite number of possible sets of estimates of the structural parameters of a particular structural equation that are consistent with a given set of reduced form estimates, thus implying no unique way of deriving structural parameter estimates of the particular structural equation from the reduced form estimates;

(ii) when there is exactly one set of possible estimates of the structural parameters of a particular structural equation consistent with a given set of reduced form estimates, thus implying a unique way of deriving structural parameter estimates of the particular structural equation from the reduced form estimates;

(iii) when there is no set of estimates of the structural parameters of a particular structural equation consistent with a given set of reduced form estimates.

In case (i) the structural equation is said to be *under-identified*. (or not identified).

In case (ii) the structural equation is said to be *exactly identified*.

In case (iii) the structural equation is said to be *over-identified*.

Exact conditions under which a structural equation is under-, exactly or over-identified can be found in any econometrics text. A full treatment of the identification problem is outside the scope of this book. We will just give illustrations of each case. For algebraic simplicity, we will use the framework of the demand and supply model developed in the introduction for illustrative purposes. Consider first the model:

$$Q_S = \alpha + \beta P + u \tag{9.55}$$

$$Q_D = \gamma - \delta P + v \tag{9.56}$$

$$Q_S = Q_D . \tag{9.57}$$

(We omit the subscripts to avoid confusion.)

We can simplify the following algebra by applying (9.57) and replacing Q_S and Q_D by Q in (9.55) and (9.56), which then become

$$Q = \alpha + \beta P + u \tag{9.58}$$

$$Q = \gamma - \delta P + v. \tag{9.59}$$

The reduced form of the model is given by

$$P = a + w \tag{9.60}$$

$$Q = b + w' \tag{9.61}$$

where

$$a = \frac{\gamma - \alpha}{\beta + \delta} \qquad b = \frac{\alpha\delta + \beta\gamma}{\beta + \delta} \qquad w = \frac{-u + v}{\beta + \delta} \qquad w' = \frac{\delta u + \beta v}{\beta + \delta}. \tag{9.62}$$

Suppose we estimate the reduced form equations by least squares. In this case, with no exogenous variables, the least squares estimators $\hat{a}$ and $\hat{b}$ are given by

$$\hat{a} = \bar{P} \quad \hat{b} = \bar{Q}. \tag{9.63}$$

Can we derive (indirect least squares) estimators of the structural parameters which satisfy (9.62)? that is, such that

$$\hat{a} = \frac{\gamma^* - \alpha^*}{\beta^* + \delta^*} \quad \hat{b} = \frac{\alpha^* \delta^* + \beta^* \gamma^*}{\beta^* + \delta^*} \tag{9.64}$$

We have two equations to determine four unknowns. Obviously there is no unique set of α^*, β^*, γ^* and δ^* which satisfy (9.64). Consider, for example, if $\hat{a} = 10$, $\hat{b} = 20$. Then one possible set of structural estimates that satisfies (9.64) is $\alpha^* = 10, \beta^* = 1, \gamma^* = 30, \delta^* = 1$. A second set, which also satisfies (9.64), is $\alpha^* = -10, \beta^* = 3, \gamma^* = 40, \delta^* = 2$. Obviously there are an infinite number of such sets. Hence for the model as given by (9.58) and (9.59) there is no unique way of solving for the structural estimates given the reduced form estimates. Both structural equations are under-identified.

Consider now the model:

$$Q = \alpha + \beta P + u \tag{9.65}$$
$$Q = \gamma - \delta P + \theta Y + v \tag{9.66}$$

where Y is exogenous.
The reduced form of this model is:

$$P = a + bY + w \tag{9.67}$$
$$Q = c + dY + w' \tag{9.68}$$

where

$$\left. \begin{array}{lll} a = \dfrac{\gamma - \alpha}{\beta + \delta} & b = \dfrac{\theta}{\beta + \delta} & w = \dfrac{-u + v}{\beta + \delta} \\[3mm] c = \dfrac{\alpha \delta + \beta \gamma}{\beta + \delta} & d = \dfrac{\beta \theta}{\beta + \delta} & w' = \dfrac{\delta u + \beta v}{\beta + \delta}. \end{array} \right\} \tag{9.69}$$

Given estimates of the reduced form parameters, can we derive indirect least squares estimates of the structural parameters? Suppose

$$\hat{a} = 2 \quad \hat{b} = 1 \quad \hat{c} = 8 \quad \hat{d} = 2.$$

Examining (9.69) we can define β^* by

$$\beta^* = \frac{\hat{d}}{\hat{b}} = 2.$$

We can define α^* by $\alpha^* = \hat{c} - \beta^*\hat{a}$ (from (9.69) $\alpha = c - \beta a$)

$$= 8 - 4$$

$$\alpha^* = 4.$$

α^* and β^* are uniquely defined by these two conditions — no other α^* and β^* would satisfy them. Thus we can uniquely find estimates of the supply curve (9.65) from the reduced form estimates. The supply curve is exactly identified. However the demand curve (9.66) is under-identified, since there is an infinite number of sets of γ^*, δ^* and θ^* which satisfy (9.69). Consider:

$$\gamma^* = 12 \quad \delta^* = 2 \quad \theta^* = 4.$$

These satisfy (9.69), but so do:

$$\gamma^* = 10 \quad \delta^* = 1 \quad \theta^* = 3.$$

There is an infinite number of such sets, and thus the demand curve is under-identified.

Finally consider the model:

$$Q = \alpha + \beta P + u \qquad (9.70)$$

$$Q = \gamma - \delta P + \theta Y + \lambda Z + v \qquad (9.71)$$

where Y and Z are exogenous.
The reduced form of this model is:

$$P = a + bY + cZ + w \qquad (9.72)$$

$$Q = d + eY + fZ + w' \qquad (9.73)$$

where

$$\left.
\begin{aligned}
a &= \frac{\gamma - \alpha}{\beta + \delta} \quad b = \frac{\theta}{\beta + \delta} \quad c = \frac{\lambda}{\beta + \delta} \quad w = \frac{-u + v}{\beta + \delta} \\[2mm]
d &= \frac{\alpha\delta + \beta\gamma}{\beta + \delta} \quad e = \frac{\beta\theta}{\beta + \delta} \quad f = \frac{\beta\lambda}{\beta + \delta} \quad w' = \frac{\delta u + \beta v}{\beta + \delta}.
\end{aligned}
\right\} \qquad (9.74)$$

Suppose we found the reduced form estimates to be

$$\hat{a} = 2 \quad \hat{b} = 1 \quad \hat{c} = 2$$
$$\hat{d} = 8 \quad \hat{e} = 2 \quad \hat{f} = 3.$$

Examining (9.74) we could define β^* by

$$\beta^* = \frac{\hat{e}}{\hat{b}} = 2$$

but we could also define it by

$$\beta^* = \frac{\hat{f}}{\hat{c}} = \frac{3}{2}.$$

In this model there is no reason why $\hat{f}/\hat{c}$ should be equal to $\hat{e}/\hat{b}$. (You should verify this by deriving explicit expressions for the reduced form estimators in terms of the observations.)

Thus we have two different values for β^* depending on which way we 'work back' from the reduced form estimates. There is no β^* that satisfies all the conditions (9.74). Similarly there is no α^* that satisfies all the conditions. The supply curve (9.70) is over-identified. (You should check that the demand curve (9.71) is still under-identified — there is an infinite number of sets γ^*, δ^*, θ^* and λ^* that satisfy (9.74).)

In this over-identified case (9.70) we have no α^* or β^* that satisfy all the conditions (9.74). Some α^* and β^* satisfy some of these conditions. We have to decide which of these α^* and β^* to select as our estimates. (Notice there are only two β^*; $\beta^* = 2$, which satisfies $\beta^* = \hat{e}/\hat{b}$, and $\beta^* = 3/2$, which satisfies $\beta^* = \hat{f}/\hat{c}$. The problem of choosing between these conflicting estimates or of reconciling them constitutes a large part of the subject area of econometrics. Several alternative estimation techniques have been proposed for estimation of structural parameters in an over-identified equation. The study of these is outside the scope of this book; the reader who wishes to pursue the problem further should consult one of the recommended econometrics texts.

9.5 Summary

This chapter extended the methods of the previous two chapters to the investigation of economic theories whose conclusions are expressed in terms of a set of simultaneous equations. We saw that, in such cases, the least squares estimators of the structural parameters are both biased and inconsistent, owing to the violation of the assumption of independence between the deviation term and the 'independent' variables. However the least squares estimators of the reduced form parameters *are* best linear unbiased and consistent. These results led us to consideration of the method of indirect least squares estimation of the structural parameters. We saw that these indirect least squares estimators, although biased, have the important property of consistency, and may thus, at least asymptotically, be considered better estimators than the ordinary least squares estimators.

However we showed that indirect least squares estimation provides unique estimators of the parameters of a particular structural equation only if that equation is exactly identified. If it is under-identified there is no way of estimating its parameters; if it is over-identified there are several conflicting estimates.

This chapter presented some of the problems raised when we wish to investigate a simultaneous model. We have managed to resolve only some of these problems; the resolution of the others is attempted in the more advanced study of economic statistics, namely econometrics.

9.6 Exercises

9.1. Using the data in section 9.3, re-estimate the reduced form equations for the model (9.18) and (9.19) taking into account the dependence between successive deviations as evidenced by the low Durbin-Watson statistics in (9.51) and (9.52) (use the methods of section 8.6). Hence find estimates of the parameters of (9.18). Compare with those of section 9.3 and comment.

9.2. Consider the model as given by (9.55) to (9.57). For $\hat{a} = 10$, $\hat{b} = 20$, find other sets of $(\alpha^*, \beta^*, \gamma^*, \delta^*)$ that satisfy (9.64).

9.3. Consider the model as given by (9.65) and (9.66). For $\hat{a} = 2$, $\hat{b} = 1$, $\hat{c} = 8$, and $\hat{d} = 2$, find other sets of $(\gamma^*, \delta^*, \theta^*)$ which satisfy (9.69).

9.4. Consider the model as given by (9.70) and (9.71). For $\hat{a} = 2$, $\hat{b} = 1$, $\hat{c} = 2$, $\hat{d} = 8$, $\hat{e} = 2$, $\hat{f} = 3$, find sets of $(\gamma^*, \delta^*, \theta^*, \lambda^*)$ which satisfy (9.74).

9.5. A supply and demand model of a market consists of the equations:

$$Q_t = \alpha + \beta P_t + u_t \quad \text{(supply)}$$
$$Q_t = \gamma - \delta P_t + \theta Y_t + v_t \quad \text{(demand)}$$

where Q is quantity, P is price, Y is income; Q and P are endogenous and Y is exogenous; u and v obey assumptions (A1), (A2), (A3) and (A5); u and v are independent. Discuss the properties of the ordinary least squares estimators of α and β. Discuss the properties of the indirect least squares estimators of α and β.

You are given the following data from 20 observations:

$$\sum_t (P_t - \bar{P})(Y_t - \bar{Y}) = 10 \quad \sum_t (Y_t - \bar{Y})^2 = 100 \quad \bar{P} = 5 \quad \bar{Y} = 42.5.$$

Suppose you know what $\gamma = 1$, $\delta = 1$ and $\theta = 0.2$; derive the indirect least squares estimates of α and β. Discuss how you would form confidence intervals for α and β.

Suppose γ, δ and θ were unknown, what further information would you need to find the indirect least squares estimates of α and β? Still supposing the parameters of the demand equation unknown, would you be able to estimate them by indirect least squares?

9.6. A simple national income model consists of the following three equations:

$$C_t = \alpha + \beta Y_t + u_t$$
$$I_t = \gamma + \delta r_t + v_t$$
$$Y_t = C_t + I_t.$$

C is aggregate consumption, I aggregate investment, Y aggregate income and r is the rate of interest. C, Y and I are endogenous, r is exogenous, and u and v are independent and obey assumptions (A1) (A2) (A3) and (A5). You are given the following data from 20 observations:

$$\sum_t (C_t - \bar{C})(r_t - \bar{r}) = -12 \qquad \sum_t (Y_t - \bar{Y})(r_t - \bar{r}) = -16 \qquad \sum_t (r_t - \bar{r})^2 = 4$$
$$\bar{C} = 55 \qquad\qquad\qquad \bar{Y} = 60 \qquad\qquad\qquad \bar{r} = 3.$$

Estimate the parameters of the reduced form equations for C and Y. Hence find the indirect least squares estimates of the parameters of the consumption function. How would you estimate the parameters of the investment function? What additional information would you need, if any?

9.7. A simple national income model consists of the following three equations:

$$C_t = \alpha + \beta Y_t + u_t$$
$$I_t = \gamma + \gamma r_t + \theta Y_t + v_t$$
$$Y_t = C_t + I_t.$$

C is aggregate consumption, I aggregate investment, Y aggregate income and r is the rate of interest. C, Y and I are endogenous, r is exogenous, and u and v are independent and obey assumptions (A1) (A2) (A3) and (A5). You are given the following data from 20 observations:

$$\sum_t (C_t - \bar{C})(r_t - \bar{r}) = -12 \qquad \sum_t (Y_t - \bar{Y})(r_t - \bar{r}) = -16 \qquad \sum_t (r_t - \bar{r})^2 = 48$$
$$\bar{C} = 26 \qquad\qquad\qquad \bar{Y} = 30 \qquad\qquad\qquad \bar{r} = 12.$$

Estimate the parameters of the reduced form equations for C and Y. Hence find the indirect least squares estimates of the parameters of the consumption function. Discuss the problems of estimating the investment function.

9.8. Construct a simultaneous economic model of your choice. Discuss the estimation problems involved. For the exactly identified equations of your model, estimate the parameters using the appropriate data.

9.9. Using the data of table 9.2 and a corresponding set of observations on an appropriate interest rate variable investigate which of the following three models appears to be 'best':

(a) the model given by equations (9.18) and (9.19);
(b) the model of exercise 9.6;
(c) the model of exercise 9.7?

Discuss what you mean by 'best'.

Appendix

A1 Summation

The use of the sigma notation (Σ) enables us to abbreviate expressions involving the sum of several terms. For example, if we wish to express the sum of the four terms X_1, X_2, X_3 and X_4, instead of writing $X_1 + X_2 + X_3 + X_4$, we can simply write $\Sigma_{i=1}^{4} X_i$, that is,

$$\sum_{i=1}^{4} X_i \equiv X_1 + X_2 + X_3 + X_4$$

Similarly, the sum of the n terms X_1, X_2, ..., X_n can be written as $\Sigma_{i=1}^{n} X_i$, that is,

$$\sum_{i=1}^{n} X_i \equiv X_1 + X_2 + \ldots + X_n$$

The expression '$\Sigma_{i=1}^{n} X_i$' can be read as 'the sum of X_i from i equals 1 to n'.

The index (i) is arbitrary since both $\Sigma_{i=1}^{n} X_i$ and $\Sigma_{j=1}^{n} X_j$ are ways of expressing $X_1 + X_2 + \ldots + X_n$. When the index is obvious, we omit it; that is $\Sigma_{i=1}^{n} X_i$ is written simply $\Sigma_{1}^{n} X_i$. Now, since, for example,

$$\sum_{1}^{4} X_i = X_1 + X_2 + X_3 + X_4 \text{, then, putting}$$

$X_1 = X_2 = X_3 = X_4 = a$, it follows that

$$\sum_{1}^{4} a = a + a + a + a = 4a$$

that is,

$$\sum_{1}^{4} a = 4a$$

Generalising, it follows that, $\sum_{1}^{n} a = na$

Consider now $\sum_{1}^{4} bX_i = (bX_1 + bX_2 + bX_3 + bX_4) = b(X_1 + X_2 + X_3 + X_4)$

Thus,

$$\sum_{1}^{4} bX_i = b\sum_{1}^{4} X_i$$

Generalising, it follows that, $\sum\limits_{1}^{n} b X_i = b \sum\limits_{1}^{n} X_i$

Consider now $\sum\limits_{1}^{4} (X_i + Y_i) = (X_1 + Y_1) + (X_2 + Y_2) + (X_3 + Y_3) + (X_4 + Y_4)$

$$= (X_1 + X_2 + X_3 + X_4) + (Y_1 + Y_2 + Y_3 + Y_4)$$

$$= \sum\limits_{1}^{4} X_i + \sum\limits_{1}^{4} Y_i$$

Thus

$$\sum\limits_{1}^{4} (X_i + Y_i) = \sum\limits_{1}^{4} X_i + \sum\limits_{1}^{4} Y_i$$

Generalising, it follows that, $\sum\limits_{1}^{n} (X_i + Y_i) = \sum\limits_{1}^{n} X_i + \sum\limits_{1}^{n} Y_i$

Note very carefully that

$$\sum\limits_{1}^{n} X_i Y_i \neq \left(\sum\limits_{1}^{n} X_i\right)\left(\sum\limits_{1}^{n} Y_i\right)$$

since the left hand side $= X_1 Y_1 + X_2 Y_2 + \ldots + X_n Y_n$

while the right hand side $= (X_1 + X_2 + \ldots + X_n)(Y_1 + Y_2 + \ldots + Y_n)$

thus, the right hand side includes terms of the form $X_i Y_j$ for $i \neq j$ while the left hand side does not.

An obvious extension of the sigma notation is to double summation. Consider, for example the six terms $X_{11}, X_{12}, X_{13}, X_{21}, X_{22}, X_{23}$, or more succinctly X_{ij} for $i = 1, 2$, and $j = 1, 2, 3$. We can express the sum

$$X_{11} + X_{12} + X_{13} + X_{21} + X_{22} + X_{23}$$

by

$$\sum\limits_{i=1}^{2} \sum\limits_{j=1}^{3} X_{ij}$$

To see the equivalence of these two expressions we first note that

$$\sum\limits_{j=1}^{3} X_{ij} = X_{i1} + X_{i2} + X_{i3}$$

Thus

$$\sum\limits_{i=1}^{2} \sum\limits_{j=1}^{3} X_{ij} = \sum\limits_{i=1}^{2} \left(\sum\limits_{j=1}^{3} X_{ij}\right) = \sum\limits_{i=1}^{2} (X_{i1} + X_{i2} + X_{i3})$$

$$= (X_{11} + X_{12} + X_{13}) + (X_{21} + X_{22} + X_{23})$$

Similarly the sum

$$(X_{11} + X_{12} + \ldots + X_{1m}) + (X_{21} + X_{22} + \ldots + X_{2m})$$
$$+ \ldots + (X_{n1} + X_{n2} + \ldots + X_{nm})$$

can be written as

$$\sum_{i=1}^{n} \sum_{j=1}^{m} X_{ij}$$

You should verify that

$$\sum_{i=1}^{n} \sum_{j=1}^{m} (X_{ij} + Y_{ij}) = \sum_{i=1}^{n} \sum_{j=1}^{m} X_{ij} + \sum_{i=1}^{n} \sum_{j=1}^{m} Y_{ij}$$

Consider now,

$$\sum_{i=1}^{2} \sum_{j=1}^{3} X_i Y_j = X_1 Y_1 + X_1 Y_2 + X_1 Y_3 + X_2 Y_1 + X_2 Y_2 + X_2 Y_3$$
$$= (X_1 + X_2)Y_1 + (X_1 + X_2)Y_2 + (X_1 + Y_2)Y_3$$
$$= (X_1 + X_2)(Y_1 + Y_2 + Y_3)$$

Thus $\displaystyle\sum_{i=1}^{2} \sum_{j=1}^{3} X_i Y_j = \left(\sum_{i=1}^{2} X_i \right)\left(\sum_{j=1}^{3} Y_j \right)$

Generalising, it follows that,

$$\sum_{i=1}^{n} \sum_{j=1}^{m} X_i Y_j = \left(\sum_{i=1}^{n} X_i \right)\left(\sum_{j=1}^{m} Y_j \right)$$

If $Y_j = 1$ for all j, it follows that

$$\sum_{j=1}^{m} Y_j = m$$

and so

$$\sum_{i=1}^{n} \sum_{j=1}^{m} X_i = m \sum_{i=1}^{n} X_i$$

Finally, putting $X_i = a$ for all i, it follows that

$$\sum_{i=1}^{n} X_i = na$$

and so

$$\sum_{i=1}^{n} \sum_{j=1}^{m} a = mna$$

A2 Independence of Uncorrelated Normal Variables

We present here only an outline of the proof that uncorrelated normally distributed random variables are independent. We omit the evaluations of the integrals. The reader familiar with the integral calculus may like to verify the integrations. To do this, the following standard result will be needed:

$$\int_{-\infty}^{\infty} e^{-t^2/2} dt = \sqrt{2\pi}.$$

If two random variables X_1 and X_2 are jointly normally distributed, then their joint probability density function is:

$$f(X_1, X_2) = \frac{1}{2\pi\sigma_1\sigma_2\sqrt{(1-\rho^2)}} \exp\left\{-\frac{1}{2(1-\rho^2)}\left[\left(\frac{X_1-\mu_1}{\sigma_1}\right)^2\right.\right.$$
$$\left.\left.-2\rho\left(\frac{X_1-\mu_1}{\sigma_1}\right)\left(\frac{X_2-\mu_2}{\sigma_2}\right)+\left(\frac{X_2-\mu_2}{\sigma_2}\right)^2\right]\right\}$$

$-\infty \leqslant X_1, X_2 \leqslant \infty.$

Now, using (3.33) (modified to be applicable for continuous random variables), we have:

$$f_1(X_1) = \int_{-\infty}^{\infty} f(X_1, X_2)dX_2$$

where $f_1(X_1)$ is the probability density function of X_1 (by itself), and

$$f_2(X_2) = \int_{-\infty}^{\infty} f(X_1, X_2)dX_1$$

where $f_2(X_2)$ is the probability density function of X_2 (by itself). Carrying out the integrations (see Hoel (13) p. 199), we find

$$f_1(X_1) = \frac{1}{\sqrt{(2\pi)}\sigma_1} \exp\left\{-\frac{1}{2}\left(\frac{X_1-\mu_1}{\sigma_1}\right)^2\right\}$$

and

$$f_2(X_2) = \frac{1}{\sqrt{(2\pi)}\sigma_2} \exp\left\{-\frac{1}{2}\left(\frac{X_2-\mu_2}{\sigma_2}\right)^2\right\}.$$

Thus the distribution of X_1 (by itself) is normal with mean μ_1 and variance σ_1^2 (cf. (4.16)).

Similarly, the distribution of X_2 (by itself) is normal with mean μ_2 and variance σ_2^2.

Further, from (3.39) (modified to be applicable for continuous random variables), the covariance between X_1 and X_2 is given by:

$$\mathrm{cov}(X_1, X_2) = \int_{-\infty}^{\infty}\int_{-\infty}^{\infty} (X_1-\mu_1)(X_2-\mu_2) f(X_1, X_2)dX_1 dX_2.$$

Evaluating this double integral, we find

$$\mathrm{cov}(X_1, X_2) = \rho\sigma_1\sigma_2$$

and so

$$\rho = \frac{\mathrm{cov}(X_1, X_2)}{\sigma_1\sigma_2}.$$

Thus ρ is the correlation coefficient between X_1 and X_2.

Now, if X_1 and X_2 are uncorrelated, then $\rho = 0$ and their joint probability density function reduces to

$$f(X_1, X_2) = \frac{1}{2\pi\sigma_1\sigma_2} \exp\left\{-\frac{1}{2}\left[\left(\frac{X_1 - \mu_1}{\sigma_1}\right)^2 + \left(\frac{X_2 - \mu_2}{\sigma_2}\right)^2\right]\right\}$$

$$= \frac{1}{\sqrt{(2\pi)}\,\sigma_1} \exp\left\{-\frac{1}{2}\left(\frac{X_1 - \mu_1}{\sigma_1}\right)^2\right\} \frac{1}{\sqrt{(2\pi)}\sigma_2} \exp\left\{-\frac{1}{2}\left(\frac{X_2 - \mu_2}{\sigma_2}\right)^2\right\}$$

$$= f_1(X_1)\, f_2(X_2).$$

Thus, if X_1 and X_2 are uncorrelated, their joint probability density function can be expressed as the product of their individual probability density functions. Hence, from (3.38), if X_1 and X_2 are uncorrelated normally distributed random variables, then they are independent.

A3 Table of the Unit Normal Distribution

An entry in the table is the proportion under the entire curve which is between $z = 0$ and a positive value of z. Areas for negative values of z are obtained by symmetry.

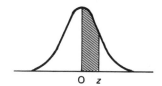

z	0.00	0.01	0.02	0.03	0.04	0.05	0.06	0.07	0.08	0.09
0.0	0.0000	0.0040	0.0080	0.0120	0.0160	0.0199	0.0239	0.0279	0.0319	0.0359
0.1	0.0398	0.0438	0.0478	0.0517	0.0557	0.0596	0.0636	0.0675	0.0714	0.0753
0.2	0.0793	0.0832	0.0871	0.0910	0.0948	0.0987	0.1026	0.1064	0.1103	0.1141
0.3	0.1179	0.1217	0.1255	0.1293	0.1331	0.1368	0.1406	0.1443	0.1480	0.1517
0.4	0.1554	0.1591	0.1628	0.1664	0.1700	0.1736	0.1772	0.1808	0.1844	0.1879
0.5	0.1915	0.1950	0.1985	0.2019	0.2054	0.2088	0.2123	0.2157	0.2190	0.2224
0.6	0.2257	0.2291	0.2324	0.2357	0.2389	0.2422	0.2454	0.2486	0.2517	0.2549
0.7	0.2580	0.2611	0.2642	0.2673	0.2704	0.2734	0.2764	0.2794	0.2823	0.2852
0.8	0.2881	0.2910	0.2939	0.2967	0.2995	0.3023	0.3051	0.3078	0.3106	0.3133
0.9	0.3159	0.3186	0.3212	0.3238	0.3264	0.3289	0.3315	0.3340	0.3365	0.3389
1.0	0.3413	0.3438	0.3461	0.3485	0.3508	0.3531	0.3554	0.3577	0.3599	0.3621
1.1	0.3643	0.3665	0.3686	0.3708	0.3729	0.3749	0.3770	0.3790	0.3810	0.3830
1.2	0.3849	0.3869	0.3888	0.3907	0.3925	0.3944	0.3962	0.3980	0.3997	0.4015
1.3	0.4032	0.4049	0.4066	0.4082	0.4099	0.4115	0.4131	0.4147	0.4162	0.4177
1.4	0.4192	0.4207	0.4222	0.4236	0.4251	0.4265	0.4279	0.4292	0.4306	0.4319
1.5	0.4332	0.4345	0.4357	0.4370	0.4382	0.4394	0.4406	0.4418	0.4429	0.4441
1.6	0.4452	0.4463	0.4474	0.4484	0.4495	0.4505	0.4515	0.4525	0.4535	0.4545
1.7	0.4554	0.4564	0.4573	0.4582	0.4591	0.4599	0.4608	0.4616	0.4625	0.4633
1.8	0.4641	0.4649	0.4656	0.4664	0.4671	0.4678	0.4686	0.4693	0.4699	0.4706
1.9	0.4713	0.4719	0.4726	0.4732	0.4738	0.4744	0.4750	0.4756	0.4761	0.4767
2.0	0.4772	0.4778	0.4783	0.4788	0.4793	0.4798	0.4803	0.4808	0.4812	0.4817
2.1	0.4821	0.4826	0.4830	0.4834	0.4838	0.4842	0.4846	0.4850	0.4854	0.4857
2.2	0.4861	0.4864	0.4868	0.4871	0.4875	0.4878	0.4881	0.4884	0.4887	0.4890
2.3	0.4893	0.4896	0.4898	0.4901	0.4904	0.4906	0.4909	0.4911	0.4913	0.4916
2.4	0.4918	0.4920	0.4922	0.4925	0.4927	0.4929	0.4931	0.4932	0.4934	0.4936
2.5	0.4938	0.4940	0.4941	0.4943	0.4945	0.4946	0.4948	0.4949	0.4951	0.4952
2.6	0.4953	0.4955	0.4956	0.4957	0.4959	0.4960	0.4961	0.4962	0.4963	0.4964
2.7	0.4964	0.4966	0.4967	0.4968	0.4969	0.4970	0.4971	0.4972	0.4973	0.4974
2.8	0.4974	0.4975	0.4976	0.4977	0.4977	0.4978	0.4979	0.4979	0.4980	0.4981
2.9	0.4981	0.4982	0.4982	0.4983	0.4984	0.4984	0.4985	0.4985	0.4986	0.4986
3.0	0.4987	0.4987	0.4987	0.4988	0.4988	0.4989	0.4989	0.4989	0.4990	0.4990

A4 Table of the chi-square Distribution

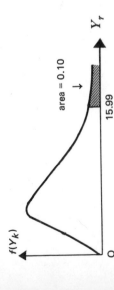

area = 0.10

Example:

for $k = 10$ degrees of freedom

$P[Y_k > 15.99] = 0.10$

k \ P	0.995	0.99	0.975	0.95	0.90	0.75	0.50	0.25	0.10	0.05	0.025	0.01	0.005
1	0.0⁴393	0.0³157	0.0³982	0.0²393	0.0158	0.102	0.455	1.323	2.71	3.84	5.02	6.63	7.88
2	0.0100	0.0201	0.0506	0.103	0.211	0.575	1.386	2.77	4.61	5.99	7.38	9.21	10.60
3	0.0717	0.115	0.216	0.352	0.584	1.213	2.37	4.11	6.25	7.81	9.35	11.34	12.84
4	0.207	0.297	0.484	0.711	1.064	1.923	3.36	5.39	7.78	9.49	11.14	13.28	14.86
5	0.412	0.554	0.831	1.145	1.610	2.67	4.35	6.63	9.24	11.07	12.83	15.09	16.75
6	0.676	0.872	1.237	1.635	2.20	3.45	5.35	7.84	10.64	12.59	14.45	16.81	18.55
7	0.989	1.239	1.690	2.17	2.83	4.25	6.35	9.04	12.02	14.07	16.01	18.48	20.3
8	1.344	1.646	2.18	2.73	3.49	5.07	7.84	10.22	13.36	15.51	17.53	20.1	22.0
9	1.735	2.09	2.70	3.33	4.17	5.90	8.34	11.39	14.68	16.92	19.02	21.7	23.6
10	2.16	2.56	3.25	3.94	4.87	6.74	9.34	12.55	15.99	18.31	20.5	23.2	25.2
11	2.60	3.05	3.82	4.57	5.58	7.58	10.34	13.70	17.28	19.68	21.9	24.7	26.8
12	3.07	3.57	4.40	5.23	6.30	8.44	11.34	14.85	18.55	21.0	23.3	26.2	28.3
13	3.57	4.11	5.01	5.89	7.04	9.30	12.34	15.98	19.81	22.4	24.7	27.7	29.8
14	4.07	4.66	5.63	6.57	7.79	10.17	13.34	17.12	21.1	23.7	26.1	29.1	31.3
15	4.60	5.23	6.26	7.26	8.55	11.04	14.34	18.25	22.3	25.0	27.5	30.6	32.8
16	5.14	5.81	6.91	7.96	9.31	11.91	15.34	19.37	23.5	26.3	28.8	32.0	34.3
17	5.70	6.41	7.56	8.67	10.09	12.79	16.34	20.5	24.8	27.6	30.2	33.4	35.7
18	6.26	7.01	8.23	9.39	10.86	13.68	17.34	21.6	26.0	28.9	31.5	34.8	37.2
19	6.84	7.63	8.91	10.12	11.65	14.56	18.34	22.7	27.2	30.1	32.9	36.2	38.6
20	7.43	8.26	9.59	10.85	12.44	15.45	19.34	23.8	28.4	31.4	34.2	37.6	40.0

k	-2.58	-2.33	-1.960	-1.645	-1.282	-0.674	0.000	0.674	1.282	1.645	1.960	2.33	2.58
21	8.03	8.90	10.28	11.59	13.24	16.34	20.3	24.9	29.6	32.7	35.5	38.9	41.4
22	8.64	9.54	10.98	12.34	14.04	17.24	21.3	26.0	30.8	33.9	36.8	40.3	42.8
23	9.26	10.20	11.69	13.09	14.85	18.14	22.3	27.1	32.0	35.2	38.1	41.6	44.2
24	9.89	10.86	12.40	13.85	15.66	19.04	23.3	28.2	33.2	36.4	39.4	43.0	45.6
25	10.52	11.52	13.12	14.61	16.47	19.94	24.3	29.3	34.4	37.7	40.6	44.3	46.9
26	11.16	12.20	13.84	15.38	17.29	20.8	25.3	30.4	35.6	38.9	41.9	45.6	48.3
27	11.81	12.88	14.57	16.15	18.11	21.7	26.3	31.5	36.7	40.1	43.2	47.0	49.6
28	12.46	13.56	15.31	16.93	18.94	22.7	27.3	32.6	37.9	41.3	44.5	48.3	51.0
29	13.12	14.26	16.05	17.71	19.77	23.6	28.3	33.7	39.1	42.6	45.7	49.6	52.3
30	13.79	14.95	16.79	18.49	20.6	24.5	29.3	34.8	40.3	43.8	47.0	50.9	53.7
40	20.7	22.2	24.4	26.5	29.1	33.7	39.3	45.6	51.8	55.8	59.3	63.7	66.8
50	28.0	29.7	32.4	34.8	37.7	42.9	49.3	56.3	63.2	67.5	71.4	76.2	79.5
60	35.5	37.5	40.5	43.2	46.5	52.3	59.3	67.0	74.4	79.1	83.3	88.4	92.0
70	43.3	45.4	48.8	51.7	55.3	61.7	69.3	77.6	85.5	90.5	95.0	100.4	104.2
80	51.2	53.5	57.2	60.4	64.3	71.1	79.3	88.1	96.6	101.9	106.6	112.3	116.3
90	59.2	61.8	65.6	69.1	73.3	80.6	89.3	98.6	107.6	113.1	118.1	124.1	128.3
100	67.3	70.1	74.2	77.9	82.4	90.1	99.3	109.1	118.5	124.3	129.6	135.8	140.2
Z_α	-2.58	-2.33	-1.960	-1.645	-1.282	-0.674	0.000	0.674	1.282	1.645	1.960	2.33	2.58

For $k > 100$ take $\chi^2 = \frac{1}{2}(Z_\alpha + \sqrt{(2k-1)})^2$. Z_α is the standardised normal deviate corresponding to the α level of significance, and is shown in the bottom of the table.

Source: This table is abridged from 'Table of percentage points of the χ^2 distribution' by Catherine M. Thompson, *Biometrika* 32 (1941) pp. 187–91, and is published here by permission of the author and editor of *Biometrika*.

A5 Table of the t-distribution

Example:
for $k = 10$ degrees of freedom
$P[t_k > 1.812] = 0.05$
$P[t_k < -1.812] = 0.05$

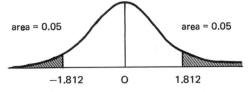

area = 0.05   area = 0.05

−1.812   O   1.812

k	0.25	0.20	0.15	0.10	0.05	0.025	0.01	0.005	0.0005
1	1.000	1.376	1.963	3.078	6.314	12.706	31.821	63.657	636.619
2	0.816	1.061	1.386	1.886	2.920	4.303	6.965	9.925	31.598
3	0.765	0.978	1.250	1.638	2.353	3.182	4.541	5.841	12.941
4	0.741	0.941	1.190	1.533	2.132	2.776	3.747	4.604	8.610
5	0.727	0.920	1.156	1.476	2.015	2.571	3.365	4.032	6.859
6	0.718	0.906	1.134	1.440	1.943	2.447	3.143	3.707	5.959
7	0.711	0.896	1.119	1.415	1.895	2.365	2.998	3.499	5.405
8	0.706	0.889	1.108	1.397	1.860	2.306	2.896	3.355	5.041
9	0.703	0.883	1.100	1.383	1.833	2.262	2.821	3.250	4.781
10	0.700	0.879	1.093	1.372	1.812	2.228	2.764	3.169	4.587
11	0.697	0.876	1.088	1.363	1.796	2.201	2.718	3.106	4.437
12	0.695	0.873	1.083	1.356	1.782	2.179	2.681	3.055	4.318
13	0.694	0.870	1.079	1.350	1.771	2.160	2.650	3.012	4.221
14	0.692	0.868	1.076	1.345	1.761	2.145	2.624	2.977	4.140
15	0.691	0.866	1.074	1.341	1.753	2.131	2.602	2.947	4.073
16	0.690	0.865	1.071	1.337	1.746	2.120	2.583	2.921	4.015
17	0.689	0.863	1.069	1.333	1.740	2.110	2.567	2.898	3.965
18	0.688	0.862	1.067	1.330	1.734	2.101	2.552	2.878	3.922
19	0.688	0.861	1.066	1.328	1.729	2.093	2.539	2.861	3.883
20	0.687	0.860	1.064	1.325	1.725	2.086	2.528	2.845	3.850
21	0.686	0.859	1.063	1.323	1.721	2.080	2.518	2.831	3.819
22	0.686	0.858	1.061	1.321	1.717	2.074	2.508	2.819	3.792
23	0.685	0.858	1.060	1.319	1.714	2.069	2.500	2.807	3.767
24	0.685	0.857	1.059	1.318	1.711	2.064	2.492	2.797	3.745
25	0.684	0.856	1.058	1.316	1.708	2.060	2.486	2.787	3.725
26	0.684	0.856	1.058	1.315	1.706	2.056	2.479	2.779	3.707
27	0.684	0.855	1.057	1.314	1.703	2.052	2.473	2.771	3.690
28	0.683	0.855	1.056	1.313	1.701	2.048	2.467	2.763	3.674
29	0.683	0.854	1.055	1.311	1.699	2.045	2.462	2.756	3.659
30	0.683	0.854	1.055	1.310	1.697	2.042	2.457	2.750	3.646
40	0.681	0.851	1.050	1.303	1.684	2.021	2.423	2.704	3.551
60	0.679	0.848	1.046	1.296	1.671	2.000	2.390	2.660	3.460
120	0.677	0.845	1.041	1.289	1.658	1.980	2.358	2.617	3.373
∞	0.674	0.842	1.036	1.282	1.645	1.960	2.326	2.576	3.291

Source: This table is abridged from table III of Fisher and Yates *Statistical Tables for Biological, Agricultural and Medical Research* London, Longman Group Ltd. (previously published by Oliver & Boyd, Edinburgh) and is reproduced by permission of the authors and publishers.

A6   Table of the Durbin-Watson Statistic

Upper and lower critical values for 5 per cent significance level

n	$k' = 1$		$k' = 2$		$k' = 3$		$k' = 4$		$k' = 5$	
	d_L	d_U	d_L	d_U	d_L	d_U	d_L	d_U	d_L	d_U
15	1.08	1.36	0.95	1.54	0.82	1.75	0.69	1.97	0.56	2.21
16	1.10	1.37	0.98	1.54	0.86	1.73	0.74	1.93	0.62	2.15
17	1.13	1.38	1.02	1.54	0.90	1.71	0.78	1.90	0.67	2.10
18	1.16	1.39	1.05	1.53	0.93	1.69	0.82	1.87	0.71	2.06
19	1.18	1.40	1.08	1.53	0.97	1.68	0.86	1.85	0.75	2.02
20	1.20	1.41	1.10	1.54	1.00	1.68	0.90	1.83	0.79	1.99
21	1.22	1.42	1.13	1.54	1.03	1.67	0.93	1.81	0.83	1.96
22	1.24	1.43	1.15	1.54	1.05	1.66	0.96	1.80	0.86	1.94
23	1.26	1.44	1.17	1.54	1.08	1.66	0.99	1.79	0.90	1.92
24	1.27	1.45	1.19	1.55	1.10	1.66	1.01	1.78	0.93	1.90
25	1.29	1.45	1.21	1.55	1.12	1.66	1.04	1.77	0.95	1.89
26	1.30	1.46	1.22	1.55	1.14	1.65	1.06	1.76	0.98	1.88
27	1.32	1.47	1.24	1.56	1.16	1.65	1.08	1.76	1.01	1.86
28	1.33	1.48	1.26	1.56	1.18	1.65	1.10	1.75	1.03	1.85
29	1.34	1.48	1.27	1.56	1.20	1.65	1.12	1.74	1.05	1.84
30	1.35	1.49	1.28	1.57	1.21	1.65	1.14	1.74	1.07	1.83
31	1.36	1.50	1.30	1.57	1.23	1.65	1.16	1.74	1.09	1.83
32	1.37	1.50	1.31	1.57	1.24	1.65	1.18	1.73	1.11	1.82
33	1.38	1.51	1.32	1.58	1.26	1.65	1.19	1.73	1.13	1.81
34	1.39	1.51	1.33	1.58	1.27	1.65	1.21	1.73	1.15	1.81
35	1.40	1.52	1.34	1.58	1.28	1.65	1.22	1.73	1.16	1.80
36	1.41	1.52	1.35	1.59	1.29	1.65	1.24	1.73	1.18	1.80
37	1.42	1.53	1.36	1.59	1.31	1.66	1.25	1.72	1.19	1.80
38	1.43	1.54	1.37	1.59	1.32	1.66	1.26	1.72	1.21	1.79
39	1.43	1.54	1.38	1.60	1.33	1.66	1.27	1.72	1.22	1.79
40	1.44	1.54	1.39	1.60	1.34	1.66	1.29	1.72	1.23	1.79
45	1.48	1.57	1.43	1.62	1.38	1.67	1.34	1.72	1.29	1.78
50	1.50	1.59	1.46	1.63	1.42	1.67	1.38	1.72	1.34	1.77
55	1.53	1.60	1.49	1.64	1.45	1.68	1.41	1.72	1.38	1.77
60	1.55	1.62	1.51	1.65	1.48	1.69	1.44	1.73	1.41	1.77
65	1.57	1.63	1.54	1.66	1.50	1.70	1.47	1.73	1.44	1.77
70	1.58	1.64	1.55	1.67	1.52	1.70	1.49	1.74	1.46	1.77
75	1.60	1.65	1.57	1.68	1.54	1.71	1.51	1.74	1.49	1.77
80	1.61	1.66	1.59	1.69	1.56	1.72	1.53	1.74	1.51	1.77
85	1.62	1.67	1.60	1.70	1.57	1.72	1.55	1.75	1.52	1.77
90	1.63	1.68	1.61	1.70	1.59	1.73	1.57	1.75	1.54	1.78
95	1.64	1.69	1.62	1.71	1.60	1.73	1.58	1.75	1.56	1.78
100	1.65	1.69	1.63	1.72	1.61	1.74	1.59	1.76	1.57	1.78

Note: k' = number of explanatory variables *excluding* the constant term.

Source: J. Durbin and G. S. Watson, 'Testing for Serial Correlation in Least Squares Regression' *Biometrika* 38 (1951) pp. 159–77. Reprinted with the permission of the authors and the *Biometrika* trustees.

Bibliography and References

This book has introduced the reader to the methods and concepts used in empirical economics. We have attempted to provide a solid base of understanding of these methods, while at the same time covering the main statistical techniques that the student is likely to encounter during his reading in economics. Inevitably, however, much material still remains uncovered. There are two main branches that the reader may like to follow to broaden his knowledge. The first leads to a more rigorous study of mathematical statistics; the second to a more detailed study of the problems raised in the final two chapters of the book.

Although we have attempted, within the limitations of the mathematical techniques that we have employed, to prove as many of the important results as possible, we have had to leave some results unproved. Where we have had to do this we have usually tried to give intuitive understanding of these assertions by means of 'illustrations'. For the reader unconvinced by these 'illustrations', we recommend a study of mathematical statistics. There he will find proofs of results like the Central Limit Theorem, the independence of $\bar{X}$ and s^2, the result that the sum of independent normally distributed random variables is also normally distributed, and so on. In the general field of mathematical statistics we suggest the books by Hoel (12), Mood and Graybill (20) and Feller (9). Kendall and Stuart (14) provides a useful reference book.

In chapter 8 we indicated how certain problems arising in the study of single equation models may be solved. However, as the reader probably noticed, other problems, for which no solutions were given in this book, may arise. Hopefully, in such cases the reader may be guided by the general methods discussed in this book towards a solution. The subject matter of econometrics is concerned partially with the investigation in more detail of the single equation model and the problems that may arise in its confrontation with evidence. Econometrics is also concerned with a fuller study of the problems encountered in the investigation of simultaneous models, in particular with the identification problem, and with estimation methods for over-identified equations. There are several good texts on econometric theory; Wonnacott and Wonnacott (25), Kmenta (16) and Koutsoyiannis (17) may be recommended as relatively straightforward introductory texts. At a higher level of mathematical rigour and sophistication are Christ (2), Goldberger (11), Johnston (13) and Klein (15). Dhrymes (5), Malinvaud (18) and Theil (24) are considerably more advanced texts which require a fairly advanced knowledge of mathematics. Texts on applied econometrics are somewhat rarer, and the student may well have to refer to journal articles for good applied econometric studies, but the books by Bridge (1) and Cramer (3) are well worth investigating.

Finally, we recommend the pamphlet by Streissler (21) as an indispensable part of the education of any numerate economist.

1. J. L. Bridge *Applied Econometrics* Amsterdam, North-Holland (1971).
2. C. F. Christ *Econometric Models and Methods* Chichester, Wiley (1966).
3. J. S. Cramer *Empirical Econometrics* Amsterdam, North-Holland (1969).
4. *Current Population Reports—Consumer Income* (series P-60) and *Consumer Buying Indicators* (series P-65), U.S. Bureau of the Census.
5. P. J. Dhrymes *Econometrics* New York, Harper & Row (1970).
6. J. Duesenberry *Income, Saving, and the Theory of Consumer Behaviour* Oxford, Oxford University Press (1967).
7. J. Durbin and G. S. Watson 'Testing for Serial Correlation in Least-Squares Regression' *Biometrika*, vol 37 (1961) pp. 409–28 and vol 38 (1951) pp. 159–78.
8. *Family Expenditure Survey Report for 1971* (Department of Employment) London, H.M.S.O.
9. W. Feller *An Introduction to Probability Theory and its Applications* vol 1 3rd edition Chichester, Wiley (1968).
10. M. Friedman *A Theory of the Consumption Function* Princeton, New Jersey, Princeton University Press (1957).
11. A. S. Goldberger *Econometric Theory* Chichester, Wiley (1964).
12. P. G. Hoel *Introduction to Mathematical Statistics* 3rd edition Chichester, Wiley (1962).
13. J. Johnston *Econometric Methods* 2nd edition New York, McGraw-Hill (1972).
14. M. G. Kendall and A. Stuart *The Advanced Theory of Statistics* (three volumes) London, Griffin (various dates).
15. L. R. Klein *A Textbook of Econometrics* Row-Peterson (1953).
16. J. Kmenta *Elements of Econometrics* London, Collier-Macmillan (1971).
17. A. Koutsoyiannis *Theory of Econometrics* London, Macmillan (1973).
18. E. Malinvaud *Statistical Methods in Econometrics* 2nd revised edition Amsterdam, North-Holland (1970).
19. F. Modigliani and R. Brumberg 'Utility Analysis and the Consumption Function' in *Post Keynesian Economics* ed. K. K. Kurihara London, Allen & Unwin (1955).
20. A. M. Mood and F. A. Graybill *Introduction to the Theory of Statistics* New York, McGraw-Hill (1963).
21. E. W. Streissler *Pitfalls in Econometric Forecasting* London, Institute for Economic Affairs (1970).
22. *Survey of Consumer Finances*, Survey Research Center, Institute for Social Research, The University of Michigan.
23. *Survey of Current Business*, Department of Commerce, U.S. Bureau of Economic Analysis.
24. H. Theil *Principles of Econometrics* Amsterdam, North-Holland (1971).
25. R. J. Wonnacott and T. H. Wonnacott *Econometrics* Chichester, Wiley (1970).

Index